The Physics of
Amorphous Solids

The Physics of Amorphous Solids

RICHARD ZALLEN
Xerox Webster Research Center
Webster, New York
now at Virginia Polytechnic Institute
Blacksburg, Virginia

A Wiley-Interscience Publication

John Wiley & Sons

New York • Chichester • Brisbane • Toronto • Singapore

Library of Congress Cataloging in Publication Data:

Zallen, Richard.
  The physics of amorphous solids.

  "A Wiley-Interscience publication."
  Includes index.
  1. Amorphous substances.   2. Solid state physics.
I. Title.

QC176.8.A44Z34   1983   530.4'1   83-3595
ISBN 0-471-01968-2

Printed in the United States of America

10 9 8 7 6 5 4 3 2

for Avram and Jennifer

PREFACE

Traditionally, solid-state physics has meant crystal physics. Solidity and crystallinity are treated as synonymous in the standard texts on condensed matter. Yet, one of the most active fields of solid-state research in recent years has been the study of solids that are *not* crystals, solids in which the arrangement of the atoms lacks the slightest vestige of long-range order. The advances that have been made in the physics and chemistry of these materials, which are known as amorphous solids or as glasses, have been widely appreciated within the research community and have contributed to the Nobel awards earned by three individuals (N. F. Mott, P. W. Anderson, and P. J. Flory). Excellent reviews of recent research on these solids have appeared and continue to appear, but relatively little has been written to guide the nonexpert and the student. My aim in writing this book has been to contribute a tutorial presentation of parts of this fascinating subject, a presentation which I hope will be of use as an introduction to newcomers to the field. The level of treatment should make it possible for most of the topics covered to be included in a first-year graduate course on the physics of solids.

Much of the intellectual fascination about the amorphous solid state arises from the fact that scientific insight must be achieved without the help of the mathematical amenities (Brillouin zones, Bloch states, group-theoretical selection rules, etc.) which accompany periodicity in the crystalline solid state. While some old approaches remain useful for amorphous solids (most notably the chemical-bonding viewpoint, which focuses on the short-range order), this challenge has been met mainly by new approaches such as localization theory and percolation. From another viewpoint, much of the intense research interest in amorphous solids is driven by the technological importance of these materials. Examples include the use of ultratransparent optical fibers in telecommunications, the use of amorphous semiconductors in xerography and solar cells, and the ubiquitous everyday uses of organic glasses as structural materials. Applications are discussed in the first and last chapters of this book.

Chapter One deals with certain general aspects, including the *inessentialness* of long-range order in solids, ways in which amorphous solids are formed, the phenomenology of the liquid $\leftrightarrow$ glass transition, and a brief survey of technological applications. Chapters Two and Three deal extensively with structural issues in the various classes of amorphous solids. The atomic-scale structure of an amorphous solid is one of its key mysteries, and structural information must be won with great effort. The stochastic-geometry models which have been found to successfully describe the topologically disordered structures of these solids include the random-close-packing model for simple metallic glasses, the continuous-random-network model for covalently bonded glasses, and the random coil model for organic polymers. Chapter Four presents a treatment of percolation theory, carried to some depth. The percolation model is a powerful unifying construct and an outstanding vehicle for exhibiting many of the modes of thought characteristic of theoretical approaches to strongly disordered systems: emphasis on statistical distributions, localization ↔ delocalization critical points, scaling behavior, and dimensionality dependences. It is thus a great help in discussions of physical phenomena in amorphous solids, and is made use of in this way in the last part of the book. Percolation also happens to be a lot of fun, which is another reason why Chapter Four is not short. As a case in point, the scaling approach of phase-transition theory becomes extremely transparent within the context of the percolation model. Scaling ideas recur several times in this book, in connection with polymer structure (Section 3.10), percolation (Section 4.5), politics (Section 4.8), and electron localization (Section 5.7). The final two chapters of the book deal with physical properties of amorphous solids. Topics include theories of the glass transition, Anderson localization and the mobility edge in amorphous semiconductors, optical properties associated with electronic and vibrational excitations, electrical properties of metallic and semiconducting glasses, and chemical-bonding approaches to electronic structure and to the technological control of properties.

My interest in the physics of amorphous solids developed over a period of years within the stimulating environment of the Xerox Webster Research Center, a laboratory that has produced many important contributions to this field. I am indebted to many of my colleagues at Xerox, but I am especially grateful to Harvey Scher, Michael Slade, and Bernard Weinstein, with whom I have had the pleasure of working on diverse aspects of research on amorphous solids. Elsewhere, I wish to thank David Adler, William Paul, and Jan Tauc for many stimulating discussions over the years. The very able and amicable help of Theresa Kusse and Nancy MacDonald in the preparation of the manuscript is gratefully acknowledged. Many of the figures in this book have benefitted from the excellent artwork of Lance Monjé. The patient encouragement of Beatrice Shube at John Wiley is also appreciated. Most of all, however, I wish to express my deep appreciation to my wife, Doris, whose indispensable support made the book possible.

Richard Zallen

Rochester, New York
April 1983

CONTENTS

CHAPTER 1   **The Formation of Amorphous Solids**   **1**

1.1   Freezing into the Solid State: Glass Formation versus Crystallization   1
1.2   Preparation of Amorphous Solids   5
1.3   Structure, Solidity, and Respectability   10
1.4   The Glass Transition   16
1.5   Applications of Amorphous Solids   23

CHAPTER 2   **Amorphous Morphology: The Geometry and Topology of Disorder**   **33**

2.1   Introduction: Geometry, Chemistry, and the Primacy of Short-Range Order   33
2.2   Review of Crystalline Close Packing   35
2.3   Partial Characterizations of Structures   38
     2.3.1   Coordination Number   38
     2.3.2   Radial Distribution Function   40
     2.3.3   EXAFS   43
     2.3.4   Froth—The Honeycomb of Aggregated Atomic Cells   45
     2.3.5   Atomic Polyhedra versus Polyhedral Holes   47
2.4   Random Close Packing   49
     2.4.1   Empirical rcp Structure   49
     2.4.2   Theoretically Derived rcp   51
     2.4.3   Characterizations of the rcp Structure   54
     2.4.4   Peas in a Pot   56
     2.4.5   Dimensionality Considerations and the Extendability of Local Close Packing   58
2.5   Continuous Random Network   60
     2.5.1   The Simplicial Graph   60
     2.5.2   Mathematical Bonds and Chemical Bonds: The Covalent Graph   60
     2.5.3   The Continuous-Random-Network Model of Covalent Glasses   63
     2.5.4   Prototype Elemental crn: Amorphous Silicon   67

2.5.5   Prototype Binary crn: Fused Silica   72

2.6   Experimental RDFs versus rcp and crn Models   73

CHAPTER 3   **Chalcogenide Glasses and Organic Polymers**   **86**

3.1   Molecular Solids and Network Dimensionality   86
3.2   One- and Two-Dimensional-Network Solids   90
3.3   Compositional Freedom in Chalcogenide
        Glasses and in Oxides   97
3.4   The $8 - n$ Rule and the ''Ideal Glass''   101
3.5   Topological Defects and Valence Alternation   104
3.6   The Random Coil Model of Organic Glasses   107
3.7   Random Walks, Drunken Birds, and
        Configurations of Flexible Chains   113
3.8   SAWs, Mean Fields, and Swollen Coils
        in Solution   120
3.9   Why Overlapping Coils are ''Ideal''   127
3.10 Scaling Exponents and Fractal Dimensions   129

CHAPTER 4   **The Percolation Model**   **135**

4.1   Introduction   135
4.2   An Example: The Vandalized Grid   136
4.3   The Percolation Path   139
4.4   Applications to Phase Transitions   146
4.5   Close to Threshold: Critical Exponents, Scaling,
        and Fractals   153
4.6   Trees, Gels, and Mean Fields   167
4.7   Continuum Percolation and the Critical
        Volume Fraction   183
4.8   Generalizations and Renormalizations   191

CHAPTER 5   **Localization ↔ Delocalization Transitions**   **205**

5.1   Localized-to-Extended Transitions in
        Amorphous Solids   205
5.2   Dynamic Modeling: Monte Carlo
        Simulations of the Glass Transition   206
5.3   The Free-Volume Model of the Glass Transition   212
5.4   Free Volume, Communal Entropy, and
        Percolation   218
5.5   Electron States and Metal ↔ Insulator
        Transitions   223

5.6   Disorder-Induced Localization: The Anderson
      Transition   231
5.7   Scaling Aspects of Localization   242

CHAPTER 6   **Optical and Electrical Properties**   **252**

6.1   Local Order and Chemical Bonding   252
6.2   Optical Properties   260
6.3   Electrical Properties   274
6.4   Native Defects and Useful Impurities   289

Index   **297**

The Physics of
Amorphous Solids

CHAPTER ONE

The Formation of Amorphous Solids

1.1 FREEZING INTO THE SOLID STATE: GLASS FORMATION VERSUS CRYSTALLIZATION

To begin with, let us suppose that we all know what is meant by the term "solid." (This innocent assumption is less harmless than it appears, and it calls for a bit of discussion, which is to be supplied in Section 1.3.) In a familiar type of thought experiment, often invoked to conceptually analyze the energetics involved in the formulation of a solid, a large collection of initially isolated atoms is gradually brought together "from infinity" until the actual interatomic spacings of the solid are attained. The actual experiment that most closely corresponds to this gedanken experiment involves cooling the vapor of the material until it condenses into the liquid state, and then further gradual cooling of the liquid until it solidifies. Results of such an experiment, for a given quantity of the material, may be represented on a volume-versus-temperature $V(T)$ plot such as the one schematically shown in Fig. 1.1.

Figure 1.1 should be read from right to left, since time runs in that direction during the course of the quenching (temperature-lowering) experiment. A sharp break or bend in $V(T)$ marks a change of phase occurring with decreasing temperature. The first occurs when the gas (whose volume is limited only by the dimensions of the experimental enclosure) condenses to the liquid phase (of well-defined volume, but shape enclosure-determined) at the boiling temperature T_b. Continued cooling now decreases the liquid volume in a continuous fashion, the slope of the smooth $V(T)$ curve defining the liquid's volume coefficient of thermal expansion $\alpha = (1/V)\,(\partial V/\partial T)_P$. (The experiment is assumed to be taking place at low pressure, $P \approx 0$.) Eventually, when the temperature is brought low enough, a liquid→solid transition takes place (with the exception of liquid helium, which remains liquid as $T \to 0$ in the absence of pressure). The solid then persists to $T = 0$, its signature in terms of $V(T)$ being a small slope corresponding to the low value (relative to that of the liquid phase) of the expansion coefficient α which characterizes a solid.

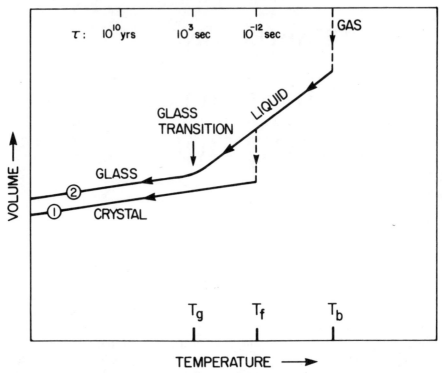

Figure 1.1   The two general cooling paths by which an assembly of atoms can condense into the solid state. Route ① is the path to the crystalline state; route ② is the rapid-quench path to the amorphous solid state.

A liquid may solidify in two ways:

1. *discontinuously* to a crystalline solid or
2. *continuously* to an amorphous solid (glass).

The two solids resulting from these two quite different solidification scenarios are labeled, correspondingly, ① and ② in Fig. 1.1. Scenario ① occurs in Fig. 1.1 at temperature T_f, the freezing (or melting) point. The liquid→crystal transition is marked by a discontinuity in $V(T)$, an abrupt contraction to the volume of the crystalline solid. In a quenching experiment carried out at a sufficiently low cooling rate, this is usually the route taken to arrive at the solid state. But at sufficiently high cooling rates, it is found that most materials alter their behavior and follow route ② to the solid phase. T_f is bypassed without incident, and the liquid phase persists until a lower temperature T_g is reached. Here the second solidification scenario is realized. The liquid→glass transition occurs in a narrow temperature interval near T_g, the *glass transition* temperature. There is no volume discontinuity, instead V(T) bends over to acquire the small slope (similar to that of the crystal) characteristic of the low thermal expansion of a solid.

Both crystals and glasses are bona fide solids and share the essential attributes of the solid state (Section 1.3). Their fundamental difference is in the basic nature of their microscopic, atomic-scale structure. In crystals, the equilibrium positions of the atoms form a translationally periodic array. The atomic positions exhibit long-range order. In amorphous solids, long-range order is absent; the array of equilibrium atomic positions is strongly disordered. For crystals, the atomic-scale structure is securely known at the outset from the results of diffraction experiments, and it provides the basis for the analysis of such properties as electronic and vibrational excitations. For amorphous solids, the atomic-scale structure is itself one of the key mysteries. Several chapters of this book are devoted to the structure of glasses. A brief preview is given in Section 1.3 and Fig. 1.6.

A note on terminology is in order at this point. The term *amorphous solid* is the general one, applicable to any solid having a nonperiodic atomic array as outlined above. The term *glass* has conventionally been reserved for an amorphous solid actually prepared by quenching the melt, as in ② of Fig. 1.1. Since, as discussed in Section 1.2, there are other ways to prepare amorphous solids than by melt-quenching, glass (in the conventional usage) is the more restrictive term. In this book, that historical distinction will not be adhered to, and both terms will be used synonymously. (''Historical'' is used here in two senses, since the distinction itself refers to the history, i.e., method of preparation, of the solid.) Not only does this lubricate the discussion because ''glass'' is one word while ''amorphous solid'' is two, it is also convenient to have ''glass'' to set in opposition to ''crystal'' (instead of ''amorphous solid'' versus ''crystalline solid''). Other terms, sometimes used in the literature in place of amorphous solid, are noncrystalline solid and vitreous solid.

A detailed view of the vicinity of the liquid→glass solidification transition is shown in Fig. 1.2 for the case of the organic glass polyvinylacetate ($CH_2CHOOCCH_3$). The data show results for $V(T)$ obtained at two different cooling rates, and reveal that the observed transition temperature T_g depends upon the cooling rate at which the experiment is carried out. This is a characteristic *kinetic* dimension of the glass transition. The two $V(T)$ curves in Fig. 1.2 are labeled by two experimental time scales, 0.02 hr for the upper curve and 100 hr for the lower curve. In these particular experiments, the stated times are the times elapsed in quenching the specimen to temperature T from a fixed initial temperature well above T_g. Note that the effect of changing this time by a factor of 5000 is to shift T_g by only 8°K. Thus this effect, while quite real, is small.

Denoting the average cooling rate $-\overline{dT/dt}$ by $\dot{T}$, the mild influence that the time scale of the measurement exerts on the experimentally observed liquid→glass temperature may be indicated by writing T_g as $T_g(\dot{T})$. The weak functional dependence may be approximated as a logarithmic one. Typically, changing $\dot{T}$ by an order of magnitude causes T_g to shift by a few degrees kelvin.

The reason that T_g shifts to lower temperatures when the cooling process is extended over longer times resides in the temperature dependence of a typical

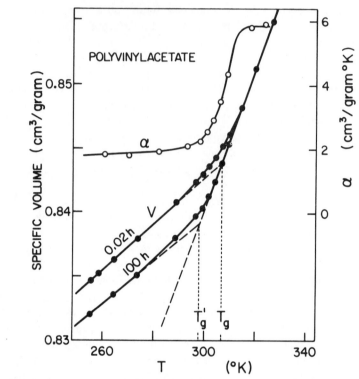

Figure 1.2 Volume-versus-temperature cooling curves for an organic material in the neighborhood of the glass transition. $V(T)$ is shown for two greatly different cooling rates, as is the coefficient of thermal expansion $\alpha(T)$ for the fast-cooling curve (0.02 hr). The break in $V(T)$, and the corresponding step in $\alpha(T)$, signal the occurrence of the liquid→glass transition (after Kovacs, Hutchinson, and Aklonis, 1977).

molecular relaxation time τ. (The adjective "typical" reflects the fact that there is actually a spectrum of relaxation times; τ may be regarded as the geometric mean of that distribution.) The quantity $1/\tau$ characterizes the rate at which the molecular configuration (atomic-scale structure) of the condensed system adapts itself to a change in temperature. This quantity varies enormously during the cooling process. An indication of this dramatic variation is given at the top of Fig. 1.1 where, in crude order-of-magnitude terms, values of τ are associated with three temperatures: T_f, T_g, and a temperature well below T_g (say, $T_g - 50°K$). The structural-rearrangement response time may increase from the order of 10^{-12} sec at T_f to 10^{10} years (age of the universe) at $T_g -$ $50°K$. (Experimentally, τ is accessible at high temperatures where it roughly scales inversely as the viscosity of the liquid. At low temperatures, in the solid, τ is inferred indirectly.)

The 30 orders of magnitude in τ, spanned between the liquid near the melting point and the glass well below the "glass point," are swept through swiftly and continuously at temperatures in between. As T traverses the region near T_g, $\tau(T)$ becomes comparable to the time scale of the measurement (typically 10^3 sec, give or take an order of magnitude or two). As T is lowered below T_g, τ becomes much larger than any experimentally accessible times, so that the material loses its ability to rearrange its atomic configuration in harmony with the imposed decline of temperature. The atoms get frozen into well-defined positions (equilibrium positions, about which they oscillate), which correspond to the configuration they had at T_g. It is now easy to understand why, in Fig. 1.2, expanding the experimental time scale (slowing the cooling rate $\dot{T}$) lowers the observed glass point T_g: If a longer experimental time t is available, then a lower temperature T is needed to achieve the condition $\tau(T) > t$ which freezes the atoms into the configuration that they maintain in the amorphous solid state. Note that the mildness of the t dependence of T_g is simply the other side of the coin with respect to the severity of the exceedingly steep function $\tau(T)$.

While kinetic effects clearly play a role in the operational definition of T_g, it is generally believed that the observed liquid $\leftrightarrow$ glass transition is a manifestation of an underlying thermodynamic transition viewed as corresponding to the limit $t \to \infty$, $\dot{T} \to 0$. Some of the experimental evidence of this is given in Section 1.4, and theories of the glass transition—which has been one of the knottiest problems in condensed matter physics—will be discussed in Chapter Four.

In addition to showing what happens to the specific volume (inverse of the density) at temperatures near T_g, Fig. 1.2 also includes a related thermodynamic variable, the expansion coefficient α. This quantity experiences a well-defined "step" near T_g, corresponding to the slope change in $V(T)$. Other thermodynamic aspects of T_g are discussed in Section 1.4.

A comment should be made about the terms "freezing" and "melting." These two terms are conventionally reserved for the two directions ($\leftarrow$ and $\rightarrow$) in which a material may traverse the crystal $\leftrightarrow$ liquid transition, the event which occurs at T_f along route ① of Fig. 1.1. This usage is usually adhered to here. But it should be realized that the same terms *also* describe the event that occurs at T_g along route ② of Fig. 1.1. For the glass $\leftrightarrow$ liquid transition, T_g denotes the temperature at which (in direction $\leftarrow$) the undercooled liquid freezes. In the other direction ($\rightarrow$, increasing temperature), T_g denotes the temperature *at which the glass melts*.

1.2 PREPARATION OF AMORPHOUS SOLIDS

For a long time it was thought that only a relatively restricted number of materials could be prepared in the form of amorphous solids, and it was common to refer to these "special" substances (e.g., oxide glasses and organic polymers) as "glass-forming solids." This notion is wrong, and it is now realized that "glass-forming ability" is almost a universal property of condensable matter. The amorphous solid state is ubiquitous. Table 1.1 presents a list of

TABLE 1.1   Representative amorphous solids, their bonding types, and their glass-transition temperatures

Glass	Bonding	T_g (°K)
SiO_2	Covalent	1430
GeO_2	Covalent	820
Si, Ge	Covalent	—
$Pd_{0.4}Ni_{0.4}P_{0.2}$	Metallic	580
BeF_2	Ionic	570
As_2S_3	Covalent	470
Polystyrene	Polymeric	370
Se	Polymeric	310
$Au_{0.8}Si_{0.2}$	Metallic	290
H_2O	Hydrogen bonded	140
C_2H_5OH	Hydrogen bonded	90
Isopentane	van der Waals	65
Fe, Co, Bi	Metallic	—

amorphous solids in which every class of bonding type is represented. The glass-transition temperatures span a wide range.

The correct viewpoint (expressed, for example, in D. Turnbull's 1969 review paper) is the following: *Nearly all materials can, if cooled fast enough and far enough, be prepared as amorphous solids.* ("Fast" and "far" are explained below.) This viewpoint has been abundantly supported in recent years by the preparation of an enormous variety of amorphous solids. Prominent among these, and providing one of the most striking demonstrations of the ubiquity of this state of condensed matter, are the metallic glasses. Because metals tend to be structurally simple materials (many form, in the crystalline state, close-packed structures), the proliferation of glassy metals is a very significant development. Traditional "glass formers" have been materials associated with considerable complexity on a molecular scale, such as organic glasses composed of polymer chains having bulky sidegroups dangling from them. Metals had been thought to be too simple to form glasses.

Figure 1.3 displays an effective technique, known as *melt spinning*, for achieving the very high rate of cooling needed to form a metallic glass. A jet of hot molten metal is propelled against the surface of a rapidly rotating copper cylinder, which is kept cool (room temperature or below). The liquid metal is drawn into a thin film, roughly 50 microns thick (50 μm = 0.05 mm). Since the film is so thin, since it is in intimate contact with a large heat sink, and since metals have high thermal conductivity, the liquid cools and solidifies extremely fast. A temperature drop of about 1000°K is accomplished in about a millisecond, i.e., $\dot{T} \approx 10^6$ °K/sec. The solid film of metallic glass is spun off the rotor, as a continuous ribbon, *at a speed exceeding 1 kilometer per minute*.

Thus the name of the game—the essential ingredient in the preparation of an amorphous solid—is *speed*. A given material may solidify via either of the

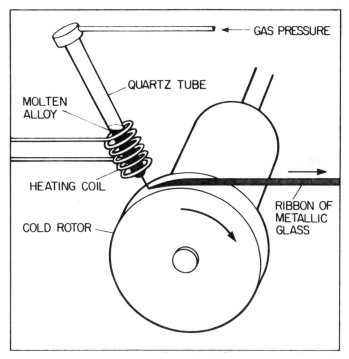

Figure 1.3 Melt spinning of metallic glass. The solid ribbon of amorphous metal is spun off at speeds that can exceed 1 kilometer per minute (from Chaudhari, Giessen, and Turnbull, 1980, copyright 1980 by Scientific American, Inc., all rights reserved, used by permission).

two routes indicated in Fig. 1.1. As soon as the temperature of the liquid is lowered to T_f, it may take route ① to the solid state and crystallize. But crystallization takes time. Crystalline centers must form (a process called nucleation) and then grow by outward propagation of the crystal/liquid interfaces. With the liquid being cooled at a finite rate, the liquid may be taken below T_f along the $V(T)$ trajectory which smoothly continues the curve from higher temperatures. In the temperature interval between T_f and T_g, the liquid is referred to as the undercooled or supercooled liquid. (The undercooled liquid is still unambiguously *liquid* and must not be confused with the glass, as is mistakenly done in a few texts.) If its temperature can be taken below T_g before crystallization has had time to occur, the undercooled liquid solidifies as the glass and remains in this form essentially indefinitely.

Glass formation, therefore, is a matter of *bypassing crystallization*. The channel to the crystalline state is evaded by quickly crossing the dangerous regime of temperature between T_f and T_g and achieving the safety of the amorphous solid state below T_g. Throughout the temperature interval $T_g < T < T_f$, the liquid is "at risk" with respect to nucleation and growth of crystallites. Earlier it was stated that, for a material to be prepared as an amorphous solid, cooling must proceed "fast enough and far enough." "Far enough" is seen to mean that the quench must be taken to $T < T_g$, and "fast enough" means that $T_g < T < T_f$ must be crossed in a time too short for crystallization to occur. In contrast to crystallization, which is heterogeneous (pockets of the solid phase appear abruptly within the liquid and then grow at its expense), the liquid→glass transformation occurs homogeneously throughout the material. This transformation would be observed for any liquid when sufficiently undercooled (i.e., all liquids would form glasses), except for the intervention of crystallization.

"Fast enough" can be, for many materials, very much slower than the quenching rate ($\dot{T} \approx 10^6$ °K/sec) quoted in connection with Fig. 1.3. Unlike the single millisecond taken to quench a metallic glass, the time taken to quench the silicate glass that forms the rigid ribbed disk of the Mt. Palomar telescope was eight months, corresponding to a leisurely $\dot{T}$ of 3×10^{-5} °K/sec. It is much easier to prepare a glass for which a low $\dot{T}$ suffices than it is to prepare one for which a high $\dot{T}$ is needed. Thus, while it is not meaningful to speak of glass-forming solids (since this classification encompasses virtually all materials), it is certainly valid to refer to *glass-forming tendency*. This attribute is correlated with $1/\dot{T}$, and is obviously much greater for oxide glasses than for metallic glasses.

Figure 1.4 schematically illustrates four techniques for preparing amorphous solids that span the range of quenching rates. These techniques are not fundamentally different from those used for preparing crystalline solids; the point is simply that care is taken to quench fast enough to form the glass rather than slow enough to form the crystal.

For materials with very high glass-forming tendency, the melt can be allowed to cool slowly by simply turning off the furnace or by bringing it down in a programmed manner (Fig. 1.4a). Typical cooling rates are in the range from 10^{-4} to 10^{-1} °K/sec. Glasses in this category, among those listed in Table 1.1, are SiO_2, As_2S_3, and polystyrene. Thus, although the crystalline form of As_2S_3 is abundant in nature (which had a long time to produce it) as the mineral orpiment, synthetic crystals cannot be prepared from the melt on any experimentally reasonable time scale. The melt always solidifies as the amorphous solid.

Somewhat faster rates are needed to quench a glass such as amorphous selenium, an elemental glass composed of long-chain polymeric molecules. Using an ice-water bath to quench modest volumes of the melt, as indicated in Fig. 1.4b, yields rates in the range 10^1–10^2 °K/sec. Se glass can be prepared by this method, as can the Pd-Ni-P metallic glass included in Table 1.1. This metallic glass has a glass-forming tendency high enough to allow it to be prepared in bulk form, rather than the thin-film form characteristic of the other metallic glasses listed in the table.

The technique sketched in Fig. 1.4c is another of the melt-quenching methods (of which the melt-spinning method of Fig. 1.3 is the most spectacular example) developed specifically for metallic glasses. These methods are collectively called *splat-quenching* techniques, and achieve $\dot{T}$ values in the range 10^5–10^8 °K/sec. The hammer-and-anvil drop-smasher method of Fig. 1.4c

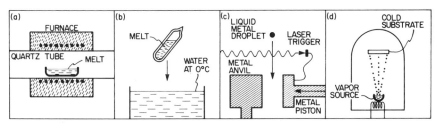

Figure 1.4   Four methods of forming amorphous solids: (*a*) slow cooling, (*b*) moderate quenching, (*c*) rapid "splat-quenching", and (*d*) condensation from the gas phase.

cools the liquid droplet from two sides at once, and is used to produce milligram-size laboratory specimens of metallic glasses such as the Au-Si alloy listed in Table 1.1.

Before going on to discuss condensation from the vapor phase (Fig. 1.4*d*), it is interesting to take note of a final method for quenching the liquid, one which is even faster than splat quenching. The technique is called *laser glazing*, and it begins with the material in crystalline form. A very short and very intense single laser pulse is focused onto a very small spot on the crystal surface, with the laser wavelength selected so that the light energy is absorbed in an extremely thin (~ 100 Å) layer of the solid. The large energy dumped into this tiny volume is sufficient to melt it, but it is swiftly quenched and resolidified by the surrounding crystal. The small, very thin, melted-and-quenched region has been found to be amorphous in the case of silicon, a material normally preparable in amorphous form only by vapor-condensation techniques. Amorphous metals can also be prepared by laser glazing. The quench rate can only be roughly estimated by highly approximate calculations; these yield towering $\dot{T}$ values in the range 10^{10}–10^{12} °K/sec.

All of the glass-forming methods discussed thus far rely upon speed-induced access to route ② of Fig. 1.1. In Fig. 1.4*d*, we show a representative of a class of techniques that bypasses the liquid phase completely and constructs the amorphous solid in atom-by-atom fashion from the gas phase. These techniques possess the highest effective quench rates ($\dot{T}$ is probably too high to be any longer a meaningful parameter in its original sense), and they are widely used to prepare glasses which have not been obtained by melt-quenching methods.

Figure 1.4*d* shows the simplest of these *vapor-condensation* techniques. A vapor stream, formed within a vacuum chamber by thermal evaporation of a sample ("source") of the material in question, impinges upon the surface of cold substrate. As the atoms condense on the surface, the as-deposited amorphous structure is quenched in if conditions are arranged so that their thermal energy is extracted from them before they can migrate to the crystalline configuration. Variations of the method involve vaporizing the source by the use of an electron beam, or the use of ion bombardment to drive atoms from it. Another method involves the plasma-induced decomposition of a molecular species, a technique employed to deposit amorphous silicon from silane (SiH_4) vapor.

Vapor-condensation techniques produce amorphous solids in the form of thin films, typically 5–50 μm thick. Among the amorphous solids listed in Table 1.1, those which normally demand vapor-condensation methods for their preparation are Si, Ge, H_2O, and the elemental metallic glasses Fe, Co, and Bi. For the pure metals, the substrate must be kept very cold ($<20°K$). It is usually difficult to define T_g for such glasses, since they are not prepared by a liquid→glass quenching process and, when subjected to a heating cycle after preparation, they often crystallize before there is a chance for the glass→liquid transition to occur.

Many glasses which may be formed by melt quenching, such as Se and As_2S_3, are often prepared instead by vapor deposition when thin films are desired (as in applications such as xerography, Section 1.5). Some differences between melt-quenched and vapor-quenched material can be detected, but these normally disappear when the latter is allowed to anneal (Section 3.1.2). It is correct to regard both techniques as producing essentially the same condensed phase.

A trend exists for glass-forming tendency to be greater for a binary material (say, a silicon–gold alloy) than for an elemental one (say, pure silicon). This has to do with the relation between T_g and T_f. Figure 1.5 shows the relevant aspect of the temperature-versus-composition phase diagram for the binary system $Au_{1-x}Si_x$. For the alloy ($0 < x < 1$), the liquid is stabilized and thus the melting point T_f is lowered, relative to that of the single-component endpoints ($x = 0$ or 1), by the entropy of mixing and the attractive interaction between the two components. There is seen to be a eutectic composition $x = 0.2$ at which the melting point is minimized at a deep cusp in $T_f(x)$. At $x = 0.2$, T_g/T_f takes on its largest value, which is about 0.5 for this system. (For comparison, in excellent glass formers such as As_2S_3 and SiO_2, T_g/T_f is about 0.7.)

Near the eutectic composition, as at a in the figure, a liquid is much more readily quenched to the glass than is a liquid at a distant composition such as b. The treacherous territory between T_f and T_g, within which the melt is both thermodynamically ($T < T_f$) *and* kinetically ($T > T_g$) capable of crystallizing, is much broader and more forbidding at b than at a. Thus the eutectic composition is favored for glass formation, a conclusion consistent with the observation that $Au_{0.8}Si_{0.2}$ can be splat quenched to the glassy state while Au and Si cannot. Pure silicon can be vapor quenched to form the amorphous form, while pure gold has yet to be prepared as an amorphous solid. The latter eventuality is, to end on a note in keeping with the theme of this section, *simply a matter of time.*

1.3 STRUCTURE, SOLIDITY, AND RESPECTABILITY

The title of this section mentions three attributes of amorphous solids, each of which will be discussed in turn. The first two are physical attributes; the third is

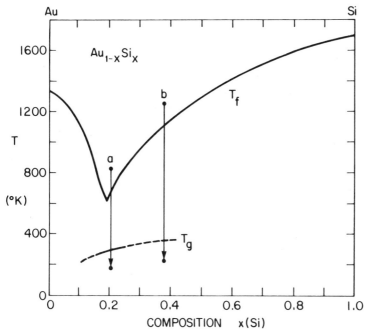

Figure 1.5   Glass formation in the gold–silicon system. Two quenches from the liquid state, at two compositions, are indicated. Glasses can be prepared much more readily in quench *a* than in quench *b*, since the latter must cross a greater temperature range between T_f and T_g in which it is "at risk" vis-à-vis crystallization. (The T_f curve is from the work of Predel and Bankstahl, 1975; the T_g curve is from the work of Chen and Turnbull, 1968.)

a different type of quality, only recently attributed to glasses in conventional attitudes about what constitutes the discipline of solid-state physics.

The subject of structure dominates the following two chapters, but it seems advisable to insert a brief preview at this point. Figure 1.6 presents, schematically and in a nutshell, the salient characteristics of the atomic arrangements in glasses as opposed to crystals. Also included, as an additional and useful reference point, is a sketch of the arrangement in a gas. Of necessity, two-dimensional crystals, glasses, and gases are represented, but the essential points to be noted carry over to their actual, three-dimensional, physical counterparts. For the two sketches representing ideal crystal (*a*) and glass (*b*) lattices, the solid dots denote the equilibrium positions about which the atoms oscillate; for the gas (*c*), the dots denote a snapshot of one configuration of instantaneous atomic positions.

For an amorphous solid, the essential aspect with which its structure differs with respect to that of a crystalline solid is *the absence of long-range order*. There is no translational periodicity. This fundamental difference is evident at a glance in Figs. 1.6*a* and 1.6*b*.

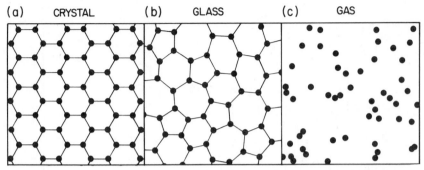

Figure 1.6   Schematic sketches of the atomic arrangements in (*a*) a crystalline solid, (*b*) an amorphous solid, and (*c*) a gas.

On the other hand, the atomic positions in the glass are *not* randomly distributed in space. Randomness is a trait more properly associated with Fig. 1.6*c*, at least in the low-density limit in which the atoms comprising the gas may be viewed as point particles. For such a dilute gas (the ideal gas of the kinetic theory), the particle positions are totally uncorrelated. Each atom may be located anywhere, independent of the positions of all other atoms. But in Fig. 1.6*b*, there is seen to be a high degree of *local* correlation. Each atom has (in the example used here for illustration) three nearest neighbors at nearly the same distance from it. Nearest-neighbor atoms are connected by lines in the figure, and the "bond angles"—formed where these lines meet at an atomic position—are also nearly equal.

In the crystalline case of Fig. 1.6*a*, the nearest-neighbor separations and bond lengths are *exactly* equal (remember that we are dealing with the equilibrium positions), rather than *nearly* equal as in the glass. The degree of local correlation in amorphous solids is quantitatively described in the following chapters; it suffices here to say that this local order is quite high. Thus glasses have, in common with crystals, *a high degree of short-range order*. As in crystals, this is a consequence of the chemical bonding responsible for holding the solid together.

Thus, while the lack of long-range order in glasses implies randomness at large separations (knowing the positions of a few atoms does *not* help to locate, as it *does* in a crystal, the positions of distant atoms), the atomic-scale structure is highly nonrandom for a few interatomic distances about any given atom. A simple thought experiment serves as one way of demonstrating (other than by just looking) the presence of local order in Figs. 1.6*a* and 1.6*b* and its absence in Fig. 1.6*c*. Suppose a single atom is plucked out of each panel of the figure by a man with a bad memory. If he later wished to reinsert each atom in its original position, he would have no difficulty doing so for Figs. 1.6*a* and 1.6*b*. Not so, however, for Fig. 1.6*c*; since it is completely random, the remaining atomic positions provide no clue about the missing one.

Since Fig. 1.6*c* is a genuinely random array while Fig. 1.6*b* is not, it may seem surprising that all three main categories of amorphous-solid structure discussed in the next chapters (random close packing, continuous random network, random-coil model) include the term random in their names. This can be accepted as a historical circumstance, but we may also agree that random applies in the limited sense of referring to *statistical distributions*, which describe quantities (such as bond angles) in the glass structure that would take on a *single* fixed value in the crystal.

The difference between Figs. 1.6*c* and 1.6*b* can be characterized as a case of *unconstrained* versus *constrained chance*. A random array of point particles, unconstrained by any particle–particle correlations, is a suitable model for a gas at low density, but the disorder in a glass is constrained at short distances by the physics and chemistry of the atom–atom bonding interactions. This dichotomy evokes a philosophical analogy which I cannot resist mentioning. Physics and mathematics can be viewed, respectively, as constrained and unconstrained logical systems. In mathematics the system is largely unconstrained by considerations other than logical consistency, but physical theory is additionally, and essentially, disciplined by the *experimental* requirement that it be (in Einstein's words) "of value for the comprehension of reality."

Before leaving Fig. 1.6*b*, we should note the term *topological disorder* (treated further in Chapter Two) in connection with the glass structure schematically represented here. *This disorder is intrinsic to the lattice structure itself.* It is a much more severe class of disorder than that mentioned in the next paragraph.

There are certain types of crystalline systems that are sometimes classified as disordered solids because, while their crystal lattices remain intact and fundamentally resemble Fig. 1.6*a*, their translational symmetry is broken by the chemical (for a "mixed crystal") or orientational (for a "plastic crystal") identity of the objects which occupy the lattice sites. In a mixed crystal such as Ge_xSi_{1-x}, we know *where* each atom is (on a site of the crystalline diamond lattice, shown later in Fig. 2.10), but we do not know *what* it is (Ge or Si). Each lattice site is occupied at random by either a Ge atom (with probability x) or a Si atom (with probability $1 - x$). In a "plastic crystal," symmetric molecules set in different orientations sit on the sites of a periodic lattice. The disorder in such systems (*compositional disorder* for a mixed crystal, *rotational disorder* for a plastic crystal) is very mild compared to the topological disorder characteristic of amorphous solids, because there remains an underlying crystal lattice with its periodicity preserved, and it is often possible to deal with such solids by conventional methods of crystal physics. Thus, the electronic and optical properties of Ge_xSi_{1-x} mixed crystals may be adequately approximated by the "virtual crystal approximation," in which the solid is viewed as a perfectly periodic crystal composed of a single type of fictitious atom that is intermediate in behavior to germanium and silicon. Such mildly-disordered essentially-crystalline solids are not treated in this book.

To move on to the second topic of this section, the subject of solidity, we now consider what *time* does to the configurations represented in Fig. 1.6. For

Fig. 1.6*c*, the effect of letting the clock run is to completely overturn the particular instantaneous structure shown here for the gas. The motion of the atoms takes the gas into other random arrays that, on the scale of atomic dimensions, are totally different from the specific arrangement of Fig. 1.6*c*. (On a macroscopic scale, of course, the effect is that of a small statistical fluctuation on, say, the density.) That atomic motion, for a dilute gas, consists of straight-line trajectories which are punctuated occasionally by sharp deflections corresponding to collisions of the atoms with each other or with the walls of the container.

Time has no such drastic effect on the structures represented in Figs. 1.6*a* and 1.6*b*. While plenty of motion is going on (even at 0°K, the zero-point motion remains), this motion does *not* overthrow the structure in a crystal or a glass. Viewing a given atom as a classical particle with a definite trajectory, the situation is as sketched in Fig. 1.7*a*. The atom stays close to a well-defined equilibrium position, and executes oscillatory motion about it. This persistent aspect is in marked contrast to the fluid scene in Fig. 1.6*c*, in which the motion is translational on an atomic scale.

An essential distinction is thereby drawn, with respect to the nature of the microscopic *motion* taking place in the material, between Figs. 1.6*a* and 1.6*b*, on the one hand, and Fig. 1.6*c*, on the other hand. In Figs. 1.6*a* and 1.6*b*, the atoms are immobilized except for *vibrational* motions (Fig. 1.7*a*) about their average positions. In Fig. 1.6*c*, the atoms are free to make long, uninhibited, *translational* excursions. Macroscopically, this is nothing less than the distinction between *solidity*, on the one hand, and *fluidity* (in the extreme form exhibited by a dilute gas), on the other hand. In a solid, the atoms oscillate about equilibrium positions, which constitute a durable structure. No such enduring structure exists in a fluid, in which the atomic motion is characterized by extensive translational movements.

The fluid of Fig. 1.6*c* is a gas. In a dense fluid, that is, a liquid, translational movement is likewise an essential characteristic of the atomic motion. Since each atom is now hemmed in to a substantial extent by nearby atoms,

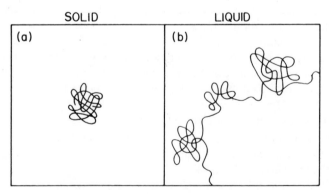

Figure 1.7   Sketches of the state of atomic motion (*a*) in a solid and (*b*) in a liquid.

the motion is also partly oscillatory. This is schematically represented in Fig. 1.7*b*, which is intended to convey a sense of the atomic movement in a liquid. (Trajectories similar to the one sketched here appear in computer simulations of the liquid state, such as those discussed in Chapter Four.) In spite of the presence of an oscillatory component, the key feature is the presence of a translational component of motion. A liquid, like a gas, possesses no enduring arrangement of atoms. Each atom in a liquid is mobile and wanders through the material, changing neighbors continually during its diffusive meandering. Atoms in a solid retain their neighbors (aside from rare events such as vacancy jumps) as all remain near fixed positions.

Moving from these atomistic descriptions to more standard macroscopic definitions: a liquid flows, lacks a definite shape (though its volume is definite), and cannot withstand a shear stress; a solid does not flow, has a definite shape, and exhibits elastic stiffness against shear. The distinction is usually quite clear. A glass (tumbler) of water consists of two transparent substances, one liquid and one solid. Water is our most familiar liquid, and the ease with which it flows is its most familar property. The rigid, brittle container in this example is an amorphous solid, an oxide glass related to fused silica (SiO_2). The structure (static for the glass, everchanging for the liquid) is of the type indicated in Fig. 1.6*b* for both materials, but the motion is as in Fig. 1.7*a* for the solid container and as in Fig. 1.7*b* for its fluid contents.

Amorphous solids are bona fide solids, having all of the requisite elastic properties (shear stiffness, etc.). There is no need to belabor this point, many of the applications of these materials (look ahead to Table 1.2) rely explicitly on properties such as rigidity and strength. Note that neither the macroscopic (rigidity, etc.) nor the atomic (Fig. 1.7*a*) description of solidity makes any reference to the presence or absence of structural long-range order, indeed, solidity is *not* synonymous with crystallinity.

It should not be necessary any longer to emphasize that solidity $\neq$ crystallinity. Unfortunately, it is, in fact, necessary to do so. The reason for this necessity arises from the following circumstance: Amorphous solids are rarely included in textbooks on solid-state physics. If one were tempted to define the subject of "solid-state physics" by the content of current textbooks with that title, one might erroneously conclude that solid-state physics is synonymous with crystal physics. So much is this the case, in fact, that it is standard procedure for a course on the physics of solids to begin with a discussion of crystal lattices and translational periodicity *as if periodicity were a prerequisite for solidity.* Since it blithely ignores an entire important class of solids, this attitude is completely wrong.

This mistaken premise, the exclusive association of solidity with crystallinity, brings us to the third in the troika of topics listed in the title of this section. This topic has to do with the issue of the *respectability* of amorphous solids as proper inhabitants of the solid state. This issue arises as a historical legacy, and it is likely to fade with time into a non-issue as it becomes impossible for new (or revised) solid-state-physics texts to ignore the scientific and technological significance of glasses. Symptomatic of the inevitable respectabilization (and helping to hasten the process along) was the 1977 Nobel Prize in Physics shared by P. W. Anderson, N. F. Mott, and J. H. Van Vleck. Anderson and Mott were recognized, in part, for their deep contributions to the theory of amorphous solids, some of which are described in Chapter Five.

Although reluctance to accept amorphous solids as an integral part of solid-state physics is an attitude that will (hopefully) disappear before long, it is instructive to consider the factors which contributed to that misguided attitude. There is no mystery here. Solid-state physicists have traditionally been raised on the mathematical amenities of translational periodicity. Much of the machinery of familiar solid-state theory explicitly depends on and exploits the presence of long-range order in the crystalline solid state. This theoretical machinery includes: Brillouin zones, Bloch functions, $\mathbf{k}$-space, $E(\mathbf{k})$ electronic band structures, $\omega(\mathbf{k})$ phonon dispersion curves, and elegant uses of symmetry and group theory for the labeling of eigenstates and the elucidation of selection rules. In the amorphous solid state, the loss of long-range order severely reduces, and possibly eliminates, the validity and utility of the above-mentioned mathematical tools. This must seem like Paradise Lost to many theorists, and it accounts for the past reluctance of some to face the reality of noncrystalline solids. But, however essential it may be to many standard theoretical techniques, long-range order is simply *inessential* to an entire class of solids—which do very well without it.

It is sometimes emphasized that an amorphous solid is metastable with respect to some crystalline phase that forms the thermodynamic equilibrium state of lowest energy. While this statement itself is correct (though no general proof of it exists), the emphasis is misplaced because experience teaches that the crystalline ground state is normally kinetically inaccessible. Once formed, glasses can persist without practical limit ($> 10^n$ yr). The situation is similar to that of crystalline diamond. Diamond, the hardest substance known and the archetypal covalent crystal, is metastable. The lowest-energy configuration of a collection of carbon atoms is not as diamond but as graphite, which is the stable thermodynamic phase at standard temperature and pressure. Despite their metastability, "diamonds are (effectively) forever"; a diamond is in no danger, and persists indefinitely at STP. Since the same is true of a glass well below T_g, metastability becomes an academic matter.

1.4 THE GLASS TRANSITION

Phenomena associated with the liquid ↔ glass transition are macroscopic manifestations of the crossover between the two microscopic motional situations of Fig. 1.7. Some aspects of the glass transition have already been shown in Figs. 1.1 and 1.2. This section presents further phenomenological aspects associated with T_g. In particular, the question of an underlying *equilibrium* thermodynamic transition is addressed.

The basic thermodynamic response function to be experimentally examined in connection with a temperature-induced change of phase is C_P, the specific heat at constant pressure. C_P measures the heat absorption from a temperature stimulus, and is defined by $C_P \equiv (dQ/dT)_P = T(\partial S/\partial T)_P$. Here dQ is the heat absorbed by a unit mass of the material to raise its temperature by dT, and S is its entropy.

Figures 1.8 and 1.9 show $C_P(T)$ data for two very different amorphous solids, the covalent glass As_2S_3 and the metallic glass $Au_{0.8}Si_{0.1}Ge_{0.1}$. In each case, the glass transition clearly appears as a "step" in the specific heat. For As_2S_3, $C_P(T)$ can be followed continuously from low temperature up through T_g and well into the liquid regime to T_f and beyond. For $Au_{0.8}Si_{0.1}Ge_{0.1}$, a portion of the liquid curve (shown dashed in the figure) is extrapolated between measured values taken just above T_g and just below T_f, because crystal-

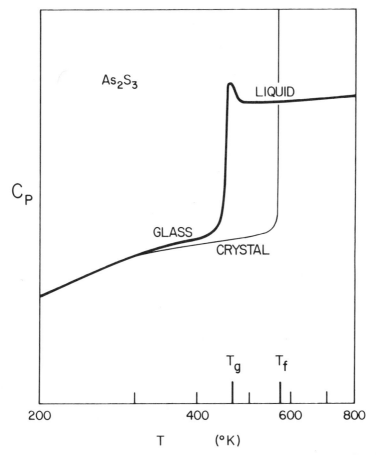

Figure 1.8 The specific heat of the crystalline, amorphous, and liquid forms of As_2S_3, a covalent material which is a prototypical glass former (after Blachnik and Hoppe, 1979).

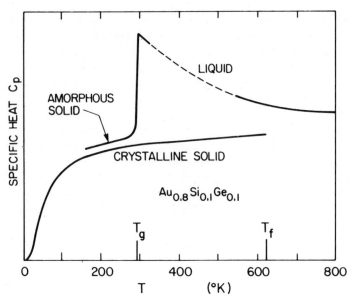

Figure 1.9 The specific-heat signature of the glass transition in a metallic glass (Chen and Turnbull, 1968).

lization intervenes shortly after passing these temperatures at the relatively slow heating or cooling rates required for measuring C_P. Figure 1.9 is of historic interest because it provided the first thermodynamic evidence of a glass transition in an amorphous metal, observed upon warming a splat-quenched sample through T_g (Chen and Turnbull, 1968).

Included in Figs. 1.8 and 1.9 are the specific heat curves for the crystalline forms. For each material, the values of C_P(glass) and C_P(crystal) are similar. The experimental curve for crystalline As_2S_3 in Fig. 1.8 is followed to the melting point T_f, where C_P diverges because of the heat of fusion—the latent heat (finite ΔQ with $\Delta T = 0$) associated with the crystal ↔ liquid transition. No such latent-heat singularity accompanies the glass ↔ liquid transition.

The behavior of $C_P(T)$ near T_g in Figs. 1.8 and 1.9 is qualitatively the same as that of $\alpha(T)$ in Fig. 1.2: Both the specific heat and the thermal expansion coefficient step up, in a narrow temperature interval, from a low value characteristic of the glass to a high value characteristic of the liquid. This behavior is very close to that expected for a second-order transition, as now discussed.

A phase transition is said to be first order if the volume and entropy (which are first-order derivatives, with respect to pressure or temperature, of the thermodynamic Gibbs function) change discontinuously. Crystallization is the prime example of a first-order transition: The volume discontinuity at T_f is illustrated in Fig. 1.1, the entropy discontinuity corresponds to the heat of fusion. A transition which is continuous in V and S is said to be second order or higher, *n*th order if the *n*th derivatives of the Gibbs function are the lowest ones to show a discontinuity. A second-order transition thus involves discontinuities in quantities such as the slopes of $V(T)$ and $S(T)$, i.e., $\alpha(T)$ and $C_P(T)$.

From the above definition and from the behavior observed near T_g in Figs. 1.1, 1.2, 1.8, and 1.9, it may be seen that characteristic properties of the glass transition closely resemble a second-order thermodynamic transition. While $V(T)$ is continuous through the vicinity of T_g (Fig. 1.1), $\alpha(T)$ and $C_P(T)$ definitely change their values upon passing through this region (Figs. 1.2, 1.8, 1.9). However, these changes are not sharp (as they should be in a true second-order transition) but instead are *diffuse*, occurring over a small temperature interval rather than at a single sharply definable temperature. Thus the kink in $V(T)$, separating the undercooled liquid from the amorphous solid in Fig. 1.1, is rounded, so that the corresponding step in $\alpha(T)$, while steep, is not a vertical discontinuity. Similar statements apply to the bend in the entropy function $S(T)$ and to the corresponding step in the specific heat $C_P(T)$. Nevertheless, the steps exhibited near T_g in Figs. 1.8 and 1.9 are certainly quite clear and pronounced. It is therefore convenient, in order to further distinguish the solidification transition which occurs at T_g from that which occurs via crystallization (in a true first-order transition) at T_f, to phenomenologically characterize the liquid $\leftrightarrow$ glass transition as an *apparent, diffuse, second-order transition*. This characterization is a useful mnemonic that helps to flesh out the continuous/discontinuous dichotomy already noted with respect to the phase-change events at T_g and T_f. It also emphasizes the fact that, in thermodynamic measurements such as those shown in Figs. 1.2, 1.8 and 1.9, the glass transition is well defined.

The above discussion dwells on thermodynamic aspects of the glass transition. Kinetic aspects of this transition are also important. The influence of $\dot{T}$ on the observed transition, illustrated previously in Fig. 1.2, is the clearest proof that the event actually observed near T_g differs from a strict second-order transition. The shift of T_g with $\dot{T}$, understood (as discussed in Section 1.1) in terms of the interplay between the time scale of the experiment and the kinetics of molecular recovery, is the main manifestation of the kinetic dimension of the glass transition. But there are other effects, some of which can be quite subtle. For example, it is typically found that bringing a glass to a temperature just below T_g—and holding it there for differing lengths of time— produces definite changes in the $C_P(T)$ anomaly observed on subsequent heating. Thus the small peak that is superimposed on the step seen in C_P for As_2S_3 in Fig. 1.8 can be made more pronounced, in this way, by long annealing times near T_g. Also, the $V(T)$ characteristics represented in Figs. 1.1 and 1.2 follow the contraction of the material upon cooling. There is a small but real asymmetry between contraction and expansion, so that a slightly different shape (describable as an "overshoot" effect) is sometimes seen upon heating as $V(T)$ turns the corner at T_g.

The confluence of both the thermodynamic and the kinetic dimensions of the liquid ↔ glass transition presents one of the most formidable problems in condensed matter physics, and it cannot be said that there yet exists a completely satisfying theoretical description of the glass transition. The two best-known theoretical treatments, both of which adopt an equilibrium thermodynamic viewpoint, are the polymer-configuration model of Gibbs and DiMarzio (1958) and the free-volume model of Turnbull and Cohen (1961, 1970). A discussion of the free-volume model (which may be more general, and is certainly simpler, than the configurational-entropy theory designed specifically for polymers) is presented in Chapter Five, in a recent format that makes use of percolation theory.

Although kinetic effects are important in all practical observations of the transition between the amorphous solid and the liquid phase, both the Gibbs–DiMarzio and the Turnbull–Cohen theories previously cited reflect the following prevalent view of the glass transition: There exists an *underlying thermodynamic transition* that is intimately related to T_g. While kinetics intervenes to influence the placement of T_g in a particular set of experimental circumstances, T_g never strays very far from the value corresponding to the underlying transition.

Strong empirical support for the idea that the observed glass transition is the kinetically modified reflection of an underlying equilibrium transition (which is presumed to be close to second order in character) is contained in data of the type displayed in Fig. 1.10. The curve shown here, obtained by integrating $C_P(T)$ specific heat data (such as that shown in Figs. 1.8 and 1.9) for the hydrogen-bonded glass-forming liquid $H_2SO_4 \cdot 3H_2O$, tracks the entropy $S(T)$ of the liquid from the melting point down to T_g. The quantity plotted is the excess entropy $S_{ex} = S_{liq} - S_{xtl}$, the amount by which the entropy exceeds that of the crystal at the same temperature. As the temperature decreases from T_f to T_g, S_{ex} drops sharply, which comes from the fact that, just as in Fig. 1.8, the specific heat (and thus dS/dT) of the liquid is substantially larger than that of the crystal. Data are not shown here for the glass ($T < T_g$), but since we know that both the crystalline and amorphous solids have nearly equal specific heats (e.g., Fig. 1.8), it follows immediately that S_{ex} levels off at low temperatures to a nearly constant value close to $S_{ex}(T_g)$.

The significant point about Fig. 1.10, the key feature which bears materially on the question of an underlying thermodynamic transition, is the dashed line that extrapolates the liquid curve below T_g. This part of the curve, of course, was not measured, solidification having occurred when the temperature decreased past T_g. However, were we to adopt a view of the glass transition as a purely kinetic phenomenon, then it would be possible to probe the dashed curve by shifting T_g to lower temperature and thereby extending the life of the liquid phase. To do this, one would choose a slower cooling rate so that a lower temperature would be reached before the microscopic time scale (τ, as schematically at the top of Fig. 1.1) became commensurate with the ex-

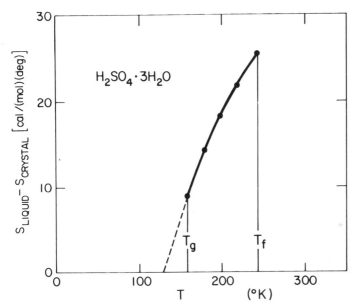

Figure 1.10 The excess entropy of a glass-forming liquid in the region between T_f and T_g, showing the extrapolation to zero excess entropy at a temperature near T_g (Angell and Sichina, 1976).

perimental time scale. Assuming infinite time at one's disposal (as well as an infinite barrier to crystallization, so that this alternative channel to the solid state is closed off), it would be possible to continue this procedure indefinitely, downshifting T_g as much as one liked and extending the liquid line down along the dashed curve of Fig. 1.10. But Fig. 1.10 shows that, even "in principle," T_g *cannot* be downshifted indefinitely. At a temperature not far from the observed T_g in Fig. 1.10, and still quite far from $T = 0$, the extrapolated entropy of the liquid becomes less than that of the crystalline solid: $S_{ex} < 0$. Since the existence, at the same temperature, of an amorphous phase (here, the liquid) with lower entropy than the stable crystalline phase is physically implausible, the extrapolated curve *cannot* be followed beyond the temperature $T(S_{ex} = 0)$ at which the excess entropy vanishes. Let us denote the vanishing-excess-entropy temperature, $T(S_{ex} = 0)$, by T_0. As the liquid is cooled, something "has to give" before this temperature is reached; the liquid phase cannot exist below T_0. What "gives" is the glass transition; the liquid solidifies to the glassy phase, and the plummeting entropy is arrested at a small positive value vis-à-vis the crystal.

Recognition of the liquid ↔ glass transition as an *entropy crisis* was long ago appreciated by Kauzmann (1948), and the low-temperature ($T < T_0$) extrapolation of the liquid to the physically unacceptable condition of negative excess entropy ($S_{liq} < S_{xtl}$) was later mislabeled as Kauzmann's "paradox."

Nature averts the "paradox" by interceding with solidification at T_g, thereby putting an end to the liquid state at a temperature above T_0. T_0 may be regarded as a limiting value that sets a lower bound (no matter how slow the cooling rate) on T_g. Thus do thermodynamic constraints strongly limit the influenced of kinetic effects upon T_g since $T_g \rightarrow T_0$ as $\dot{T} \rightarrow 0$. (As Gibbs and DiMarzio put it in 1958: "The existence of glasses is *not* dependent on kinetic phenomena.") Moreover, since the excess entropy of the glass must itself remain nonnegative at low temperature, the $S_{ex}(T)$ characteristic must turn *sharply* from its steep behavior above T_g (the liquid characteristic illustrated in Fig. 1.10) to its flat behavior below T_g. In other words, the glass transition cannot be *arbitrarily* gradual; it must be sharp enough to turn the corner formed by the intersection of the dashed curve with the horizontal axis in Fig. 1.10. This is why the steps [in $C_P(T)$ at T_g] in Figs. 1.8 and 1.9 are so well defined.

To summarize the argument, the presence of $T(S_{ex} = 0) \equiv T_0$ just below T_g implies the existence of a thermodynamic transition underpinning the observed liquid $\leftrightarrow$ glass transition. The liquid phase simply cannot survive below T_0, and the inevitable solidification takes place at T_g which is shifted upward from T_0 by the mediation of kinetic effects. Note that the residual entropy of the glass is small, not substantially larger than that of the crystal. Although the configurational arrest that occurs at T_g freezes the liquid into one specific amorphous structure selected from a large number of similar structures, these structures are mutually inaccessible below T_g. Upon unlocking the structure of the glass by heating back into the liquid state above T_g, the equivalent disordered structures recover their mutual accessibility (via the diffusive motion of Fig. 1.7*b*), and the entropy increases accordingly.

In this discussion, entropy has been taken to be the determining thermodynamic function. Earlier, volume (or density) was used to introduce T_g in Figs. 1.1 and 1.2, because V is a much simpler quantity than S and is more easily accessible experimentally. An extrapolated temperature $T(V_{ex} = 0)$, a temperature of vanishing excess volume defined in analogy to $T(S_{ex} = 0) \equiv T_0$, is a less suitable benchmark than T_0 for the underlying transition. Nevertheless, there is a connection between V_{ex} and S_{ex}, as can be seen by the following crude argument. Model the condensed system by N atoms distributed among $N + m$ identical cells in the usual combinatorial shell game. The number of ways of placing the atoms among the cells (one or none in a cell) is $W = (N + m)!/(N!m!) = N(N - 1)(N - 2) \cdots (N - m)$. Very crudely, for $N \gg m \gg 1$, $W \sim N^{m+1}$ so that $S = k \ln W \sim k(m + 1) \ln N. \sim km \ln N$. Thus S is roughly proportional to m. If we assume that thermal expansion increases the volume by increasing m (i.e., by adding more "holes"), then this proportionality relates S to the increase in V. Interpreting $m = 0$ as the high-density low-entropy locked-in state, this relation becomes $S_{ex} \propto V_{ex}$. Hence, in this drastically simplified picture (in which S amounts to the "entropy of mixing" of filled and empty cells), the excess entropy and the excess volume are closely connected. This connection (though not the simple proportionality) persists in more realistic models.

There is another facet to T_0, which concerns the viscosity η, a property of the liquid whose steep increase with decreasing temperature is often measured in the approach to the glass transition. The fact that η can be adjusted continuously from enormous (solidlike) values near T_g to moderate values at somewhat higher temperatures (a viscosity range inaccessible in the melting of a crystalline solid at T_f, in which the solid transforms abruptly to a mobile low-viscosity liquid) is of great practical importance in the technological applications of amorphous solids. (Glassblowing is the most familiar example of the usefulness of the ability to "tune" the viscosity of the melt.) Since such applications are about to be discussed in the following section, it is appropriate to make the phenomenological connection to viscosity at this point. For many materials, over much of the temperature range from T_f to T_g, the following relation has been found to describe the steep (by $> 10^{12}$) rise in viscosity with falling temperature: $\eta = A \exp[B/(T - T_0)]$. This empirical equation for the temperature dependence of the viscosity is known as the Vogel–Fulcher equation, and it contains a characteristic temperature T_0 (at which η diverges) that is found to correlate well with the T_0 that has been discussed in this section.

1.5 APPLICATIONS OF AMORPHOUS SOLIDS

Ever since (and even before) the Phoenicians built a large export-oriented glassworks industry on the quartz-rich sands of the Lebanese coast, technological uses of amorphous solids have been a factor in human affairs. Books can be (and have been, e.g., Rawson, 1980) written on the applications of glasses. The quick survey given in this section is intended to convey a sense of the scope and variety of these technological uses.

Table 1.2 presents a representative list of present-day applications (or potential applications, at this writing, in the case of the last two entries) of amorphous solids. The application itself is given in the third column of the table. The first two columns show the type of glass used in the application and a specific example of such a material, while the last column indicates properties of the solid that are exploited in the application.

Traditionally, the most familiar format is as a structural material, ordinary "window glass." This architectural usage of a variety of glasses that are based on fused silica (SiO_2) exemplifies two general aspects common to many applications of amorphous solids. In the first place, as alluded to at the end of the last section, these materials *harden continuously* with decreasing temperature near T_g. The ability to control the viscosity (and thereby the flow properties) of the melt provides a valuable processing advantage in the preparation of products formed from glasses. In this way *the phenomenon of the glass transition is itself technologically significant.* Secondly, when an amorphous solid can be used in place of a crystalline one in an application calling for large-area sheets or films, it is generally advantageous to do this and thereby avoid the problems associated with polycrystallinity or the expense of preparing large single crys-

TABLE 1.2 Some examples of applications of amorphous solids

Type of Amorphous Solid	Representative Material	Application	Special Properties Used
Oxide glass	$(SiO_2)_{0.8}(Na_2O)_{0.2}$	Window glass, etc.	Transparency, solidity, formability as large sheets
Oxide glass	$(SiO_2)_{0.9}(GeO_2)_{0.1}$	Fiber optic waveguides for communications networks	Ultratransparency, purity, formability as uniform fibers
Organic polymer	Polystyrene	Structural materials, "plastics"	Strength, light weight, ease of processing
Chalcogenide glass	Se, As_2Se_3	Xerography	Photoconductivity, formability as large-area films
Amorphous semiconductor	$Te_{0.8}Ge_{0.2}$	Computer-memory elements	Electric-field-induced amorphous $\leftrightarrow$ crystalline transformation
Amorphous semiconductor	$Si_{0.9}H_{0.1}$	Solar cells	Photovoltaic optical properties, large-area thin films
Metallic glass	$Fe_{0.8}B_{0.2}$	Transformer cores	Ferromagnetism, low loss, formability as long ribbons

tals. Thus it would be prohibitively expensive to fabricate large windows out of crystalline SiO_2 (quartz), while it is eminently feasible to do so using SiO_2-based silicate glasses.

Of course, besides the above general advantages, which have to do with practicality of preparation, it is also true that silicate glass is far superior to crystalline quartz in this application, which is the first entry in Table 1.2. Although both materials are transparent to visible light, the glass is optically isotropic while quartz is optically anisotropic—a drawback in a window material. Moreover, the glass is a far better thermal insulator, and a window should keep heat and cold out as well as let light in. Both the optical isotropy and the low thermal conductivity are the consequences of disorder in a nonmetallic amorphous solid. Isotropy results because no special (i.e., symmetry) directions survive the loss of long-range order, while low thermal conductivity results because of the disorder-induced scattering of thermal phonons.

The Phoenicians are credited with being the first to fabricate transparent glass, and the second application entered in Table 1.2 represents a modern development of their achievement which carries it to a phenomenal level. The transparency of the glasses developed for fiber-optic communications is so great that, at certain wavelengths, light can pass along a path within the material that is a kilometer long and yet retain over 90% of its intensity.

Copper wires transmitting electrical signals presently carry most of the telephone messages transmitted around our planet, but it seems likely that this function will someday be carried out largely by glass fibers transmitting optical signals. Bell himself, in 1880, demonstrated a "photophone" in which he used the voice-induced motion of a reflecting diaphragm to modulate a light beam and transmit a signal through the air to a photoreceptor 200 meters away. At that time, the absence of a convenient low-loss conduit for the light (the atmosphere is unreliable as an optical-transmission medium) prevented light-wave communications from progressing further. What has made this idea practical now is the availability of glass fibers of such purity and homogeneity that their optical attenuation is so low (<0.3 decibels/kilometer) as to make them ideal conduits for the light.

Essential aspects of the use of glass fibers in telecommunications networks are schematically shown in Fig. 1.11. Electrical pulses produced by pulse-code modulation from speech in a telephone system (or digitally encoded pulses originating from a computer in other kinds of information-transmitting systems) are converted into a similar sequence of light pulses by a semiconductor laser or light-emitting diode coupled to one end of the optical fiber. The signal is transmitted along the fiber as a stream of light pulses, and at the other end it is converted back into electrical pulses and then finally into the desired form (speech, printed output, computer file, etc.).

The hairlike glass fiber, about 100 μm in diameter, functions as a light-guide—a waveguide in the optical portion of the electromagnetic spectrum. The simplest type is illustrated at the upper left of Fig. 1.11. A central core of ultratransparent glass is sheathed by a coaxial cladding of a glass having a

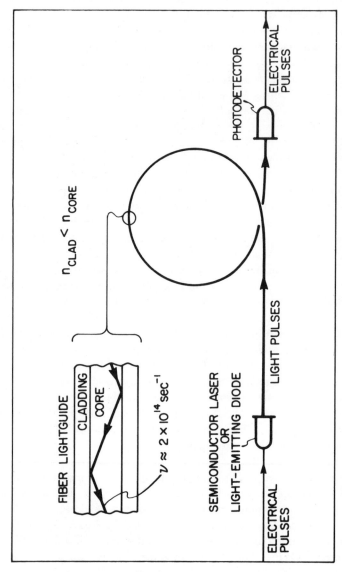

Figure 1.11   The use of ultratransparent glass fibers in telecommunications networks.

refractive index lower than that of the core. Because $n_{\text{cladding}} < n_{\text{core}}$, light rays propagating within the core at small angles relative to the axis are totally internally reflected at the core/cladding interface and remain confined within the lightguide core. Typical materials used for the core are silica–germania glasses such as the one listed in Table 1.2, and the cladding can be a similar glass of somewhat different composition. Careful control of composition makes it possible to construct "graded-index" fibers in which the refractive index falls off radially from the core axis in a predetermined profile designed to keep the light pulses from spreading out as they travel down the guide. In the simple stepped-index configuration shown in Fig. 1.11, different optical modes would propagate with different velocities along the fiber (sharply angled zigzag paths are longer, and travel slower, than zigzag paths more parallel to the axis). Graded-index fibers are tailored to elegantly cancel out this "mode dispersion," so that the light pulses are transmitted down the guide without distortion.

The substitution in communications networks of optically transmitting glass wires for electrically conducting crystalline wires is driven by several advantages of the fiber lightguides. Light losses are so low that "repeater" spacings can be much larger than for metal-conductor systems. ("Repeaters" are receiver/transmitter elements that periodically rejuvenate the pulses along long lines.) More fibers (even clothed in their protective polymer packaging) than copper wires fit into a cable of given size. Optical fibers also have negligible crosstalk and are immune to electromagnetic interference. But by far the most significant advantage of an optical communications system is its information-transmitting capacity, which is far higher than that of a conducting system. This high capacity derives from the very high frequency (about 2×10^{14} sec^{-1} for operation at a wavelength of 1.5 μm, a typical wavelength in the near-infrared ultratransparent "window" of the glass) of the lightwave carrier, which allows it to be modulated at very high frequencies. Much more traffic (e.g., many more simultaneous conversations) can be borne by a glass fiber than by its copper counterpart.

The applications discussed thus far involve oxide glasses in a classic tradition. But the most ubiquitous amorphous solids in present-day society are organic glasses, polymeric solids composed of intermeshed long-chain organic molecules. These materials are represented by the third row in Table 1.2. Many of the innumerable "plastic" products in everyday use belong to this class of amorphous solid, which is now the most technologically pervasive type of glass. (On a volume basis, more plastic is now produced than steel.) Among the reasons for this enormous prominence of organic polymers as structural materials are ease of processing on a large scale (again capitalizing on thermal behavior near T_g), low cost, light weight, and high mechanical strength.

In the communications link sketched in Fig. 1.11, the role played by the glass element is one of transmission. The photoelectronic elements that bracket the optical fiber, the semiconductor-laser transmitter at one end and the photodetector receiver at the other, are both based on crystalline semiconductors. In the next group of three applications listed in Table 1.2, semiconducting amorphous solids play an electronically active role.

Figure 1.12 displays the process of xerography, the most widely used form of electrophotography or electrostatic imaging. This process is the basis of most plain-paper copying machines used in offices, libraries, schools, etc., and at the heart of the process is a large-area thin-film photoconducting element (shown shaded in Fig. 1.12), which is an amorphous solid. Typical materials employed as the photoconductor are chalcogenide glasses such as Se or As_2Se_3, formed by vapor condensation on a metal substrate. These glasses are semiconductors with bandgaps of about 2 eV, transparent in the infrared but highly absorbing for visible light (photon energy $h\nu$ in the range from 2 to 3 eV).

The steps in the xerographic (literally "dry writing") process are schematically represented in the four-part sequence of Fig. 1.12. In the dark, the photoconductor is a good insulator, and while in this state its surface is uniformly charged by ions from a corona discharge, as indicated in Fig. 1.12a. An equal and opposite induced charge develops at the metal–photoconductor interface. The chalcogenide film, about 50 μm thick, supports an electric field in excess of 10^5 V/cm.

The imaging step is shown in Fig. 1.12b. The photoconductor plate is exposed to a pattern of visible light in the form of an image reflected from the document being copied. Where light strikes the photoconductor, photons are absorbed and electron–hole pairs are created by excitation across the energy gap. Photogeneration of mobile charge carriers is assisted by the large electric field present, which helps to pull apart the mutually attracting electron and hole of each light-created pair so that each charge is free to move separately. The electrons then move under the influence of the field to the surface, where they neutralize positive charges located there, while the holes move to the photoconductor–substrate interface and neutralize negative charges located there. Where intense light strikes the photoconductor, the earlier charging step of Fig. 1.12a is totally undone; where weak light strikes it, the charge is partially reduced; and where no light strikes, the original electrostatic charge remains on the surface. The optical image has been converted into a latent image consisting of an electrostatic potential distribution that replicates the light and dark pattern of the original document.

To develop the electrostatic image, fine negatively charged pigmented particles are brought into contact with the plate. These "toner" particles are attracted to positively charged surface regions, as shown in Fig. 1.12c. The toner is then transferred, in a step illustrated in Fig. 1.12d, to a positively charged sheet of paper. Brief heating of the paper fuses the toner (an organic glass) to it and produces a permanent photocopy. To prepare the photoreceptor for the next copy, remaining toner is cleaned off and the residual electrostatic image is erased (i.e., discharged) by flooding with light. In high-speed duplicators the photoconductor layer is usually in the form of a moving continuous drum or belt, around the perimeter of which are located stations for performing the functions of Fig. 1.12. The advantage of an amorphous semi-

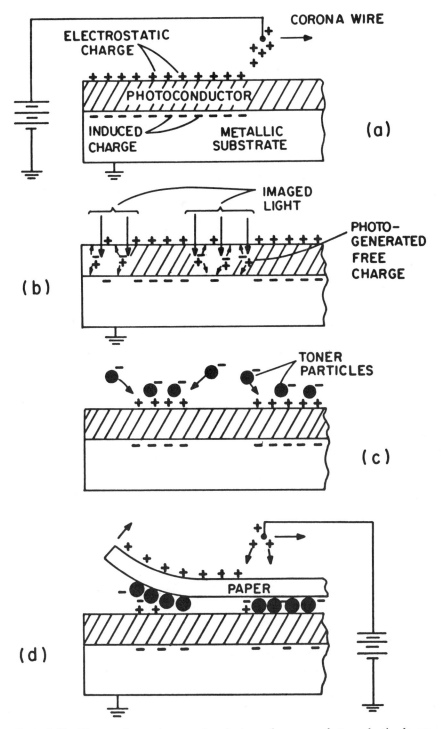

Figure 1.12 The use of amorphous semiconductors as large-area photoconducting layers in xerography. Steps in the xerographic process include (a) charging, (b) exposure, (c) development, and (d) transfer.

conductor for the photoconductor layer in this application resides largely in the ability to prepare uniform large-area thin films of amorphous materials. The speed of xerographic printing technology is presently on the order of several copies per second.

The application of tellurium-rich semiconducting glasses as computer-memory elements, the fifth entry in Table 1.2, exploits a phenomenon specific to amorphous materials. This is electric-field-induced crystallization of the glass. For Te-rich Ge-Te glasses, the crystal→glass transformation results in a large ($\times 10^6$) increase in electrical conductivity because, although both the crystalline and amorphous forms are semiconductors, the bandgap of the crystalline form is very small in this system. A current pulse converts the low-conductivity glass to the high-conductivity crystal; in computer terms, the pulse "switches" the material from the "off" to the "on" state (Adler, 1977; Ovshinsky, 1968). What is interesting is that the "off" state may be restored by another short pulse of current; the glass↔crystal transition is electrically reversible for these materials.

Although not yet completely understood, the reversible switching phenomenon may be visualized in the following way. The off→on glass→crystal transition results from heating the glass to just above T_g, where crystallization occurs readily for this system, and produces filamentary conducting paths of crystalline material which span the film thickness. The current pulse which triggers the on→off crystal→glass reverse transition does this by melting the crystalline filaments (taking them above T_f). The molten filaments are then quenched (taken below T_g) so quickly by the surrounding material that the glass is re-formed, returning the film to the low-conductivity off state. (This description suggests the possibility of using short laser pulses in place of current pulses, and such optically driven switching is indeed an interesting variation of this phenomenon.) Devices based on these films are used as electronically alterable ("read-mostly") memories in computer applications.

The ultimate application of large-area photoreceptor films is clearly in the field of solar energy. If solar-cell technology is ever to contribute an appreciable fraction to worldwide energy production, many hundreds of square miles of the planet's land area will need to be turned over to such "sun-harvesting" activity (Mort, 1980). Amorphous materials, such as polymers, immediately suggest themselves for an application calling for large-area films in such quantity. The example given in Table 1.2 is amorphous silicon, an interesting long-term candidate for solar-cell technology.

Crystalline silicon, of course, is of overwhelming importance in the electronics industry. While crystalline silicon photovoltaic cells are used to power equipment on space-probe vehicles, this material is too expensive for use on the scale envisaged here. Amorphous silicon, on the other hand, can be prepared in large-area films at a far lower cost than its crystalline counterpart. Moreover, a very thin film (~ 1 μm) of a-Si suffices to absorb most of the solar-spectrum light, while a much thicker film (> 50 μm) of c-Si is required for this purpose. (*Note*: When the amorphous and crystalline forms of a given solid are discussed side by side, the prefixes *a-* and *c-* will often be used to distinguish them.) The reason for the much higher optical absorption of *a*-Si is the disorder-induced breakdown of the optical selection rules that control interband transitions in the crystal. (This consequence of disorder is described in Chapter Six.)

As noted in Table 1.2, the material of interest with respect to solar-cell technology is not pure amorphous silicon but is instead an amorphous alloy containing appreciable hydrogen (usually written *a*-Si:H). The superiority, as an electronic material, of hydrogenated amorphous silicon over the pure material, is an interesting scientific story—to be discussed in Chapter Six. In brief, the role of hydrogen is to eliminate electronic defects ("states in the gap") which are intrinsic to elemental *a*-Si.

The final entry in Table 1.2 is one of the applications of metallic glasses formed by the melt-spinning technique of Fig. 1.3. One of the attractions of these metal glass ribbons is that they are cast directly from the melt in a swift, single-step process. Conventional methods needed to manufacture crystalline metal in thin-strip form call for a complex sequence of steps involving casting, hot-rolling, cold-rolling, and annealing, and consume about five times as much energy as melt spinning of glassy ribbon. One of the earliest (circa 1978) commercial uses of these ribbons was in the form of a nickel-based brazing alloy employed in building aircraft engines. However, the main potential of these glasses, as indicated in the last line of Table 1.2, is in their use as magnetic materials.

Ferromagnetic glasses such as $Fe_{0.8}B_{0.2}$, $Fe_{0.7}P_{0.2}C_{0.1}$, and $Co_{0.8}Fe_{0.1}B_{0.1}$ combine high saturation magnetization with the useful property of being "magnetically soft" (low coercivity, easily magnetized by small magnetic fields). Crystalline ferromagnets that are magnetically soft are also, unfortunately, mechanically soft, but the glass ferromagnets mentioned here are found to be quite hard. Also, the magnetic glasses are isotropic, and the absence of a crystalline axis of easy magnetization allows the magnetization direction to be rotated at a much smaller energy cost than in crystals. The high (relative to crystalline metals) electrical resistivity of amorphous metals is also helpful in this regard. For these reasons, ferromagnetic glasses are under active development for use in the magnetic cores of power transformers, where their low-loss properties are very important. Other potential applications of amorphous metal magnets are in magnetic disk memories and read/write recorder heads.

It is a mainstream tradition in solid-state physics that much scientific insight follows in the wake of research motivated by the technological importance of a class of materials (as with crystalline semiconductors after the invention of the transistor), and amorphous solids—which had practical utility long before they acquired scientific respectability—fit well into that tradition. The applications listed in Table 1.2, as well as others omitted for lack of room, have contributed greatly to progress in the physics of amorphous solids.

This survey of applications of glasses has contained within it a brief overview of physical properties: mechanical, optical, electronic, and magnetic. It is noteworthy that the paragraph before last casually began with the term "ferromagnetic glasses." The occurrence of *ferromagnetism* among amorphous solids comes as a shock to any physicist who insists upon identifying this distinctive solid-state property as an exclusive consequence of crystalline order. There are also *superconducting* glasses; examples are bismuth, gallium, $Pb_{0.9}Cu_{0.1}$, and $La_{0.8}Au_{0.2}$. Superconductivity and ferromagnetism in the amorphous solid state exemplify a main theme of this book. Mentioned earlier (Section 1.3) but worth repeating, this is the *inessentialness of long-range order in solids.* Having done away with long-range order and translational periodicity, we now go on to analyze the disorder (*and* the remaining order) that characterizes the structure of amorphous solids.

REFERENCES

Adler, D., 1977, *Scientific American* **236**, No. 5, 36.

Angell, C. A., and W. Sichina, 1976, *Ann. N.Y. Acad. Sci.* **279**, 53.

Blachnik, R., and A. Hoppe, 1979, *J. Non-Crystalline Solids* **34**, 191.

Chaudhari, P., W. C. Giessen, and D. Turnbull, 1980, *Scientific American* **242**, No. 4, 98.

Chen, H. S., and D. Turnbull, 1968, *J. Chem. Phys.* **48**, 2560.

Gibbs, J. H., and E. A. DiMarzio, 1958, *J. Chem. Phys.* **28**, 373.

Kauzmann, W., 1948, *Chem. Rev.* **43**, 219.

Kovacs, A. J., J. M. Hutchinson, and J. J. Aklonis, 1977, in *The Structure of Non-Crystalline Materials,* edited by P. H. Gaskell, Taylor and Francis, London, p. 153.

Mort, J., 1980, *Phys. Technol.* **11**, 134.

Ovshinsky, S. R., 1968, *Phys. Rev. Letters* **21**, 1450.

Predel, B., and H. Bankstahl, 1975, *J. Less-Common Metals* **43**, 191.

Rawson, H., 1980, *Properties and Applications of Glass,* Elsevier, Amsterdam.

Turnbull, D., 1969, *Contemp. Phys.* **10**, 473.

Turnbull, D., and M. H. Cohen, 1961, *J. Chem. Phys.* **34**, 120.

Turnbull, D., and M. H. Cohen, 1970, *J. Chem. Phys.* **52**, 3038.

CHAPTER TWO

Amorphous Morphology: The Geometry and Topology of Disorder

2.1 INTRODUCTION: GEOMETRY, CHEMISTRY, AND THE PRIMACY OF SHORT-RANGE ORDER

In traditional solid-state physics, which is to say, crystal physics, the crystal structure is the starting point. And a fine start it is, too; every honest solid-state physicist will readily admit that an enormous amount of valuable information about a given solid is handed to the physicist on a silver platter by the crystallographer who tells him or her of its crystal structure. Consider, for example, the clues that the structure provides about bonding. The coordination number z of a particular atom is the number of nearest-neighbor atoms which surround that atom in the solid. A low coordination number ($z \leq 4$) provides virtually *prima facie* evidence for a dominant role of covalent bonding (with its highly directional character) in coupling nearest-neighbor atoms. Thus for a binary compound AB, such as GaAs or CdTe, encountering the crystalline form in the zincblende structure (in which $z = 4$ for each atom) tells us that covalent bonding is primarily responsible for the cohesion of the solid. That is a simple and familiar example. An example of a more complex AB binary is crystalline AsS, which is discussed in connection with chalcogenide glasses in the next chapter. Here our crystallographer friends tell us of a structure with discrete molecules (As_4S_4 in this case) with strong covalent bonding ($z = 2$ and $z = 3$) within the molecules and weak van der Waals bonding between molecules. The latter part of the last sentence is based on the relatively large atom–atom spacings between the molecular units.

Close-packed structures are symptomatic of the nondirectional forces associated with ionic or metallic bonding (or van der Waals bonding in the case of the solid rare-gas elements of column eight in the Periodic Table). The cesium chloride structure, with $z = 8$, is the solution of the geometric problem of maximizing z for an AB compound subject to the proviso that the nearest neighbors of A atoms are all B atoms (and vice versa). This proviso is an energy-minimizing condition for an A^+B^- salt, and many ionic compounds do indeed adopt the CsCl structure. Some ordered intermetallic alloys (such as beta brass, CuZn) also adopt this structure. Thus the CsCl structure cannot, by itself, be taken to be a structural signature of ionic bonding but, to be distinguished from the metallic case, must be supplemented by a glance either at the Periodic Table (for the relative positions of A and B) or, better yet, at a sample of the solid itself.

The extreme of maximum coordination for an AB solid, in which both types of atom occupy at random the positions of a close-packed lattice ($z = 12$, as described in the next section), obtains in several metallic alloy systems such as AuCu. Each atom is now forced to accept nearest neighbors of *both* types; A–A and B–B contacts can no longer be avoided if close packing is to be achieved.

The above examples, illustrative of the intimate connection between structural and chemical-bonding considerations, provide a glimpse of one dimension of the great advantage which inheres in knowing, *at the outset*, the atomic-scale structure of the solid. Of course, another dimension of the crystallographer's invaluable contribution toward easing the travail of the condensed-matter physicist involves the aspect of symmetry. As noted earlier, the translational and the factor-group symmetries that characterize a crystalline solid introduce enormous simplifications into the theoretical treatment of its electronic and vibrational states. Such symmetry does not enter, at any stage, into the physics of amorphous solids, because it depends on the presence of long-range order.

Note that even for crystals, symmetry and group-theoretical considerations connote little about the *chemical* nature of the solid. This is precisely because the crystal symmetry reflects a long-range, and not a short-range, characterization of the solid's structure. From a chemical-bonding viewpoint, it is the short-range order that is of paramount importance. This is the reason that the short-range order, epitomized here by the coordination z, has been emphasized in the preceding introductory discussion of the structure/chemistry connection for a few crystalline cases. In now moving on to the question of the structure of glasses, the following point cannot be overemphasized. *Short-range order is very much in evidence in the structure of amorphous solids.* Thus the essential structural difference between crystal and glass consists "merely" in the latter's loss of long-range order.

From the viewpoint of the *study* of the atomic-scale structure of the condensed phase, the crystal/glass difference is much more substantial. While for crystalline solids the crystallographic data yield detailed and quite complete information about the equilibrium positions of the atoms, providing an advanced initial outpost from which to progress further in our understanding of the material, for amorphous solids the nature of the atomic arrangement is itself a difficult and hard-to-win objective. The lack of long-range order makes it very hard to extract structural information from the diffraction experiments, as will be described in the last section of this chapter. Nevertheless, much can now be said about the structure of amorphous solids; the picture is no longer as bleak as that portrayed by Zachariasen in the opening sentence of his classic 1932 paper: "It must be frankly admitted that we know practically nothing about the atomic arrangement in glasses."

The picture that has gradually emerged turns out to be concordant with the homogeneous, continuous-random model that was proposed by Zachariasen for inorganic covalent glasses. Three notable types of continuous-random models are described later in this and the following chapter:

1. *Continuous random network*, appropriate to the structure of covalent glasses.
2. *Random close packing*, appropriate to the structure of simple metallic glasses.
3. *Random coil model*, appropriate to the structure of polymeric organic glasses.

Although these models necessarily are, to some extent, idealizations, they represent the best available pictures of the atomic-scale structure of the main classes of amorphous solids. Ample evidence supports these models; some of the relevant experimental results will be shown.

The feature that is common to all of the continuous-random models is that their description involves, in an essential way, the use of *statistical distributions*. This structural aspect of amorphous solids belongs in the realm of what may properly be called *stochastic geometry*, a subject which also includes percolation theory (a subject to be introduced in Chapter Four).

Prior to presenting the structural models characteristic of the amorphous solid state, we first introduce a few of the mathematical objects (parameters, functions, etc.) that are useful in the quantitative description of atomic-scale structure. It is simplest to do this by using a familiar crystal structure as an example, and the face-centered cubic lattice is selected for this purpose. This close-packed structure is also needed to provide the crystalline analog of the first of the continuous-random models to be discussed: the random-close-packed model of simple glassy metals.

2.2 REVIEW OF CRYSTALLINE CLOSE PACKING

Elemental (or monatomic) solids are solids composed of a single type of atom. Among the 100 or so elements of the Periodic Table, approximately 55 (including most metals, as well as the five rare-gas crystals) normally form crystalline solids whose periodic lattices correspond to one or another of the solutions of the following geometric problem: How may we arrange very many identical hard spheres so as to most efficiently fill space? By "hard" spheres we mean that the spheres may touch, but not overlap; by "very many" we mean essentially indefinitely many (or, more to the point, a number comparable to the number of atoms in a "typical" macroscopic solid sample, say 10^{20} or so); and by "most efficiently" we mean an arrangement that maximizes the fraction of space contained within the spheres (or, equivalently, that maximizes the number density of sphere centers per unit volume). The basic solution of this problem (of which, as we will see shortly, there exist an infinite number of closely related variations) is the face-centered cubic (fcc) lattice discovered by Kepler in 1611.

Before describing the fcc lattice, we should describe the simpler analogs of close packing in one and two dimensions. In one dimension a "sphere" is simply a line segment, and one-dimensional (1d) space (an infinite line) is completely filled by equal line segments in end-to-end contact. The two-dimensional (2d) problem corresponds to the most efficient arrangement of equal nonoverlapping circles on a plane, the solution to which is shown by the heavy circles in Fig. 2-1. Each circle contacts six others, and the circles occupy $\pi/(2\sqrt{3})$ = 90.7% of the area of the plane. The lattice formed by the circle centers, marked in the figure by the letter A, is called the triangular lattice and is the unique solution for close packing in two dimensions.

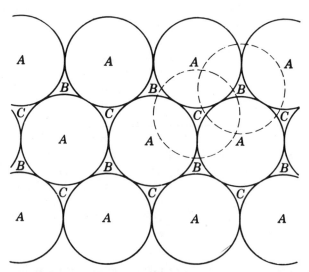

Figure 2.1   Crystalline close packing of spheres. Starting with layer A, the next layer, either B or C, nests above it as shown. The layer sequence $\cdots$ABCABCABC$\cdots$ corresponds to cubic close packing, with the sphere centers arrayed on a face-centered cubic (fcc) lattice.

Close packing in three dimensions can be built up as follows. An indefinitely extended layer of equal spheres, with their centers coplanar, is arranged with each sphere in contact with six neighbors, so that an equatorial section corresponds to the 2d close packing of Fig. 2-1. Now a second similar layer is placed in contact with the first. There are *two* ways in which to obtain the closest approach of the two layers. If the solid circles centered on the A lattice denotes the first layer, then the second layer may approach most closely by positioning its sphere centers over either the B or C set of interstitial sites (interstitial with respect to A) shown in Fig. 2.1. In both cases, each sphere of one layer contacts three of the adjacent layer. If the first layer corresponds to (i.e., centers lie over the sites of) the A lattice and the second layer corresponds to the B lattice, then a third layer can similarly be closely stacked by positioning its centers over the sites of lattice A or C. Continuing this procedure leads to a packing in which each sphere contacts 12 others. This is the most efficient three-dimensional packing. The filling factor for 3d close packing, defined as the fraction of space occupied by the spheres, is $\pi/(3\sqrt{2}) = 74.0\%$. Unlike the 2d case, a "rigorous" proof that there exists no denser 3d packing has not yet been given to the satisfaction of all mathematicians. In a formal mathematical article in which he proved an upper limit for the filling factor of about 0.78 (which still stands as the lowest "rigorous" upper bound for the packing of equal spheres), Rogers (1958) wryly remarked that "many mathematicians believe, and all physicists *know*" that 0.74048 (corresponding to crystalline close packing as described here) is the best packing possible. So we do!

Because of the choice of two possibilities for the addition of each layer in building up the structure, there are an infinite variety of close-packed structures with filling factor 74% and coordination number 12. The layer stacking sequence . . .ABCABCABC. . . has cubic symmetry and is the fcc structure; . . .ABABABAB. . . has hexagonal symmetry and is the hexagonal-close-packed (hcp) structure. Any disordered, nonperiodic stacking can be constructed which is also equivalently close packed.

Thus an unlimited number of crystalline structures can be so generated, with arbitrary repeat sequences. These possibilities are realized in nature in the enormous number of crystalline modifications (termed "polytypes") exhibited by certain layer crystals. Layer crystals are molecular crystals in which the tightly bonded molecular unit is macroscopically extended in two dimensions and of atomic dimensions in the third dimension. PbI_2 and CdI_2 are layer crystals in which the two-dimensionally-extended molecule consists of a sheet of metal atoms sandwiched between two sheets of iodine atoms. The array of iodines that constitutes the "surface" of the molecular unit is a triangular lattice. Starting with one layer (A), a second may be stacked upon it in closest approach of the iodine sheets in two ways (B or C in Fig. 2.1). The interlayer interaction is quite weak, energetic differences between different stacking arrangements are very small, and dozens of distinct polytypes have been prepared. Because the ionic radii of Pb and Cd are so much smaller than that of iodine, the iodines approximate a close-packed array in each of these crystals. Thus these layer crystals provide physical examples of the multifarious variety of crystalline close packing of equal spheres.

To return to elemental crystals, we note that at room temperature the difference in energy is very small between the fcc and the hcp forms of cobalt. Small crystals of cobalt exhibit a close-packed structure with a random stacking sequence of layers. It should be noted that even a random, nonperiodic, stacking sequence is quite highly ordered: It possesses trigonal C_{3V} symmetry about axes perpendicular to the layers, as well as translational periodicity in the two dimensions parallel to the layers.

2.3 PARTIAL CHARACTERIZATIONS OF STRUCTURES

This is a convenient point to introduce, using for purposes of illustration the just-discussed fcc structure, several of the ways in which solid lattices are quantitatively or numerically characterized. In this discussion the term *lattice* will be taken to mean the geometric array of sites or points corresponding to the equilibrium positions about which the atoms of a given solid oscillate. In our usage of lattice the site array is considered in a general sense, devoid of periodicity or symmetry considerations since these aspects are absent for amorphous solids.

A full and complete specification of the lattice naturally consists of the coordinates of all sites. For a crystalline solid, this is possible to do and (assuming, as usual, a perfect crystal) is very economically accomplished by specifying the small number of locations contained within a single unit cell, along with the three translation vectors that can be repetitively employed to generate the full structure from the unit-cell basis. Analogously complete information is out of the question for an amorphous solid. We must therefore be content with incomplete information about the lattice, information which typically is statistical in character. Several such partial characterizations of structure will now be described.

2.3.1 Coordination Number

For a crystal, the simplest single numerical parameter that provides a partial characterization of the lattice structure is the coordination number z, the number of nearest neighbors. This quantity is 12 (the largest value possible for z) for the many metallic elements which crystallize in the fcc structure, as well as for solid argon, krypton, neon, and xenon, which adopt the same lattice. For crystalline compounds, the single value of z is replaced by a small set of numbers specifying the nearest-neighbor environment of each distinct type of site that occurs in the lattice. For example, $z(\mathrm{Na}) = 6$ and $z(\mathrm{Cl}) = 6$ in the rock-salt structure, and $z(\mathrm{Ca}) = 8$ and $z(\mathrm{F}) = 4$ in the fluorite structure. To ease the discussion, we shall continue to refer to "the" coordination number, in the singular.

Although z comprises a ridiculously oversimplified data set in comparison with the full crystal structure, it nevertheless conveys a great deal of chemical and topological information and is the most valuable single piece of structural information. As already mentioned at the beginning of this chapter, information about the nature of the chemical bonding is implicit in z because of its intimate connection with the nature of the short-range order. The close connection between coordination number and lattice topology will be treated in Chapter Four in a discussion of network dimensionality in covalent and molecular glasses. Here it suffices to note that $z = 1$ and $z = 2$ both lead to discrete molecular units embedded in a disconnected lattice structure: $z = 1$ implies diatomic molecules, as in solid hydrogen and crystalline iodine; while $z = 2$ implies closed rings or extended linear chains, as in both the crystalline and amorphous forms of solid sulfur and selenium.

While z is a well-defined and valuable characteristic for a crystal lattice, it might superficially appear to lack utility for characterizing a glass lattice because nearest neighbors occur at various separations. Strictly speaking, each atom in an amorphous solid has exactly one nearest neighbor because of this dispersion in distances. Moreover this does *not* mean that atoms in a glass pair off (as in crystalline H_2, with $z = 1$), because if atom b is the "nearest neighbor" of atom a, the "nearest neighbor" of b could be, say, atom c. If the sides of the triangle formed by the three atoms are ab, bc, and ac, then this situation could locally occur whenever bc < ab < ac. We need not continue this particular line of mathematical reasoning any further, however, because in actuality there *is no such qualitative difference* in the nature of the nearest-neighbor spacings between the two classes of *physical* objects we are considering: crystalline solids and amorphous solids. The experimental evidence for this is contained in the observed width of the first peak (corresponding to nearest neighbors) of the radial distribution function, a functional characterization of structure which will be discussed next. This peak width corresponds to the spread in nearest-neighbor spacings, and has finite width in crystals because of vibrational motion (even at low temperature, because of the quantum-mechanical zero-point component of vibrational energy which is present as a consequence of the uncertainty principle). The basic observation is that this first-peak width in an amorphous solid is very similar indeed to that observed for the corresponding crystal. The basic reason for this insubstantial difference in width between crystal and glass is that the bonding interactions responsible for the existence of a solid provide a powerful determinant of the short-range order. Thus the short-range order is essentially as well defined in the glass as it is in the crystal. In both cases the coordination number may be measured by determining the integrated area under the first peak.

The principal qualitative differences between nearest-neighbor aspects of crystals and glasses were schematically illustrated earlier in Fig. 1.6 and the accompanying discussion. Quantitative characterizations of these differences are developed in the present chapter, the last section of which contains representative experimental results dealing with this important issue.

2.3.2 Radial Distribution Function

For concreteness, let us again consider an elemental solid. By generalizing the idea of a single coordination number to a sequence of numbers embracing "shells" of neighbors at distances beyond the nearest ones, we are led to a more substantial structural characterization called the radial distribution function. To illustrate, we return to the fcc crystal lattice. Let D denote the diameter of the balls which we imagine to be arranged in cubic close packing. D then specifies the nearest-neighbor spacing of the site positions (ball centers) of the corresponding fcc lattice. We now imagine the construction of a very small sphere of variable radius r centered on a particular lattice site, which has been arbitrarily singled out to serve as the origin. As r is increased from 0, no neighboring sites are encountered until $r = D$, at which radius the sphere surface intersects the 12 nearest neighbors. This first *coordination shell* is characterized by the pair of values z_1 and r_1, where z_1 is the number of neighboring sites it intersects ($z_1 = z$, 12 in this case) and r_1 is its radius. Increasing r further leads to an encounter with a second set of sites. For the fcc lattice there are six such next-nearest neighbors at a distance from the origin site of $D\sqrt{2}$; that is, the second coordination shell is characterized by $z_2 = 6$, $r_2 = D\sqrt{2}$. Going on to higher r next picks up the third coordination shell containing 24 ($= z_3$) third-nearest neighbors at a spacing of $D\sqrt{3}$ ($= r_3$) from the origin. Continuing on in this fashion defines a discrete sequence of coordination shells, with the ith shell specified by (z_i, r_i). Figure 2.2 shows a graph of z_i versus r_i for the fcc lattice for the first 15 coordination shells, which accounts for 356 neighboring sites out to a distance of $4D$. (It happens that, up through $i = 13$, $r_i = D\sqrt{i}$ for the fcc lattice. There are no neighbors at $D\sqrt{14}$, and $D\sqrt{30}$ is also missing.) In general, the spacing between shells ($r_{i+1} - r_i$) decreases with increasing r.

The radial distribution function (henceforth RDF), defined for an average atom in a solid, is the generalization of Fig. 2.2. From a random atom as the origin, $\rho(r)dr$ gives the probability of finding a neighboring atom at a distance between r and $r + dr$. For a crystal lattice with each atomic nucleus regarded as clamped to its equilibrium position (lattice site), the RDF $\rho(r)$ is a sum of delta functions, with each term corresponding to a coordination shell:

$$\rho_{\text{clamped xtl}}(r) = \sum_i z_i(r)\delta(r - r_i) \qquad (2.1)$$

As already mentioned, for an actual crystal the spikes in $\rho(r)$ [many researchers use $J(r)$ for the RDF] corresponding to the coordination shells are broadened by thermal and zero-point motion. Despite this blurring, well-defined peaks can be seen in the experimentally determined RDFs for crystals out to about a dozen coordination shells. Beyond that, the shells become too closely spaced (Fig. 2.2) to be resolved.

Direct evidence of the existence of short-range order in glasses, in the form of well-defined nearest-neighbor and next-nearest-neighbor coordination shells, is provided by the presence of the clearly seen first and second peaks in

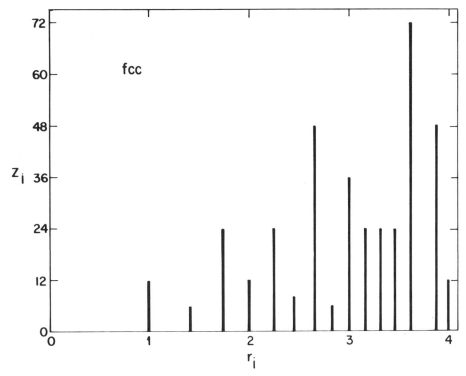

Figure 2.2 Coordination shells for the fcc structure. z_i is the number of atoms centered at a distance r_i from a given atom.

the radial distribution function. On the other hand, the absence of long-range order manifests itself in the fact that, for glasses, discernible peaks in the RDF rarely occur beyond third-nearest neighbors. Similar remarks apply to the RDF of liquids.

For point particles distributed randomly in space with an average number density of $\bar{n}$ particles per unit volume, the RDF follows directly from the fact that the expected number of particles in a shell of volume $4\pi r^2 dr$ is just that volume element times $\bar{n}$:

$$\rho_{\text{random}}(r) = 4\pi\bar{n}r^2. \tag{2.2}$$

This corresponds physically to the RDF of a dilute gas (Fig. 1.6c), for which the mean spacing $\bar{n}^{-1/3}$ dwarfs the atomic size. For condensed phases, interatomic spacings and atomic dimensions are comparable. Thus $\rho(r)$ is essentially zero, $\rho(r) \ll 4\pi\bar{n}r^2$, within a sphere (of radius approximately one atomic diameter) from which neighboring atom centers are excluded because of the repulsion between the core-electron shells. This close-in hiatus is followed by the well-defined nearest-neighbor peak that is centered near r_1, has an integrated area equal to the coordination z, and possesses a peak value $\rho(r_1)$ lying well *above* the smeared-background value of $4\pi\bar{n}r_1^2$. Here $\bar{n}$ is the number density (atoms/cm^3) of the solid. With increasing r, $\rho(r)$ exhibits damped oscillations about $4\pi\bar{n}r^2$, with which it eventually merges. In amorphous solids the convergence of $\rho(r)$ to $4\pi\bar{n}r^2$ is rapid. The utility of the RDF for structural characterization comes down to the amount of information which is extractable from oscillations in $\rho(r)$ before it merges with $4\pi\bar{n}r^2$. A schematic comparison of RDF's representative of crystals, glasses, and gases (in correspondence with the structural sketches shown previously in Fig. 1.6) is contained in Fig. 2.3.

The information about the average density appears in the RDF in the form of the large-r asymptote, the light dashed curve of Fig. 2.3. Another form of this function that is sometimes used is the reduced RDF $g(r)$, which is equal to $(1/r)\rho(r) - 4\pi\bar{n}r$. For $g(r)$ the large-r asymptote is zero, and the average density appears in the form of the initial negative slope.

The radial distribution function is widely used to characterize, albeit incompletely, the structure of glasses. The utility of this characterization derives from the fact that it is derivable, via Fourier transformation, from the results of diffraction experiments. A sketch of the geometry of such experiments is indicated in Fig. 2.4. The incident beam consists of monoenergetic electrons, X-rays, or neutrons at a selected energy E. E is chosen so that the wavelength $\lambda = hc/E$ of the incident photons (in the X-ray experiment), or the de Broglie wavelength $\lambda = h/\sqrt{2mE}$ of the incident energetic particles (in the electron or neutron experiments) is of the order of 10^{-8} cm, comparable to interatomic spacings. The measured quantity is the scattering interference function $I(k)$, where the scattering vector k is related to the observed scattering angle 2θ by $k = (4\pi/\lambda)\sin\theta$. The radial distribution function $\rho(r)$ is then obtained as the real-space transform of $I(k)$. This seemingly straightforward procedure actually involves many subtleties and technical difficulties in both the acquisition and the processing of the data.

The type of scattering experiment schematicized in Fig. 2.4 is often used as a diagnostic test for amorphicity. Crystalline solids, in powdered form or even in fine-grained polycrystalline form, display a diffraction pattern composed of sharp rings. Amorphous solids characteristically reveal rather diffuse bands.

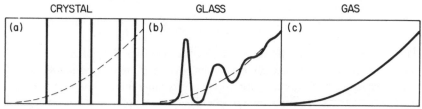

Figure 2.3 Schematic sketches of the radial distribution functions for (*a*) a crystalline solid, (*b*) an amorphous solid, and (*c*) a gas. These distributions schematically correspond to the atomic arrangements sketched in Fig. 1.6.

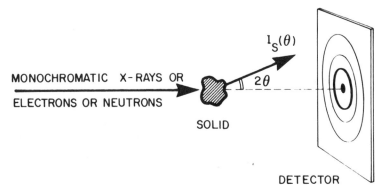

Figure 2.4   Basic geometry of structure-probing scattering experiments.

The RDF provides a structural characterization of unquestioned value. As we have already seen, it yields crucial information about the all-important short-range order and, thereby, the nature of the chemical bonding. It also serves as a key test of different structural models. In this regard, however, it has been demonstrated that a given observed RDF cannot be placed in *unique* correspondence with a definite structural model. Stated somewhat differently, it is found that we can construct a fairly wide range of topologically distinct structural models (but with the same RDF-defined short-range order, of course) of a particular amorphous solid, *all* of which generate RDF's indistinguishable from each other and from the observed RDF. This disappointing state of affairs, discussed further in Chapter Four, permits an uncomfortable amount of ambiguity. The mystery of the structure remains. Nevertheless, the use of the RDF has enormously narrowed the field of possible structures. It has been the detailed analyses of RDF data that have clearly shown the untenability of microcrystallite models for amorphous solids. Conversely, these experiments and their interpretation have demonstrated the applicability of models like those discussed in the remainder of this chapter: random-close-packing models for metallic glasses and continuous-random-network models for covalent glasses.

2.3.3   EXAFS

The availability of intense sources of radiation over broad regions of the X-ray spectrum, most notably the light emitted by the relativistic electrons rapidly circulating along the curved path of a synchrotron beam, has made possible an experimental technique that yields information that remedies a gap left by the scattering techniques discussed above. The technique is commonly referred to by the acronym EXAFS, standing for Extended X-ray-Absorption Fine Structure.

The RDF obtained from a conventional scattering experiment suffers from the limitation that it describes the environment of an *average atom* in the solid. This is adequate for an elemental solid, but for any heteroatomic solid this average aspect loses valuable information about the nature of the chemical correlations and interatomic bonding. Consider the prototype covalently bonded chalcogenide glass As_2S_3. In this material, the coordination numbers corresponding to the actual short-range order are $z(As) = 3$ and $z(S) = 2$. The result obtained as the area under the first scattering-derived RDF peak is 2.4, the weighted average. It is here that the EXAFS technique makes its special contribution; it makes it possible to directly investigate the individual contributions of each type of atom present in the solid. (We would like to be able to go even further and to directly determine, for each atomic species, not only the number of nearest neighbors and their radial location but also their chemical identity. However, it is still necessary to *infer* this type of specific information from further analysis.) By ''directly investigate'' we mean that the information of interest is obtained as a generalized Fourier transform of the experimentally measured quantity. With the EXAFS technique we can so obtain, instead of the average-atom $\rho(r)$ RDF which comes out of electron-scattering experiments or X-ray scattering experiments, the individual-component RDFs $\rho_{As}(r)$ and $\rho_S(r)$ corresponding to the separate geometric environments seen by arsenic atoms and by sulfur atoms. In the case of As_2S_3, we would find nearest-neighbor first peaks centered at the same value of r for both functions. However, the integrated area $\rho(r)dr$ under the first peak would be 3 for $\rho_{As}(r)$ and 2 for $\rho_S(r)$. As_2S_3, in both its crystalline and amorphous forms, is characterized by a short-range order in which each arsenic atom is covalently bonded to three nearest-neighbor sulfur atoms and each sulfur is bonded to three neighboring arsenics.

The EXAFS technique is simply an optical absorption experiment which is carried out in the X-ray portion of the spectrum. Optical absorption is measured as a function of photon energy. With increasing photon energy, an edge or sharp rise in absorption is observed when the incident photon has sufficient energy to ionize an atom in the solid by exciting an electron out of a deeply bound core state into an unbound free-electron-like state. In the high-absorption plateau above the edge, fine structure is seen in the form of an oscillatory behavior of absorption with varying photon energy. It is this spectral fine structure that contains the structural information, playing a role analogous to that of the interference function discussed earlier in connection with diffraction studies of structure. It is the fact that the spectral location of a particular X-ray absorption edge is specific to a particular type of atom which allows the EXAFS method to separately probe the environments of the different types of atoms in a solid.

Although the EXAFS technique is operationally rather different from the scattering techniques used for structural investigations, the underlying physics is essentially similar. This is so because the spectral fine structure seen in EXAFS is basically a manifestation of electron diffraction on an atomic scale. The electron emitted from the absorbing atom travels outward, encounters neighboring atoms, is scattered by these back toward the excited atom, and there its wave-function amplitude is modified by the interference between the outgoing and incoming waves. This modification of the excited-state wave-function amplitude at the position of the absorbing atom modifies, in turn, the magnitude of the matrix element for the photon absorption process. At a given photon energy the interference depends on the lattice structure via the path-length excursion for the returning waves, and for a given structure it varies with photon energy via the effective wavelength of the emitted electron. Thus EXAFS amounts to a diffraction technique in which the incident-wave probe is generated *inside* the sample at the sites of atoms of one selected type.

2.3.4 Froth—The Honeycomb of Aggregated Atomic Cells

The idea is known to mathematicians as a Voronoi polyhedron or a Dirichlet region and to physicists as an atomic polyhedron or a Wigner-Seitz cell. It describes the partitioning of space into cellular neighborhoods which is implied by, and corresponds to, a lattice of point sites. Thus far we have focused on the sites as the equilibrium positions of the atomic nuclei. It is well to remember that in a solid an appreciable fraction of the total volume is occupied by the core electrons, the extent of whose wave functions ordinarily provide the limit on the closest approach between atoms. The cellular characterization of a lattice structure provides a point of view that is complementary to the previously described viewpoint with its attention to attributes of the sites (e.q., coordination number, RDF). The cellular viewpoint is a natural one to adopt for close-packed structures. We introduce it here because it turns out to be conceptually useful for illustrating the topological properties of structural models for amorphous metals.

The atomic polyhedron or Wigner–Seitz (WS) cell surrounding a given site is defined as the polyhedron whose interior contains all points which are closer to that site than to any other. Since each plane surface forming a boundary of the cell is the locus of points equidistant from two lattice sites, the cell can be constructed by drawing the planes that perpendicularly bisect each line (or "bond," in a terminology to be discussed near the end of this chapter) connecting the chosen atom and its neighbors. For the fcc lattice with its 12-fold coordination, the atomic polyhedron is a rhombic dodecahedron. This particular polyhedron is probably most familiar as the first Brillouin zone for metals with the body-centered cubic (bcc) structure. (The reciprocal lattice for a bcc real-space lattice is an fcc lattice in k space, and the Brillouin zone bears the same relation with respect to a site of the reciprocal lattice as does the WS cell with respect to a site of the crystal lattice.)

The role of the atomic polyhedron in the Wigner–Seitz treatment of the cohesive energy of the alkali metals is well worth a brief digression here, because of the clear insight their simple picture provides into the nature of metallic binding. Naturally, the same sort of considerations are operative in the binding of amorphous metals. For simple one-electron atoms like the alkalis (one electron outside of closed shells), the key is the nature of the change experienced by the valence electron in going from the free atom to the solid state. For the free atom, the usual Schrödinger-equation boundary condition is the vanishing at infinity of the electronic wave function. In the crystal, for the $k = 0$ wave function at the bottom of the conduction band, this is replaced by the requirement that the normal derivative vanish on the faces of the atomic polyhedron. For the symmetric WS cells of fcc, hcp, and bcc crystals, it is a good approximation to replace the polyhedron by a sphere of equal volume. Call the sphere radius r_0. The radial derivative of the wave function is to vanish at r_0. The effect of this is to allow the amplitude of the wave function at r_0 to be larger than in the free atom (whose wave function is rapidly converging to zero because of the vanishing-at-infinity boundary condition). That, in turn, means that the crystal wave function possesses less integrated curvature than the atomic wave function between $r = 0$ and $r = r_0$. Less curvature means less kinetic energy. This pretty argument is an elegant expression of the lowering of kinetic energy which accompanies the delocalization of electronic wave functions in a solid, an energy lowering that provides the driving force for the formation of a metallic solid from the individual atoms.

More on this type of quantum-mechanical energetics later in connection with covalent bonding, but now we return to geometry, which is the main issue at hand in this chapter. It is clear that the atomic polyhedra enclosing all of the sites of an infinite lattice will completely fill space. Figure 2.5 illustrates the two-dimensional (i.e., polygonal rather than polyhedral) WS cells for a portion of a two-dimensional lattice of irregularly arranged sites. For sites arranged on a triangular lattice, the 2d close-packed lattice corresponding to the circle centers in Fig. 2.1, the WS cells are congruent regular hexagons, which themselves define a 2d lattice called the honeycomb lattice. Mathematicians refer in general to such space fillings by polygons (in two dimensions) or polyhedra (in three dimensions) or polytopes—generalized d-dimensional polyhedra (in higher dimensions)—as *honeycombs*. Other terms for a three-dimensional honeycomb are solid tesselation and (particularly appropriate for an irregular structure) *froth*.

For our standard three-dimensional example, the fcc lattice, all of the atomic polyhedra forming the cells of the honeycomb are identical (a ''regular'' honeycomb). This is a consequence of the crystallographic statement that the fcc structure has one atom per unit cell. In this case the WS cell also serves as one possible choice for the primitive unit cell of the crystal. (In mathematician's terms, the Voronoi polyhedron is here also a parallelohedron capable of filling space by means of operations of the translation group.) For crystals with several symmetrically inequivalent atoms per unit cell, there are several types of atomic polyhedra, but we shall focus for simplicity on the elemental metals. For fcc, hcp, and bcc metals (together, these three structures account for over 80% of the metallic elements), there is only a single shape of atomic polyhedron in each case. As already noted, for the fcc structure it is a rhombic dodecahedron (which will be illustrated in Fig. 2.7). For the hcp structure it is

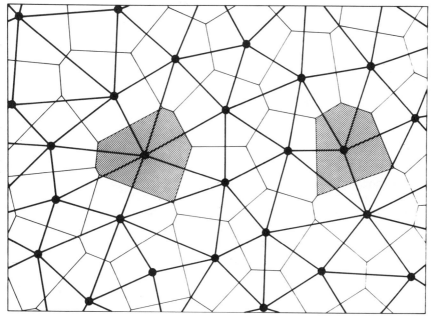

Figure 2.5 Polygonal divisions of a plane defined by an irregular array of sites. The heavy dots denote the atomic sites, the Wigner–Seitz atomic cells of the Voronoi froth are shown by the light lines (two cells are shown shaded), and the heavy lines denote the bonds of the simplicial graph.

a dodecahedron distinct from, although related to, the fcc dodecahedron, while for the bcc structure it is a 14-hedron bounded by eight hexagons and six squares (the ''truncated octahedron'' familiar as the first Brillouin zone for fcc and diamond-structure crystals).

In passing from crystalline to amorphous metals, the single atomic polyhedron characteristic of the former is replaced by a *statistical distribution* of occurrence of distinct polyhedra. The regular honeycomb is replaced by an irregular froth of cells, no two identical. The distribution of WS cells or Voronoi polyhedra can be used to mathematically characterize the topology of the structure, discussed after the following brief digression, which is believed to best represent amorphous metals.

2.3.5 Atomic Polyhedra versus Polyhedral Holes

The reader may already be aware of another set of space-filling polyhedra, distinct from the dodecahedra of the atomic-cell honeycomb, which are sometimes mentioned in connection with crystalline close packing. These are the octahedral and tetrahedral interstitial ''holes'' in the lattice. In order to avoid confusion between these polyhedra and the ones we are considering, we shall pause here to distinguish between the two complementary types of space fillings (or "tilings") in a commentary that anticipates the alternate approach to be adopted later for covalent structures. For concreteness we refer specifically to the fcc lattice, but the same statements apply to hcp and the other close-packed crystal polytypes.

Suppose we wish to insert smaller balls into the 26% of space which is left unfilled in an fcc arrangement of spheres of diameter d. Clearly the largest interstices available for small impurity atoms are to be found between the 2d-close-packed layers of the host lattice. A glance at Fig. 2.1 reveals that there are interlayer locations (which lie directly over host-lattice sites of type A, B, and C) such that: a third of these interstitial sites nest over a triangle of host sites in the lower layer and are directly under a site in the upper layer, another third nest under a triangle of host sites in the upper layer and are directly over a site in the lower layer, and the remaining third nest between a pair of triangles (rotated with respect to each other by 60°) on the two layers. The last set of interstitial locations are those known as the octahedral holes, since each is surrounded by six lattice spheres whose centers form the vertices of an octahedron. Likewise the other interstitial locations present, which are twice as numerous, are the tetrahedral holes; each of these is surrounded by four host spheres, all in contact with each other.

The octahedral holes are the more spacious, being able to accommodate spheres of diameter up to $0.41D$. Interstitial sites of this type are the ones that are occupied by the Pb or Cd ions in the layer crystals PbI_2 and CdI_2 discussed in Section 2.2. Spheres up to $0.22D$ in diameter may fit into the more cramped tetrahedral holes.

The interstitial polyhedra thus defined have edges formed by lines which link sites of the lattice. The edges correspond to the atom–atom bonds which are represented by the sticks in a conventional ball-and-stick model of the crystal lattice. Together these polyhedra—for the fcc lattice two-thirds of them being tetrahedra and the rest octahedra—fill space. (It is important to note that neither tetrahedra nor octahedra *alone* can tile space.)

The bonding network composed of these interstitial polyhedra provides a structural representation that is complementary to the Voronoi network (or cellular froth, or honeycomb) of atomic polyhedra. A bit more precisely, each is the topological inverse of the other. Justification for this assertion about their relationship will be given later, when the bonding-network characterization is applied to covalent glasses, for which it is most appropriate. For now it suffices to be aware of the clear distinction between the two separate networks, both of which are aspects of the same lattice from different viewpoints.

Our initial interest is in amorphous metals, materials with fairly dense close-packed structures. The appropriate approach for metallic glasses is the one based on the Voronoi polyhedra of the atomic-cell honeycomb. For crystalline close packing, we know that the honeycomb is composed completely of dodecahedra. We now go on to a discussion of the distribution of atomic polyhedra characteristic of noncrystalline close packing.

2.4 RANDOM CLOSE PACKING

2.4.1 Empirical rcp Structure

The geometric notion has been around for a long time. Grain provides a familiar illustration. For a few thousand years, volume has served as a useful measure of grain in barter and commerce, volume occupied by disordered arrays of macroscopic particles in contact with one another. A biblical reference (St. Luke, VI, 38) reveals the existence of an early, and evidently widespread, appreciation of the basic structural idea:

> Give, and it shall be given unto you; good measure, *pressed down and shaken together*, and running over, shall men give into your bosom. For with the same measure that ye mete withal it shall be measured to you again.

The description given here of "good measure" contains the two key ingredients of the structure that we will now discuss: randomness ("shaken together") and, in some sense, close packing ("pressed down"). In recent times (> 1970), this *random close-packed* (rcp) structure has emerged with significance on a *microscopic* scale: *It presently provides the most satisfactory model for the structure of amorphous metals* (Cohen and Turnbull, 1964; Cargill, 1970, 1975; Chaudhari and Turnbull, 1978).

The following type of experiment yields very reproducible results. A large number of hard spheres of uniform size are quickly poured and packed into a container having irregular surfaces. (Plane surfaces are avoided because they encourage the laying down of layers of balls and the consequent formation of domains of crystalline close packing.) From the disordered but stable configuration which results, the coordinates of the sphere centers are determined by direct (albeit laborious) measurement. The rcp lattice thus obtained can be characterized, with varying degrees of completeness, by the same sort of numerical and functional measures as used for crystals. For a topologically disordered lattice, the various partial characterizations generally take on the form of a statistical distribution.

Consider first the atomic polyhedron (Wigner–Seitz cell, Voronoi polyhedron). For the fcc lattice (as for any crystal structure with a single atom per unit cell), this structural parameter is single valued and consists of one lone geometric object—in this case a rhombic dodecahedron. For the rcp lattice, this single polyhedron is replaced by a melange of polyhedra of various sizes and shapes. The mean size of the rcp atomic polyhedra exceeds the size of the fcc polyhedron (for structures resulting from packing of the same size spheres) by the ratio $0.7405/0.637$. Here 0.7405 is recognized as the filling factor (fraction of space contained within the spheres) for the fcc or hcp lattice, and 0.637 is the corresponding quantity for random close packing. The latter quantity is, empirically, remarkably well defined and reproducible. It is recovered in a variety of experiments of the type in which, say, ball bearings are shaken into bumpy-walled containers, or kneaded inside rubber balloons, or settled in oil, or even in which such processes are simulated on a computer. Individuals associated with modern versions of such experiments are Bernal (1965), Scott (Scott and Kilgour, 1969), and Finney (1970). (The "Bernal structure" is sometimes used as a synonym for random close packing, as is "dense random packing.") As already indicated, venerable antecedents of these exist; one very interesting example will be discussed shortly.

The 0.637 filling factor for the rcp structure means that, for balls of the same size, random close packing is about 86% as dense (or as "efficient" a filling of space) as is crystalline close packing. For hard spheres interacting via an overall attractive potential, a rough model for rare-gas solids, with closed-shell atoms interacting via intermolecular or "van der Waals" forces, or for positive ions in a metal interacting via the sea of conduction electrons, crystalline close packing corresponds to an absolute potential-energy minimum since this arrangement provides the maximum packing density.

In its own right, random close packing comprises a *local* energy minimum in configuration space. Though it corresponds to a metastable arrangement, the minimum is very deep relative to energies of nearby configurations, since each sphere is locked in by contacts with (typically, six) neighboring spheres. There is no way to pass continuously, via increasing density, from random to crystalline close packing. The structure would have to be taken apart and then put back together (i.e., traverse configurations of low density and, consequently, quite high energy) in order to effect this metastable→stable transformation. Thus a conversion of rcp to, say, fcc, is *reconstructive*, requiring a change of topology. This is completely analogous to the breaking and reforming of bonds required for the crystallization of the continuous-random-network structure of covalently bonded glasses, to be discussed in Section 2.5.3. In both cases, the amorphous→crystal transition normally proceeds via the formation of crystalline nuclei and their subsequent growth by the outward propagation of the glass/crystal interface as it consumes the amorphous matrix.

In a macroscopic hard-sphere close-packing experiment, the overall long-range interparticle attraction is usually provided by an external compressive force (and/or gravity), while the short-range repulsion is of course provided by the impenetrability of the balls [which simulates $V(r) \to \infty$ for $r < D/2$, with D being the sphere diameter]. Clearly this is a fairly crude model of the interatomic interactions in an amorphous metal. The previous paragraph's discussion of the metastability of an rcp arrangement of hard spheres is obviously highly plausible on an intuitive basis, and on the basis of experience with macroscopic analogs. But it should also be emphasized that this conclusion has been supported by calculations for monatomic metals in which realistic interatomic potentials have been used to describe the microscopic-scale interactions. With such potentials (in which, *inter alia*, the repulsive part is softened vis-à-vis a hard-sphere repulsion), the rcp structure is shown to be in metastable internal equilibrium, that is, it is found to be stable against small displacements. A pretty picture of random close packing, obtained in such a computer study (Barker et al., 1975) is displayed in Fig. 2.6.

2.4.2 Theoretically Derived rcp

Random close packing has been defined empirically. It is natural to wonder if there exists a purely mathematical derivation of this structure, in analogy to the inevitable occurrence of the fcc and hcp structures in the systematic study of symmetric lattices. Such a deductively-arrived-at rcp lattice, besides being intellectually satisfying, would provide a theoretical basis for the observed uniqueness of the observed structure.

The possibility of a geometric derivation is suggested by an argument due to Coxeter (1958) in a paper vividly entitled "Close Packing and Froth." The WS cells generated by a random array fill space in such a way that every face is surrounded by (naturally) two cells, every edge is surrounded by three cells, and every vertex is surrounded by four cells. (The reader can confirm this by examining a soap-bubble froth at his or her next shower.) The first follows by definition, the second and third follow because the locus of positions equidistant from three (four) unrelated points is a line (point). In order for more than three cells to meet at an edge, or more than four points to meet at a vertex, there must be special symmetry present which relates the centers that generate the Voronoi polyhedra. Such symmetry is absent for the sphere centers in random packing, so that the described topology applies to the resultant "froth" of Voronoi cells. Coxeter considers the question of whether or not there exists a regular honeycomb (division of space by identical polyhedra, each polyhedron bounded by identical regular polygons) with these topological specifications. In non-Euclidean spaces, such space-fillings exist: in elliptic space, the cells of the specified honeycomb have pentagonal faces; in hyperbolic space, the cells have hexagonal faces. In ordinary space, there is no solution. At least, there is none unless one accepts the possibility of cells whose faces are polygons with precisely 5.12 sides (p-gons with $p = 5.12$)! (5.12 is the positive root of $3p^2 - 13p - 12 = 0$, which turns out to be the algebraic statement of the topological constraint.) Coxeter proposes a statistical interpretation of this result, identifying it with what we have been referring to as random close packing. In fact, the average number of faces per cell for Coxeter's "statistical honeycomb" is $F = 12/(6 - p) = 13.6$, remarkably close to the values (discussed below) observed for empirical rcp lattices.

It is instructive to see how the above expression for F comes about, as an illustration of the existence of topological constraints even for random systems. The Euler–Poincaré relation is, in three dimensions,

$$V - E + F - N = 1, \tag{2.3}$$

for a Voronoi network or froth of polyhedra in which V is the number of vertices, E is the number of edges, F is the number of faces, and N is the number

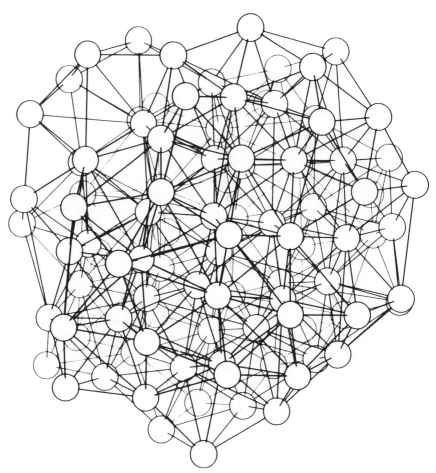

Figure 2.6 Computer-generated portrait of a 100-atom portion of a random-close-packed structure (Barker, Hoare, and Finney, 1975). The input for the calculation consisted of the atomic positions of an experimentally constructed Bernal model. These positions were then permitted to "relax" via a potential-energy minimization in which a Lennard-Jones interaction (defined in Section 5.2) was assumed to connect all atom pairs. In this picture, small spheres are shown for clarity; their diameter is roughly $0.3D$, where D is the equivalent hard-sphere diameter. Atoms whose centers are closer than about $1.4D$ are shown bonded to each other. (Reprinted by permission from *Nature* **257**, 120, copyright 1975, Macmillan Journals Ltd.)

of cells. Consider the application of Eq. (2.3) to the case of a single isolated polyhedron, $N = 1$. For a cube: $V = 8$, $E = 12$, and $F = 6$. For a rhombic dodecahedron: $V = 14$, $E = 24$, $F = 12$. For the case of interest, the "average" cell of the statistical honeycomb, we need to make use of the already-noted connectivity characteristics of froth: four cells (and edges) meet at each vertex and three cells (and faces) meet at each edge. A *single* polyhedral cell dissected from this network then has three faces (and edges) meeting at each vertex and two faces meeting at each edge. Adding the fact that each polygonal face is in contact with p edges (and vertices), all of this topological information is then embodied by the relations

$$3V = 2E = pF. \tag{2.4}$$

Substituting (2.4) into (2.3) then yields $12/(6 - p)$ for F.

It is interesting to view (2.4) in a different light. Suppose we interpret the polyhedron as representing a molecule composed of three atomic species, with the faces, edges, and vertices representing the three types of atom. The nearest neighbors of each atom are equal numbers of atoms of the other two types. The numerical coefficients in (2.4) then determine the coordination numbers of the components (e.g., each face coordinated with p vertices and p edges, etc.).

It should be noted that although a regular dodecahedron satisfies relations (2.4), with $p = 5$, space cannot be filled ("tiled") by such cells. It is impossible to construct a three-dimensional honeycomb from regular dodecahedra. In what follows, we will see that 12-faceted cells occur only rarely in the froth for random close packing.

It is also interesting to note that if we let p approach 6 in (2.4), the number of faces $12/(6 - p)$ goes to infinity. The surface of this limiting polyhedron actually corresponds to an infinite two-dimensional honeycomb in which the Voronoi polygon is a regular hexagon.

Now we move on to consider the infinite ($N \gg 1$) aggregate of polyhedra for random close packing. The relations which, for the full Voronoi network, are the analogs of relations (2.4) for an isolated polyhedron, are

$$6V = 3E = pF = (fp/2)N. \tag{2.5}$$

Summarized in (2.5) are the circumstances that each vertex is connected to 4 edges, 6 faces, and 4 cells; each edge to 3 faces, 3 cells, and 2 vertices; each face to 2 cells, p vertices, and p edges; and each cell to f faces, $fp/3$ vertices, and $fp/2$ edges. Substituting the three equations of (2.5) into Euler's relation [note that we now may set the right-hand side of Eq (2.3) equal to zero, since each quantity on the left is much greater than unity for the "infinite" froth], and solving for f in terms of p, again yields the result

$$\bar{f} = \frac{12}{6 - \bar{p}}. \tag{2.6}$$

We have now written bars over the two geometric quantities in order to emphasize their statistical interpretation: $\bar{p}$ is the average number of edges per face, and $\bar{f}$ is the average number of faces per cell.

Another way of expressing Eq. (2.6) is

$$\sum_p (6 - p)F_p = 6N, \tag{2.7}$$

where F_p is the number of faces with p sides. The sum (which starts with $p = 3$, with F_3 being the number of triangular faces) is an invariant for all random rearrangements of N centers. This topological constraint, for a three-dimensional froth of Wigner–Seitz cells, is relatively weak. In two dimensions, the corresponding constraint is much stronger: Topology requires the average number of edges per polygonal cell to be precisely six for any planar froth in which three edges meet at each vertex. Symbolically, the 2d statement analogous to Eq. (2.7) is $\Sigma_p(pF_p)/\Sigma_p F_p = \bar{p} = 6$. This statement applies to the two-dimensional Voronoi network generated by a random array of points in a plane, as in Fig. 2.5. Thus the structure of even a completely disordered network is restricted by topological requirements, which become more restrictive when combined with restricted dimensionality. In one dimension, of course, each cell is (topologically) the same—a line segment.

2.4.3 Characterizations of the rcp Structure

Finney (1970) has carried out a detailed analysis of the statistical topology of an experimentally constructed random-close-packed model containing 8000 spheres. In particular, the characteristics of the Voronoi polyhedra corresponding to the obtained array of the sphere centers were examined. Two such irregular Wigner–Seitz cells are shown in Fig. 2.7 (along with the rhombic dodecahedron, which is the quite regular Voronoi cell of cubic close packing).

Figure 2.8 shows a pair of histograms that correspond to data obtained by Finney on the statistical topology of Wigner–Seitz cells in random close packing. For the distribution in p, the various types of polygonal faces, the most common occurrence (about 41% of the faces) is $p = 5$, followed in frequency by $p = 6\,(29\%)$, $p = 4\,(19\%)$, $p = 7\,(6\%)$, $p = 3\,(4\%)$, and $p = 8\,(1\%)$. For the distribution in f, the number of faces per polyhedron, the most common occurrence is that of 14-faceted cells, followed in frequency by $f = 15$, $f = 13$, $f = 16$, $f = 12$, and $f = 17$. By comparison, for crystalline close pack-

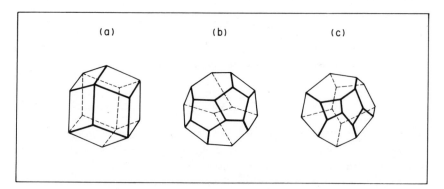

(a)   (b)   (c)

Figure 2.7 The Wigner-Seitz cell of cubic close packing (a), compared to two possible cells (b and c) that can occur in random close packing (after Finney, 1970).

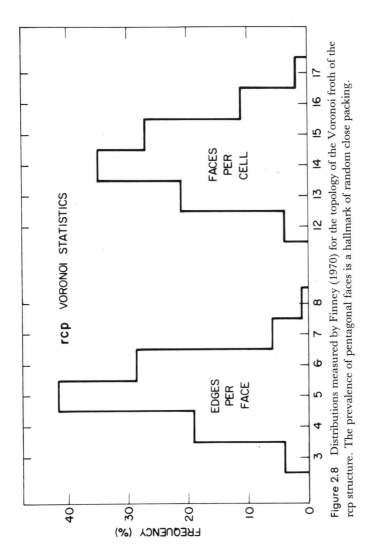

Figure 2.8 Distributions measured by Finney (1970) for the topology of the Voronoi froth of the rcp structure. The prevalence of pentagonal faces is a hallmark of random close packing.

ing with its single-valued cell topology, the distributions contract to delta functions at $p = 4$ and $f = 12$.

The distribution in f may be interpreted in a way that provides an alternative to the more famililar construct of coordination number. Instead of the idea of nearest neighbors, we use the rather different idea of *geometric neighbors.* Geometric neighbors are defined as sites whose Voronoi polyhedra are in contact. Contact means that the two atomic polyhedra share a face in common. Contiguity is another term used in this topological context, with *contiguous pair* denoting two geometric neighbors and *contiguity number* denoting the number of geometric neighbors surrounding a given cell. The contiguity number f is the cellular analog of the coordination number z. Comparing these two different characterizations of short-range order as they apply to crystal lattices, we note that for some crystal structures (e.g., fcc, with $f = z = 12$, and sc, with $f = z = 6$) the number of geometric neighbors coincides with the number of nearest neighbors, while for other structures (e.g., bcc, with $f = 14$ and $z = 8$) f exceeds z. For the 14-faceted truncated-octahedron WS cell of the bcc lattice, the eight hexagonal faces correspond to contacts with the nearest neighbors at distance d while the six square faces correspond to "contacts" with the second-nearest neighbors at distance $d(4/3)^{1/2}$.

With this interpretation in mind, we see immediately that Fig. 2.8 tells us the distribution of geometric neighbors characteristic of random close packing. The histogram specifies the frequency of occurrence of each possible number of geometric neighbors that appear in the rcp structure. The average number of geometric neighbors is 14.3. The distribution peaks near $f = 14$, and appreciable occurrences appear from $f = 12$ through $f = 17$. Sites that are geometric neighbors in the rcp structure can be as close as d or as distant as $2d$. Most geometric neighbors occur at separations near the lower end of this range, as will be seen in the RDF for rcp to be discussed below. Up to $1.5d$, all neighbors are geometric neighbors.

2.4.4   Peas in a Pot

The ancient place occupied by random close packing in human affairs (viz. in grain dealings as a measure of volume) has already been mentioned. It also happens that there exists a well-documented, more-than-two-and-a-half-centuries-old, experimental measurement of the rcp-structure Voronoi polyhedra. Although this book is not intended to provide a chronology of our subject, this particular bit of history is too tempting to resist mentioning. It also constitutes a nice example of interrelationships in science via the appearance, in totally different naturally occurring contexts, of the same mathematics (geometry in this case). *Vegetable Staticks,* subtitled *An Account of some Statistical Experiments on the Sap in Vegetables,* by Stephen Hales, Rector of Farrington and Minister of Teddington, was published in London in 1727. This early classic of what is now called plant physiology (read "the mechanics of plants" for

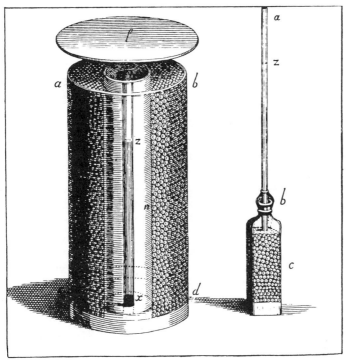

Figure 2.9 Stephen Hales (1727) was a highly gifted biologist whose interests included the uptake of water by plants. The diagram on the left shows the apparatus he used to demonstrate the substantial force exerted by dilating peas. However, when the lid (f) was covered with a weight great enough to prevent its lifting, the dilated peas deformed into the Wigner–Seitz cells of the rcp structure. The peas in the bottle at right, used by Hales in a related experiment, would not have served so well to illustrate random close packing because of the presence of the crystallinity-inducing planar walls.

"vegetable staticks") contains, *inter alia*, descriptions of many experiments by which Hales demonstrated "the great force with which vegetables imbibe moisture." In one of these, illustrated in Fig. 2.9, he demonstrated that peas in a pot, dilating via uptake of water, would lift up a heavy lead lid covering the pot. Pushing the experiment further in order to find the limit for this effect, he carried out the following extensions, described here in his own words:

> Being desirous to try whether they would raise a much greater weight, by means of a lever with weights at the end of it, I compressed several fresh parcels of Pease in the same Pot, with a force equal to 1600, 800, and 400 pounds; in which Experiments, tho' the Pease dilated, yet they did not raise the lever, because what they increased in bulk was, by the great incumbent weight, pressed into the interstices of the Pease, which they adequately filled up, being thereby formed into pretty regular Dodecahedrons.

This work is evidently a not-unsophisticated precursor of modern equivalents involving plasticene balls or lead shot. Instead of filling a volume with randomly packed spheres of given size and then decreasing the total volume by compression, the "Pease" in Hales' experiment approximate rcp spheres, which deform into Wigner–Seitz cells as they extend to fill a space of fixed volume. Now from Hales' point of view, these were negative-result experiments ("they did not raise the lever"), so that he probably was disappointed and, in contrast to his normal careful style, did not bother to numerically quantify his observations in this case. Nevertheless, it is clear that his reference to "pretty regular Dodecahedrons" is tantamount to a statement of the prevalence of pentagonal faces on his Voronoi-polyhedron Wigner–Seitz peas, since it is the presence of pentagonal faces that most distinguishes the regular dodecahedron from the other Platonic solids. Although dodecahedra per se play only a minor role, we have already noted the predominance of pentagonal faces in the cell-distribution statistics characteristic of the rcp structure. Thus, what is now known to be a typical feature of the *microscopic structure of amorphous metals* was anticipated (by about 250 years) by this observation recorded during an early example of systematic *biological* research.

2.4.5   Dimensionality Considerations and the Extendability of Local Close Packing

In their craving for simplicity and tractability, physicists habitually examine the behavior of one- and two-dimensional analogs of three-dimensional systems. This is an admirable and time-honored tradition, and has yielded (and will continue to yield) valuable insights. However, it is also true that it is frequently perilous to attempt to generalize specific results and conclusions from one dimensionality to another. We shall encounter several illustrations of this caveat throughout this book. In fact, we have already done so during the course of our discussion of random close packing, and will now make this explicit. One may reasonably adopt the viewpoint that the geometric stability and consequent physical significance of the rcp structure arises from the following circumstance: *In remarkable contrast to one and two dimensions, short-range dense packing and long-range crystalline order are not concordant with each other in three dimensions.*

The construct of an rcp structure is not meaningful in one dimension. A linear array of movable nonoverlappable line segments of equal length (the 1d equivalent of identical hard spheres) necessarily collapses under compression to a crystal lattice. Likewise, experience also indicates the absence of a stable rcp structure in two dimensions. An agitated and compressed planar array of equal coins tends to aggregate in domains of 2d crystalline close packing. In both one and two dimensions, there exists an intrinsic consistency between the crystalline close-packed (ccp) structure and the demands of the most efficient *local* packing. But in three dimensions this is *not* true. These assertions are easily demonstrated as follows. The largest number of d-dimensional spheres which can be locally packed so that each contacts every one of the others is 2 for $d = 1$, 3 for $d = 2$, 4 for $d = 3$ (and $d + 1$ for $d = d$, an example of an interdimensionality generalization which works). The resulting d-dimensional polyhedral hole formed by taking the sphere centers as vertices is a line segment, equilateral triangle, and regular tetrahedron for $d = 1$, 2, and 3, respectively. In each dimensionality the figure is a *simplex*, the simplest regular polyhedron. Now in one and two dimensions, equal simplices (plural of simplex) can tile space. But in three dimensions there is no way to arrange equal nonoverlapping regular tetrahedra so as to completely fill space. This is why the polyhedral holes in 3d crystalline close packing are not exclusively tetrahedra; the octahedra are necessary to complete the structure.

Suppose that, in two dimensions, we start with a close-packed cluster of three mutually contacting circles, and then proceed to add on circles so that each new circle contacts two mutually contacting circles already in the cluster. This procedure automatically leads to the buildup of the 2d close-packed structure, with the circle centers on a triangular lattice as in Fig. 2.1. Now let us carry out the corresponding exercise in three dimensions, starting with a tetrahedral cluster of four mutually contacting spheres. This time there is no automatically produced unique structure. It is possible to proceed for some time to build up a dense cluster based exclusively on tetrahedra, with each new sphere nesting onto an equilateral triangle of three mutually contacting older ones. Such a cluster may actually correspond to a local density which exceeds that of crystalline close packing, and its topology is obviously different from that of ccp. But we cannot keep on building up the cluster in this way, catering to a preference for tetrahedra, without soon being forced to incorporate some rather large holes in order to continue. Thus this rep-like structural "hare" is, in the long run, overtaken and passed in terms of density by the ccp "tortoise."

The above argument provides a basis for understanding the stability of the three-dimensional rcp structure in the face of the instability of any two-dimensional or one-dimensional counterpart. It also provides plausibility for the appearance of random close packing in a rapidly quenched simple solid, in which the haste to form local (i.e., microscopic, atomic scale) close-packed clusters does not give the ccp structure (which provides the maximum *macroscopic* density by virtue of its "planning ahead" long-range organization) an opportunity to organize itself. Competition between spatial short-sightedness and farsightedness is one way to view the contrast between the rcp and ccp packings. The former myopically focuses on maximum *short-range* density, while the latter patiently adopts the long view to attain maximum *long-range* density. To conclude this paragraph's anthropomorphic analogy we note that in order to be patient, one must have time, and time is the ingredient that is deliberately withheld in the formation of amorphous solids.

2.5 CONTINUOUS RANDOM NETWORK

2.5.1 The Simplicial Graph

Suppose that Fig. 2.5 represents a noncrystalline array of atoms in two dimensions. The solid circles, of diameter significantly smaller than the interatomic spacings, may be interpreted as the atomic cores. The positions of the nuclei define a set of points, and these sites in turn determine two well-defined and distinct divisions of the plane into polygonal regions separated by linear boundaries. One division, indicated by the light lines in Fig. 2.5, is the Voronoi/Wigner–Seitz/cellular network already described. This construction is uniquely defined, each boundary having the property that it is equidistant from the two nearest sites. The second polygonal division of the plane, shown by the heavy lines in Fig. 2.5, is now obtained by connecting with straight lines those sites whose Voronoi cells are contiguous. The polygons thus formed are all triangles, irregular ones. In three dimensions, the analogous procedure leads to the division of space into irregular tetrahedra whose vertices are sites and whose edges are the lines joining contiguous sites. The framework formed by the sites and the intersite lines or *bonds* is called the *simplicial graph* of the array. "Simplicial" because each cell of the resulting tesselation is a simplex, or simplest polyhedron (i.e., a triangle in two dimensions, a tetrahedron in three, a "pentatope" or five-cell in four, etc.). The partition of space into irregular simplexes is called the Delaunay division. It is the topological inverse or *dual* of the Voronoi division.

In general, a *graph* is a topological object which consists of a set of sites or vertices and a set of bonds or edges (pairwise connections between sites). In two dimensions, the Voronoi network is itself a graph and is said to be the *dual graph* of the simplicial graph. Two graphs are duals of each other if each vertex of one corresponds to a cell of the other, and if each bond of one corresponds to a bond of the other (which it crosses). In Fig. 2.10 we illustrate the dual relationship between the triangular and honeycomb lattices. A close-packed crystalline array of two-dimensional atoms has a triangular lattice as its simplicial graph and a honeycomb lattice as its Voronoi froth of polygonal atomic cells. Conversely an array of atoms on a honeycomb lattice, as in a simple covalently bonded layer in graphite, has a triangular lattice as its two-dimensional aggregate of Wigner–Seitz cells. A square lattice is said to be self-dual because its dual is another square lattice.

2.5.2 Mathematical Bonds and Chemical Bonds: The Covalent Graph

The simplicial graph is the topological construct that is the complement of the Voronoi froth used in the preceding section to discuss the rcp structure of metallic glasses. However, the reason for introducing graph concepts into the discussion of the structure of amorphous solids is not merely to complete the

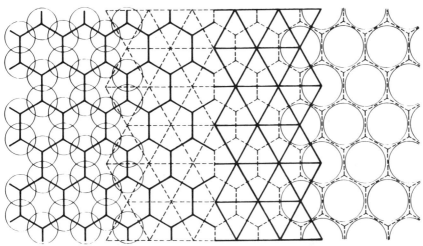

Figure 2.10   Illustration of the duality between the honeycomb and triangular lattices. Graph (heavy lines) and froth (dashed lines) representations of atomic-site lattices are appropriate for covalently bonded and metallically bonded solids, respectively. Covalent and metallic structures are schematically indicated here on the left and right sides of the diagram, with the lattice sites shown ''dressed'' with atoms (the light circles).

picture in a mathematical sense (although it does indeed happen to accomplish that esthetic, if subsidiary, purpose). Graphs provide a viewpoint which is very appropriate to the ensuing discussion of the structure of covalently bonded network glasses. Graph concepts also will be useful in our later discussions of percolation theory.

The conventional type of ''ball-and-stick'' model, which is so often used to represent the structure of crystal lattices, is equivalent to a graph with the ''balls'' (atoms) as sites and the ''sticks'' as bonds. Such models serve not only as aids in visualizing the structure of solids, but also as valuable props for guiding the thinking about them. As illustration we display, in Fig. 2.11, a familiar example of this graphic genre. Shown here is a model representing the diamond-structure lattice of crystalline C, Si, Ge, and α-Sn; the microscopic viewpoint being approximately along a (110) direction. Each atom is shown with tetrahedral coordination, $z = 4$, in a continuously connected periodic network. The skeleton of this structure is the graph consisting of sites at the atom centers and the bonds connecting each site with its four nearest-neighbor sites. This graph is *not* the simplicial graph of the diamond structure; it is actually simpler, being a subgraph of the simplicial graph. If we construct the Voronoi polyhedra corresponding to sites arranged as in diamond, we find them to be truncated tetrahedra with each sharp tetrahedral point blunted by three small triangular shavings. Thus the atomic cell has 16 faces so that the corresponding simplicial graph has $z = 16$, with each site coordinated with (bonded to) 12 next-nearest-neighbor sites in addition to its 4 nearest neigh-

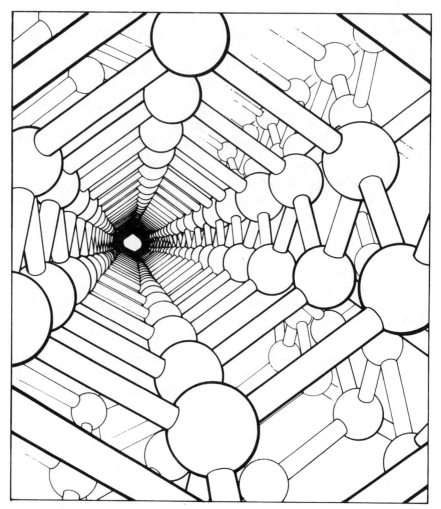

Figure 2.11   Model of the diamond structure [adapted from a drawing in Pauling and Hayward (1964)].

bors. We shall refer to a graph such as the one underlying the model shown in Fig. 2.11 as a *covalent graph*. In such a graph only those mathematical bonds of the simplicial graph are retained that have physical significance as the axis of a chemical covalent bond in the solid. Coordination numbers of $z \leq 4$ occur in covalent graphs.

In covalent bonding, as in metallic bonding, the basic mechanism for the lowering in energy relative to the free-atom state is the reduction in the kinetic energy of valence electrons achieved by their delocalization in the condensed phase. However, the geometries which result are quite different in the two cases. Since the delocalized electrons in metals spread out nondirectionally to fill the regions between spherical ion cores, close-packed structures occur as noted earlier. Because such structures are so conveniently understood in terms of dense packings of spheres, their geometry can be regarded to be primarily repulsion-determined. By contrast, the structure of covalently bonded solids is primarily attraction-determined and more quantum mechanical in character. The structure provides nearest-neighbor configurations which facilitate the directional overlap of symmetric wave-function combinations on adjacent atoms. The (mathematical) bonds retained in the topological object we will call the covalent graph are intersite lines corresponding to the interatomic covalent bonds, the regions about which the kinetic-energy-lowering spreading out of the valence-electron wave functions takes place.

Although all graphs are, strictly speaking, topologically one dimensional because they consist of a countable set of points and lines, we will be concerned with a dimensionality that is somewhat related to (although not quite the same as) the dimensionality of the space in which the covalent graph is embedded. This notion of the macroscopic dimensionality of the covalent network, embracing dimensionalities of 3, 2, 1, and even 0, is discussed in the next chapter. Here we are concerned only with three-dimensional covalent graphs, a crystalline example of which is the diamond-structure example of Fig. 2.11. We now proceed to the discussion of noncrystalline covalent graphs.

2.5.3 The Continuous-Random-Network Model of Covalent Glasses

In 1932 W. H. Zachariasen, in his classic paper entitled ''The Atomic Arrangement in Glass,'' set forth what has since become known as the continuous-random-network model (henceforth, crn model) for the structure of covalently bonded amorphous solids. Although Fig. 1*b* of Zachariasen's famous article is very familiar, it is such an important landmark in the development of ideas about the structure of glasses that it will also be reproduced here. But not quite yet. First it is useful to quickly note the difference between the pair of covalent graphs shown in Fig. 2.12. By now familiar, the honeycomb lattice of Fig. 2.12*a* is the two-dimensional covalent graph corresponding to a layer of bonded carbon atoms in graphite. Figure 2.12*b* shows a graph representing a binary compound in which each black atom is covalently bonded to three white atoms and each white atom is covalently bonded to two black ones. This is called the ''decorated'' honeycomb lattice, as it is derivable from a honeycomb lattice by replacing each bond by a pair of bonds to an intervening twofold-coordinated ''bridging'' atom. There are several interesting cases in which the structure of a class of covalent binary compounds is topologically related to that of a class of elemental solids by the type of decoration transformation illustrated in Fig. 2.12. In particular, the decorated honeycomb lattice of Fig. 2.12*b* corresponds to the covalent graph of the layers that make up crystalline As_2S_3 and As_2Se_3.

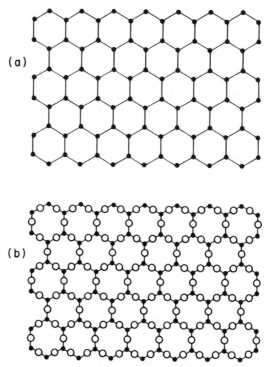

Figure 2.12 (*a*) The honeycomb lattice and (*b*) the decorated honeycomb. The topological structure of the honeycomb is the same as that of a layer in graphite or crystalline arsenic, while the topology of the decorated honeycomb occurs in the structure of the layers that make up crystalline As_2S_3 and As_2Se_3.

The two covalent graphs of Fig. 2.12 represent periodic structures. In Fig. 2.13 we show a corresponding pair of graphs which represent continuous random networks; each of these noncrystalline structures has the same short-range order as its crystalline counterpart in Fig. 2.12. Figure 2.13*b* reproduces Zachariasen's well-known diagram. The essential characteristics of crn structures are revealed by comparing Figs. 2.12 and 2.13. It is logical to begin with the simpler case, the elemental glass represented on the left of Fig. 2.13.

The elemental network glass of Fig. 2.13*a* shares the following features in common with the honeycomb crystal lattice:

1. $z = 3$, each atom is threefold coordinated.

2. Nearest-neighbor distances (i.e., bond lengths) are constant, or nearly so.

3. Both structures are "ideal" in admitting no dangling bonds.

Both networks are indefinitely extendable and no notice is taken here of surface effects. Actually statement 3 is implied by statement 1, but it is worth separate mention because of its chemical significance, as discussed in the next chapter. Statement 2, as emphasized by Zachariasen, is the condition which ensures that the energy of the covalent glass is little different from that of the

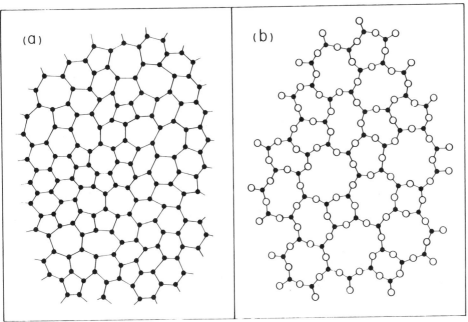

Figure 2.13 Two-dimensional continuous random networks. A sketch of a three-fold-coordinated elemental glass is presented in *a*, while Zachariasen's (1932) diagram for an A_2B_3 glass is shown in *b*.

crystal. The two fundamental ways in which the crystalline and the continuous-random networks distinctly differ from each other are:

4. A significant spread in bond angles, not permitted in the crystal, is characteristic of the crn structure.

5. Long-range order is absent, of course, for the crn glass.

For the A_2B_3 compound represented in crystalline and glassy forms by the decorated graphs in Figs. 2.12 and 2.13, respectively, similar statements as those of 1–5 apply. Now there is an additional degree of latitude in the presence of a second type of bond angle, that occurring at the bridging atom. Since the bond angle at a twofold-coordinated atom is expected to be much softer (i.e., much less costly in energy to deform) than that at a threefold-coordinated atom, all of the bond-angle leeway may be supposed to be taken up at the bridging atoms. Thus statement 4 now applies only to the soft bond angles in the network, while the stiff bond angles may be lumped together with the bond lengths in statement 2 as little different in the crn glass from their values in the crystal. This is the case in Fig. 2.13*b*. Also evident in a comparison of this structure with that of Fig. 2.13*a* is the relative ease, because of the bends permitted at the bridging atoms, of developing a covalent network without bond-length distortion in the case of the compound.

Figure 2.14, taken from a pattern used in a children's coloring book, il-

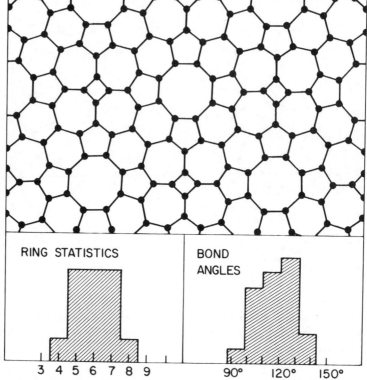

RING STATISTICS   BOND ANGLES

3 4 5 6 7 8 9   90°   120°   150°

Figure 2.14   An illustration of ring-size and bond-angle distributions.

lustrates several aspects in the context of a two-dimensional structure. The solid lines represent bonds in a threefold-coordinated covalent graph. Although the structure shown is actually a crystalline one, having high (fourfold) overall symmetry, the unit cell is large and displays a variety of closed loops or rings containing different numbers of atoms. The topology of this graph can be characterized by its "ring statistics," the distribution describing the frequency of occurrence of n-atom rings. That distribution, which is occupied in this case for $4 \le n \le 8$, is given by the histogram shown at the lower left of the figure. As mentioned earlier in Section 2.4.2, the application of Euler's law to a two-dimensional graph with $z = 3$ requires the average ring size in this example to be exactly $\bar{n} = 6$. At the lower right in Fig. 2.14 is shown a histogram corresponding to the distribution of bond angles in the network.

Besides the feature of ring statistics and of bond-angle statistics, Fig. 2.14 also illustrates a point with respect to a technique sometimes employed for calculating electronic or vibrational states in amorphous solids. This is the use of a large cluster of atoms, possessing structural features characteristic of the glass, which is periodically repeated in space to generate a crystal structure having a large unit cell. The calculation is then carried out by means of some method used for crystals, crunching it through in brute-force fashion with the aid of a powerful computer.

2.5.4 Prototype Elemental crn: Amorphous Silicon

We now consider a pair of prototypical covalently bonded amorphous solids, one an element and one a binary compound. Their covalent graphs define three-dimensional continuous random networks, and they may conveniently (albeit, schematically) be regarded as being mutually related by the same type of bridging-atom decoration transformation as that which relates the two two-dimensional structures of Fig. 2.13. The two solids are amorphous silicon (a-Si) and fused silica (a-SiO$_2$). The latter, henceforth simply silica (the amorphous form being assumed, unless otherwise noted), is the archetype among oxide glasses and probably the most familiar amorphous solid. In a historical treatment the discussion of the a-SiO$_2$ structure would come quite early, certainly preceding that of the a-Si structure, which is a more recent (and in some respects subtler) development. But in the present context it is more orderly to first introduce the elemental network.

The starting point is the three-dimensional-network crystalline graph of Fig. 2.11. The diamond structure plays the same role in relation to the structures of a-Si and a-SiO$_2$ as does the honeycomb lattice (Fig. 2.12a) in relation to the amorphous structures of Fig. 2.13. Like the crystalline diamond lattice, the ideal a-Si crn is characterized by:

1. $z = 4$, each atom is fourfold coordinated.
2. Constant bond lengths.
3. No dangling bonds.

It differs from diamond in that:

4. There is a significant spread in bond angles.
5. There is no long-range order.

Topological disorder is intrinsic to the amorphous network. After the local coordination z, the next simplest topological characterization of a covalent graph is in terms of bond loops or rings. Starting at a given atom, we trace a closed path of bonds and atoms, visiting each atom along the way only once, and returning to the initial atom. In the diamond structure of c-Si, the shortest such ring contains six atoms (and six bonds, of course). There are 12 distinct six-atom rings passing through each atom in the crystal. Each atom also belongs to many larger rings; in c-Si all rings contain an even number of atoms. By contrast, both odd- and even-membered rings appear in a-Si, and the shortest loops present contain five atoms. However, a unique characterization of the a-Si structure in terms of its ring statistics, in analogy with that shown for the two-dimensional example of Fig. 2.14, is not yet unambiguously known.

Model building, the construction by hand of macroscopic ball-and-stick-type representations of atomic-scale structures, is a craft that has historically played a useful, and occasionally even crucial, role in various fields. The most famous example of this genre is, of course, the Watson–Crick double-helix model for DNA. Although nothing quite so momentous has occurred in our subject, model building has nevertheless played an important part in extending our understanding of the structure of amorphous solids. We have already met the Bernal construction for the rcp structure of metallic glasses. Now we will describe how models have also made critical contributions in establishing the credentials of the continuous random network for covalent glasses.

In the case of tetrahedrally bonded amorphous semiconductors, it was a model built by Polk (1971) that first demonstrated the possibility of building up an extended $z = 4$ crn without developing intolerable bond-length strain. Before Polk's model, it was simply not known whether or not a space-filling tetrahedrally bonded topologically disordered structure could be constructed in consonance with conditions 1–5. Condition 2 is the critical question mark, since the length of the covalent bond must be held close to its equilibrium minimum-energy value if the glassy structure is to have an energy close to that of the crystal and be able to compete with it (at least for metastability) in nature. The indefinite extendability of such a crn structure, while maintaining compliance with condition 2, is what was established by Polk. To quote from a sequel to that work (Polk and Boudreaux, 1973), in which the original hand-built model was computer-refined:

> It has been demonstrated that it is possible to construct a tetrahedrally coordinated random-network structure in which all first-neighbor distances are equal and the bond distortions do not increase as the structure is made larger.

A photograph of the full 440-"atom" Polk model would not be terribly helpful because of its complexity. Instead, in Fig. 2.15, we show a small cluster composed of snap-together tetrahedral units of the type used in the original model. Random-network features illustrated here include a five- and a seven-membered ring, as well as six-membered rings that are somewhat twisted with respect to the "chair" (diamond structure) or "boat" (wurtzite structure) crystal configurations. (See Fig. 2.17, discussed below.)

The radial distribution function of this structure is discussed in a following section, in a comparison with experimental data on amorphous semiconductors. Other aspects of interest of the tetrahedral crn will be mentioned now. Substantial numbers of five-, six-, and seven-membered rings are present, along with larger rings. The density is within 1% of that of a diamond-structure crystal having the same bond length. As implied in the above quote, the bond-length variation can be reduced to zero, as in the crystal. *Unlike* the crystal, the Si—Si—Si bond angle takes on a *distribution* of values, which can be described by an rms spread $\Delta\theta$ of about 9° about the tetrahedral value of $\theta = \cos^{-1}(-1/3) = 109°$. Actually, although a bond-bending displacement in a covalently bonded molecule or solid is much less energetically costly than is a bond-stretching displacement, still a finite $\Delta\theta$ does cost *some* energy, which is

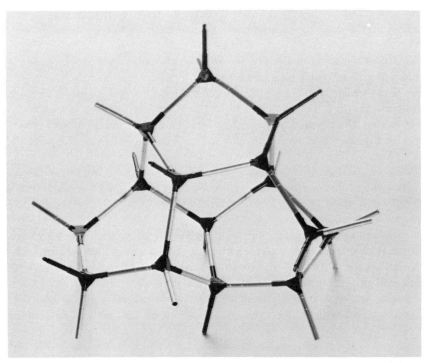

Figure 2.15 Cluster composed of snap-together units of the type used by Polk (1971) in constructing his four-coordinated continuous random network for amorphous silicon and germanium. Five-membered rings may be seen at the top and at the lower right. A six-membered ring in approximately the "chair" configuration is seen on edge at the right, while two in roughly the "boat" configuration are at the left.

proportional to $(\Delta\theta)^2$. The ratio of bond-bending to bond-stretching stiffnesses, though small, is nonzero. Consequently, in a further *elastically relaxed* computer refinement of the Polk model, a small spread Δr in bond lengths is permitted in order to reduce $\Delta\theta$ and thereby trade off a small increase in bond-length distortion energy for a larger decrease in bond-angle distortion energy. Elastic potential energy is minimized when the rms bond-angle spread is reduced to about $\Delta\theta = 7°$, with a concomitant minuscule bond-length spread of about $\Delta r/r = 1\%$ appearing in the relaxed structure. *This very sharp definition of the nearest-neighbor configuration, notably coordination and bond length, is a hallmark of covalently bonded amorphous solids.* In covalent glasses, the short-range order is little different than in the corresponding crystals. This is a much higher degree of local order than that possessed by metallic glasses.

A higher-order angular characteristic of a tetrahedrally coordinated covalent network is the distribution of dihedral angles. The dihedral angle describes the relative orientation of adjacent tetrahedra, tetrahedra which share one bond. It is a secondary bond angle. The primary bond angle (called throughout, simply, *the* bond angle) specifies the relative orientation of *adjacent* bonds, that is, nearest-neighbor bonds emanating from a single atom. By contrast, the dihedral angle pertains to the relative disposition of a pair of bonds that are *not* adjoining but that both adjoin a common intervening bond, i.e., it refers to a bond pair which consists of two next-nearest-neighbor bonds. On the plane which perpendicularly bisects the bond joining a pair of nearest-neighbor atoms, project the other three bonds that are attached to each of the two atoms. The dihedral angle is the average angle separating a bond projection belonging to one atom from the nearest bond projection belonging to the other atom. If the two sets of projections line up perfectly with each other (the "eclipsed" configuration), the dihedral angle is zero. If the two sets of projections avoid each other as much as possible (the "staggered" configuration), the dihedral angle is close to its maximum value of 60°. These two limiting situations are illustrated in Fig. 2.16.

In crystalline diamond, only the staggered configuration (60°) appears. In the related crystalline wurtzite structure, both staggered and eclipsed configurations occur. In the tetrahedral crn structure, a *continuous* distribution of dihedral angles replaces the delta-function crystal distributions. It extends smoothly and monotonically from 0° to 60°, with no minimum in between. The staggered configuration is found to be about twice as likely as the eclipsed configuration. Again, we have the theme of a discrete spectrum, describing a structural characteristic of the crystalline solid, which transforms in the corresponding amorphous solid into a continuous spectrum of statistically distributed values.

In the two crystalline tetrahedral structures mentioned above, the smallest bond circuits of the covalent graph are six-membered rings. The 6-rings in dia-

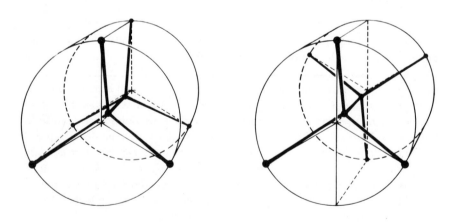

ECLIPSED   STAGGERED

Figure 2.16   The eclipsed and staggered bond configurations in tetrahedrally bonded semiconductors.

(a)

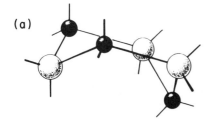

(b)

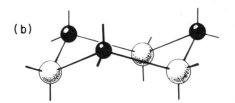

Figure 2.17   The (*a*) boat and (*b*) chair configurations of six-atom rings.

mond (an *n*-ring is a circuit having *n* bonds, *n* atoms) all have the same geometrical configuration—the "chair" configuration shown in the lower part of Fig. 2.17. Not only are all 6-rings topologically equivalent in diamond, but in this symmetric crystal structure they are also geometrically equivalent (same size *and shape*). All six dihedral angles within the chair-shaped ring correspond to the staggered bond configuration. In crystalline wurtzite, whose symmetry is not quite as high as that of diamond (hexagonal versus cubic), the six-membered rings adopt two distinct configurations: one-quarter of them are in the chair configuration, the other three-quarters are in the "boat" configuration shown at the top of Fig. 2.17. Within a boat-shaped ring, two bonds find themselves in the eclipsed configuration, the other four are in the staggered configuration. In both diamond and wurtzite, *n*-rings occur for *n* = 6, 8, 10, etc.

We have seen that in the topology of the tetrahedral continuous random network, five- and seven-membered rings occur, in addition to six-membered rings (e.g., Fig. 2.15). In fact, all sizes of *n*-rings with $n \geq 5$ appear in the covalent graph of the tetrahedral crn. From a geometric point of view, the six-membered rings which occur in the crn naturally take on a continuous spectrum of shapes. They are not limited to the two specific geometries (i.e., the crystalline "discrete spectrum") illustrated in Fig. 2.17.

The covalent graph characteristic of the three-dimensional crn structure of amorphous silicon possesses its own *ring statistics*, analogous to that illustrated earlier for the two-dimensional graph of Fig. 2.14. Just as the coordination *z* is the key topological measure of the extremely well-defined short-range order, so the ring statistics provide a measure of the less-well-defined "medium-range" topology. The histogram corresponding to the distribution of *n*-rings in *a*-Si is only crudely known. One estimate is that a network containing N sites (atoms) has about $0.3N$, $1.5N$, and $0.8N$ n-rings for n = 5, 6, and 7, respectively.

The crn structure which has been described here applies to the elemental glasses of group 4 (fourth column of the Periodic Table), notably amorphous silicon and germanium. It also serves as a model for amorphous 3-5 compounds of the GaAs family, materials which are similarly tetrahedrally coordinated and highly covalent. Note that the presence of odd-membered rings implies the presence of like-atom or "wrong" bonds, e.g., Ga—Ga or As—As. This entails an additional energy cost, reducing the stability of these semiconductor compounds in the amorphous form.

2.5.5   Prototype Binary crn: Fused Silica

Amorphous SiO_2, or fused silica, provides the archetypal example of a three-dimensional continuous random network formed by a binary compound. Extensive model-building studies of the SiO_2 crn have been carried out by Bell and Dean (1972). Rules used in building up the network (listed in a way similar to that used earlier for a-Si) specify the following:

1. $z(Si)$ = 4 and $z(O)$ = 2, fourfold and twofold coordination for silicon and oxygen, respectively, with complete chemical ordering.
2. Constant bond lengths and O—Si—O bond angles.
3. No dangling bonds in the bulk.
4. A significant spread is permitted in Si—O—Si bond angles.
5. There is no long-range order.

The first two prescriptions, describing the high degree of short-range order, are more detailed than the simpler statements that sufficed previously for the elemental a-Si structure. "Chemical ordering," in condition 1, refers to the strong requirement that each Si atom be covalently bonded to four nearest-neighbor O atoms and each O atom to two nearest-neighbor Si atoms. In condition 2, not only is the Si—O bond length to be kept constant, but so is the O—Si—O bond angle (at 109°). Each SiO_4 unit, consisting of a central silicon bonded to four oxygens, is to define the body center and vertices of a regular tetrahedron. All of the quantitative variations present in parameters of the nearest-neighbor configurations are to be assumed by the Si—O—Si bond angles. Since the oxygens are bonded to only two neighbors while the silicons are bonded to four, the angles at the oxygens are much less constrained than those at the silicons. The Si—O—Si bond angles are thus much softer than the O—Si—O bond angles, and their variation entails a much smaller energy increase. The mean Si—O—Si bond angle in the structure is about 150°, and the rms $\Delta\theta$ is about 15°.

Topologically, it is evident that condition 1 is automatically satisfied by replacing every Si—Si bond in the elemental a-Si structure by Si—O—Si. Such a "decoration transformation" relationship between the Si and SiO_2 random networks, an idea already mentioned in the previous section, is a useful conceptual device with which to connect the two structures. However, as a consequence of conditions 2 and 4, it should be viewed as a rough overall relation, and should not be taken too literally. (The idea would apply in detail only if it were the O—Si—O angle that varied, with the Si—O—Si angle being constant at 180°.) Because the Si—O—Si linkage is bent, with a bend that is variable, there is a great deal more flexibility in the SiO_2 system than in a-Si. It is consequently significantly easier to generate a lightly strained random network; unlike a-Si, for amorphous SiO_2 the feasibility of constructing an appropriate crn was never in doubt! In a-Si, there is a negligible occurrence of four-membered rings because a 90° bond angle represents too great a distortion from the tetrahedral angle of 109°. In the SiO_2 crn of Bell and Dean, however, there are a substantial number of eight-membered rings (the corresponding structural unit, containing four Si atoms) because of this greater flexibility provided by the ''bridging'' oxygens. Thus the topologies of the elemental and compound continuous random networks are not quite interconvertible via a decoration transformation, as seen by this sample from their ring statistics. Nevertheless, the topological notion of the decoration transformation provides a convenient mnemonic for categorizing random-network structures. It will appear again in connection with A_2B_3 chalcogenide glasses, which are close to a realization of the original structure sketched by Zachariasen (Fig. 2.13b).

The RDF of the crn of fused silica will be discussed in the following section on comparisons with experiment. In addition to SiO_2 glass, this structure is believed to be relevant to other amorphous AB_2 compounds with 4,2 coordination. This class of glasses includes GeO_2, GeS_2, $GeSe_2$, $GeTe_2$, $SiTe_2$, and BeF_2. All of these are covalent glasses except for BeF_2, which is an unusual highly ionic glass adopting a 4,2 coordination because the F^- ions are too fat for more than four of them to fit around a positive ion.

2.6 EXPERIMENTAL RDFs VERSUS rcp AND crn MODELS

Although microcrystalline models for glasses continue to appear periodically (pun intended), the great preponderance of evidence is in favor of continuous-random models like those described in this chapter. Some of that evidence will be described now. Material cited in this section has been selected as much for illustrative and historical significance as for technical definitiveness. Although we will emphasize the affirmative in enumerating successes of the continuous-random (or noncrystalline) picture of amorphous solids, it should be borne in mind that there exists a converse aspect which is of equal significance. This is the inability of the microcrystalline picture to achieve comparable agreement with experiment via calculations for arrays of small crystals. The inconsistency of observed scattering patterns with microcrystalline models has been demonstrated in those instances in which accurate diffraction data have been available to test detailed model calculations. That negative evidence for microcrystallinity, in combination with the positive evidence cited below, makes a convincing case for continuous-random models.

Before presenting these data it may be useful to remark again on the distinction between microcrystallinity and noncrystallinity. Fine-grained polycrystalline solids certainly exist in nature. It has been demonstrated in one study that crystal grains on a scale as small as 15 Å can be identified in diffraction studies on a very-fine-grained metal film. What has been contradicted is the universality of a crystalline picture for all solids, via the inapplicability of that picture for the amorphous solid state. Microcrystalline models are heterogeneous in that the microcrystals are supposed to be separated from each other by a disordered grain-boundary region of connective material. (The substantial volume fraction occupied by the low-density "connective tissue," which is a necessary consequence of a small crystal-grain size, implies an appreciable density deficit that is typically in disagreement with experiment and is one of the main problems associated with a microcrystalline model.) Correlations between atomic positions are very high at small separations, and then decrease abruptly when the separation traverses a grain boundary. Continuous random models, on the other hand, are homogeneous. In these models, which are concordant with the data, the fall-off in the correlation between atomic positions occurs gradually with increasing separation.

The probes are photons or electrons or neutrons of wavelength comparable to interatomic spacings, and the experiments (sketched earlier in Fig. 2.4) determine momentum distributions of elastically scattered particles (i.e., diffraction patterns). Structural information, in the form of the averaged radial distribution function $RDF(r) = 4\pi r^2 n(r)$, where $n(r)$ is the number density per unit volume for atom centers at a distance r from a given atom, is obtained as a Fourier transform of the observed momentum distribution:

$$RDF(r) = 4\pi r^2 n_0 + \frac{2r}{\pi} \int_0^\infty k[I(k) - 1]\sin(kr)dk. \qquad (2.8)$$

$I(k)$ is the normalized intensity distribution and n_0 is the average number of atoms per unit volume. The momentum transfer or scattering vector k is $(4\pi/\lambda)\sin\theta$, where λ is the wavelength of the probing radiation and 2θ is the scattering angle. Sources of error in $RDF(r)$ include the experimental uncertainty in I, the angular resolution $\Delta\theta$ governing the discernibility of details in $I(k)$, "truncation errors" arising from the finite range of k accessible to experiment [the Fourier inversion assumes $I(k)$ known for all k], and complications in the analysis for solids containing several atomic species. With much effort, these problems have successfully been overcome by various experimentalists to yield RDFs sufficiently accurate to give a substantial amount of structural information.

Among the several techniques, scattering of X-ray photons is in widest use. Neutron scattering typically requires large sample volumes (~ 1 cm^3), and it has been used for bulk glasses such as SiO_2. Electron scattering, because of the relatively low penetration depth, is typically limited to thin films (< 1000 Å). However, since several interesting classes of amorphous solids (such as metals and Ge-family semiconductors) are normally preparable only in thin-film form, electron diffraction has been a very useful tool in this field and several classic examples are included here.

Examples of metallic and covalent glasses will be discussed together in this section, along with the corresponding rcp and crn continuous-random models. Figure 2.18 displays a pair of photographic records of electron diffraction patterns from thin samples of amorphous and crystalline iron (Ichikawa, 1973). Nominally pure amorphous metals are stable only as thin films at low temperature. The pattern of diffuse concentric bands in the upper part of Fig. 2.18 was produced by diffracted electrons transmitted through a vapor-deposited film of amorphous iron prepared (by condensation from the vapor) and held at liquid-helium temperature. The pattern of sharp rings shown in the lower part

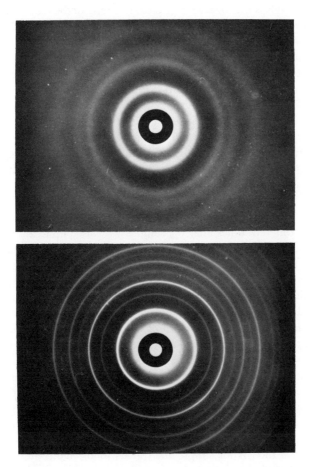

Figure 2.18 Electron diffraction patterns of amorphous (upper photograph) and crystalline (lower photograph) iron (Ichikawa, 1973).

of the figure was obtained later on the same sample after warming to room temperature; the film has crystallized. An impressive feature of Fig. 2.18 is the ease with which the sharp-ring Debye–Scherrer pattern of polycrystalline iron is discerned and recognized (and contrasted to the diffuse-halo pattern of the amorphous form), despite the very fine scale of crystallinity involved; the sample was only 100 Å thick!

Metallic glasses composed of two or more atomic species can be retained at room temperature and, like conventional covalent glasses, can be prepared by cooling (but *quickly*) the liquid. Alloy compositions favorable for glass formation usually correspond to low-lying eutectics, that is, the addition of the minority component raises the stability of the liquid and amorphous states relative to that of the crystalline state. Also, once formed, the presence in the glass of several types of atoms impedes diffusion and helps prevent crystallization. A diffraction pattern obtained at room temperature from such a melt-quenched alloy glass is shown in Fig. 2.19, this time in the form of a plot of the angular distribution of the scattered intensity observed in an X-ray experiment (Waseda and Masumoto, 1975). The metallic glass in this case was an iron alloy containing 20 atomic percent of covalent impurities, $Fe_{80}P_{13}C_7$, and the sample thickness was 0.03 mm. The sharp pattern seen after heating-induced crystallization of the sample is included in Fig. 2.19 as the analog of the lower part of Fig. 2.18.

Now we move on to a covalent glass of the simplest type. Figure 2.20 displays the experimental intensity profile for a 100-Å-thick sample of amorphous silicon, observed at room temperature in electron diffraction (Moss and Graczyk, 1969). This time the diffracted electron intensity is shown as a function of the size of the scattering vector $k = (4\pi/\lambda)\sin\theta$, in a form appropriate for the Fourier transformation needed for obtaining real-space information such as the radial distribution function. Also shown are a pair of intensity profiles taken at different times after the sample was raised up to a temperature of 600°C and kept there; these correspond to two stages in the partial consumption of the amorphous matrix by the nucleation and growth of crystalline regions. Again, the dichotomy between the observation of diffuse halos (for the glass) and sharp rings (for the crystal) is clear.

Moss and Graczyk, the experimenters responsible for the careful data of Fig. 2.20, also carried out a program of model calculations to see whether or not a mosaic of very small crystals could reproduce the scattering pattern they observed for a-Si. No satisfactory agreement with experiment could be obtained. That exercise, and others like it, have provided a damaging contradiction of the microcrystalline picture. We now proceed to the positive evidence in favor of the continuous-random picture of the amorphous solid state.

Figures 2.21 and 2.22 represent historically important (albeit relatively recent, both appeared after 1970) successes of the two basic prototypes of continuous-random models for simple amorphous solids: the continuous random network for elemental covalent glasses and random close packing for metallic glasses. Although subsequent refinements in the crn and rcp structural models

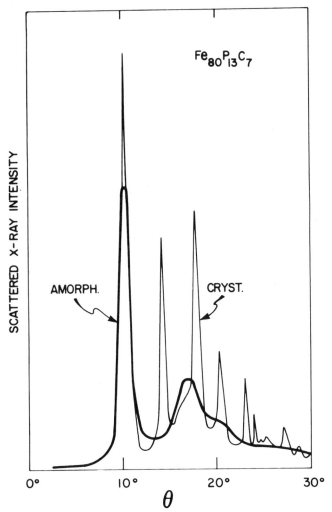

Figure 2.19 X-ray scattering data (Waseda and Masumoto, 1975) on a metallic glass (heavy curve) and on the same sample after crystallization (light curve).

have further improved the agreement with experiment, these early diagrams illustrate the main aspects.

The shaded histogram in Fig. 2.21 is the radial distribution function that Polk obtained from his original hand-built model for a tetrahedrally bonded amorphous semiconductor. It is superimposed upon the RDF of a-Si, derived by Moss and Graczyk by Fourier transformation of the momentum distribution of Fig. 2.20. The distance axis of the experimental RDF has been scaled up by a factor of order 10^8, since the scale used in Fig. 2.21 (corresponding to the model) is given in inches! The vertical lines in the figure, and their numerical labels, correspond to the discrete r_i, z_i coordination shells of the crystal. The

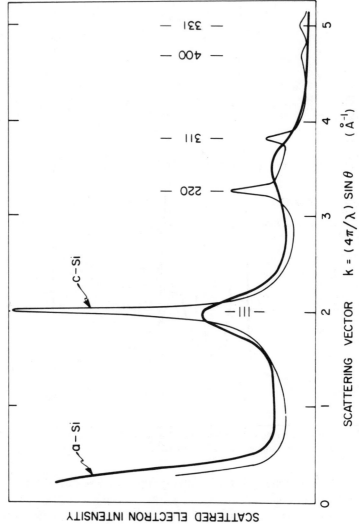

Figure 2.20 Electron diffraction pattern (Moss and Graczyk, 1969) of amorphous silicon (heavy curve) and of the same film after partial crystallization (light curve).

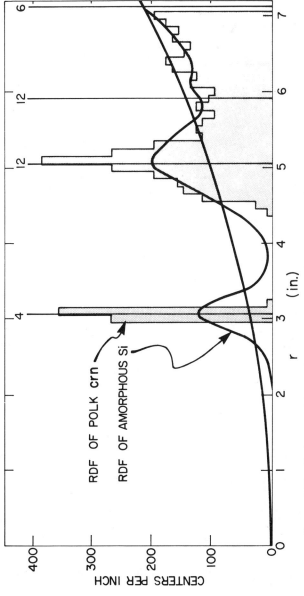

Figure 2.21 Early test of the Polk model (1971) via a comparison of its radial distribution function with that of amorphous silicon.

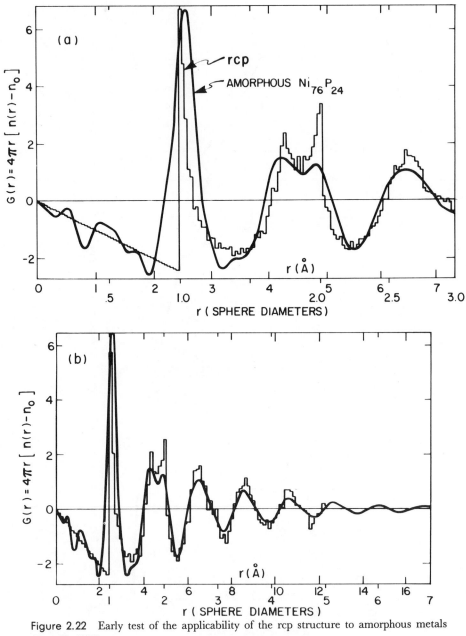

Figure 2.22 Early test of the applicability of the rcp structure to amorphous metals (Cargill, 1975).

model RDF agrees with the experimental one in exhibiting a sharp first peak, a broad second peak in the proper place, the absence of a peak in the vicinity of the crystalline third-neighbor distance, and the subsequent rapid convergence to the average-density line (the parabola in Fig. 2.21). All of these are first-order observations satisfactorily accounted for by this relatively primitive crn model. Refined versions of the Polk tetrahedral crn yield truly impressive agreement with experiments; an example will be given shortly. (This is an opportune moment to briefly mention one of the central difficulties encountered by microcrystalline models in their failure to produce comparable fits to experimental RDFs. This is their inability to generate a first peak as sharp as the observed one without *also* yielding too-sharp peaks at large separations.) Another aspect of the crn RDF deserves comment. The "bottoming out" of the pair-correlation distribution, which occurs not only before the first peak but also *between* the first- and second-neighbor peaks, is a reflection of the very high degree of short-range order which characterizes covalent glasses.

Figure 2.22, due to Cargill (1975), contains a comparison (between experiment and a structural model) which is a legitimate counterpart, for metallic glasses, of Fig. 2.21 for covalent glasses. The quantity plotted in Fig. 2.22 is the reduced radial distribution function $G(r) = 4\pi r[n(r) - n_0] = (1/r)\text{RDF}(r) - 4\pi r n_0$. The histogram is the reduced RDF for random close packing obtained from the detailed model constructed by Finney (1970). The curve is the $G(r)$ derived by Cargill from X-ray measurements on amorphous $Ni_{76}P_{24}$ foils electroplated from solution. Although an alloy was used in that study in order to have a "thick" (0.3 mm) glassy metal available for detailed investigation at room temperature, it happens that the metallic atomic radii of nickel and phosphorus are very close so that a model based on equal spheres is not inappropriate in this case. Results obtained on elemental metallic glasses at low temperature are quite similar. Cargill constructed the comparison of Fig. 2.22 after failing to obtain agreement with his nickel–phosphorus data on the basis of microcrystallite models. Random close packing yielded satisfactory overall agreement with the observed distribution function.

The first peak in Fig. 2.22, reflecting nearest-neighbor atom–atom configurations, corresponds in the model to spheres in contact with each other. It occurs, of course, at $r = D$, where D is the hard-sphere diameter. The splitting of the second peak in Fig. 2.22, a ubiquitous feature observed for metallic glasses, is a characteristic signature of the rcp structure. The geometric origins of the two split "subpeaks" are sketched in Fig. 2.23 (Bennett, 1972). The second subpeak, occurring near $r = 2D$, originates from the set of local configurations in which two second-neighbor spheres are in contact with an intervening sphere in a nearly linear array. Such closely linear three-sphere arrangements occur very frequently in close-packed structures. A sharp drop-off is expected at $2D$, the largest r possible for such a configuration, and the associated maximum is located just below $r = 2D$. The other subpeak of the splitting, occurring near $r = \sqrt{3}D$, originates from the continuum of configurations illustrated

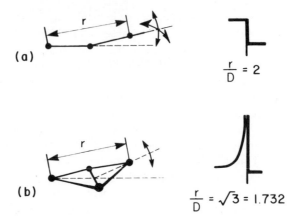

Figure 2.23 Geometric origin of the split second peak in the RDF of metallic glasses (Bennett, 1972).

in the lower part of Fig. 2.23. Two second neighbors each contact a pair of mutually touching intervening spheres.

The split second peak is a key characteristic of the RDF of amorphous metals. It is notably *absent* in the RDF of *liquid* metals. Its prediction by the rcp model is a marked success of that continuum-random structural picture for simple metals in the amorphous solid state. In Fig. 2.22, it can be seen that the subpeaks of the rcp histogram are reversed in relative height vis-à-vis the subpeaks of the experimental curve. This defect disappears in relaxed rcp models (Barker et al., 1975; Yamamoto et al., 1978). The assumption of a simple intermolecular-type potential between the sphere centers of the empirically constructed rcp structure leads to small energy-minimizing adjustments that shift the relative height of the subpeaks into agreement with observation for amorphous metals. Refinements have also been developed to deal with details of the alloy structures. A form of chemical short-range order exists in the alloy glasses in that the minority covalent atoms avoid having each other as nearest neighbors.

We conclude these case studies by returning to covalent glasses and their crn structures. Figures 2.24 and 2.25 contain a pair of comparisons involving a careful X-ray-derived RDF of amorphous germanium. The former figure shows a superposition of this RDF and the RDF of a crystalline powder of the same element. This crystal/glass comparison is an especially nice one because both curves were obtained under the same experimental conditions and by means of the same analytical transformation procedure (Temkin et al., 1973). Although the polycrystalline Ge used in this measurement was quite fine-grained, its RDF reveals the characteristic sharp crystal peaks out to distances beyond four times the bond length. The contrast to the RDF of amorphous Ge, which by this distance has settled down close to the asymptotic average-density parabola, is clear and unmistakable.

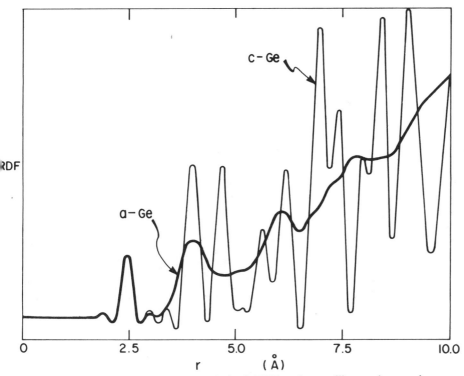

Figure 2.24 Comparison of X-ray derived RDFs of crystalline and amorphous germanium (after Temkin, Paul, and Connell, 1973).

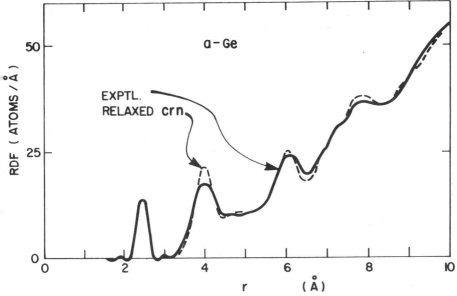

Figure 2.25 Comparison of the RDF of a refined version of the four-coordinated continuous random network with that observed for amorphous germanium (Steinhardt, Alben, and Weaire, 1974).

The *a*-Ge RDF is also relevant to the experimental/model comparison contained in Fig. 2.25. The latter compares the observed RDF with one obtained for a relaxed, broadened, Polk-type, tetrahedrally bonded crn (Steinhardt et al., 1974). Aside from the bond length, which is determined from the nearest-neighbor peak position, the model-structure RDF contained no adjustable parameters. The broadening used for the crn RDF was determined from the observed *c*-Ge RDF of Fig. 2.25, automatically correcting for the vibrational, instrumental, and analytical broadening contained in the experimental curve. The RDF thereby obtained for the tetrahedrally bonded continuous random network is seen in Fig. 2.25 to be in excellent agreement with the observed RDF of amorphous Ge.

Having focused on simple, mainly elemental, glasses, it is well to close with the classic oxide glass, *a*-SiO_2. Figure 2.26 displays a comparison of the X-ray RDF for fused silica and the corresponding 4,2-coordinated crn RDF constructed by Bell and Dean (1972). The first three peaks in the distribution function correspond to the smallest Si—O, O—O, and Si—Si interatomic spacings, in that order. Again, the crn model succeeds in accounting for the observed RDF of the covalent glass; the experimental curve is well reproduced

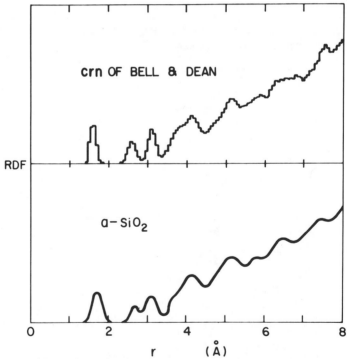

Figure 2.26 Comparison displayed by Bell and Dean (1972), between calculated and observed X-ray RDFs, supporting the applicability of their crn model to the structure of fused silica.

over its full range of definition, out to a distance of about five times the bond length.

REFERENCES

Barker, J. A., M. R. Hoare, and J. L. Finney, 1975, *Nature* **257**, 120.

Bell, R. J., and P. Dean, 1972, *Phil. Mag.* **25**, 1381.

Bennett, C. H., 1972, *J. Appl. Phys.* **43**, 2727.

Bernal, J. D., 1965, in *Liquids: Structure, Properties, Solid Interactions,* edited by T. J. Hughel, Elsevier, Amsterdam, p. 25.

Cargill, G. S., 1970, *J. Appl. Phys.* **41**, 2248.

Cargill, G. S., 1975, *Solid State Phys.* **30**, 227.

Chaudhari, P., and D. Turnbull, 1978, *Science* **199**, 11.

Cohen, M. H., and D. Turnbull, 1964, *Nature* **203**, 964.

Coxeter, H. S. M., 1958, *Illinois J. Math.* **2**, 746.

Finney, J. L., 1970, *Proc. Roy. Soc. (London) A* **319**, 479.

Hales, S., 1727, *Vegetable Staticks*, Innys and Woodward, London.

Ichikawa, T., 1973, *Phys. Status Solidi A* **19**, 707.

Moss, S. C., and J. F. Graczyk, 1969, *Phys. Rev. Letters* **23**, 1167.

Pauling, L., and R. Hayward, 1964, *The Architecture of Molecules*, Freeman, San Francisco.

Polk, D. E., 1971, *J. Non-Crystalline Solids* **5**, 365.

Polk, D. E., and D. S. Boudreaux, 1973, *Phys. Rev. Letters* **31**, 92.

Rogers, C. A., 1958, *Proc. London Math. Soc.* **8**, 609

Scott, G. D., and D. M. Kilgour, 1969, *J. Phys. D* **2**, 263.

Steinhardt, P., R. Alben, and D. Weaire, 1974, *J. Non-Crystalline Solids* **15**, 199.

Temkin, R. J., W. Paul, and G. A. N. Connell, 1973, *Adv. Phys.* **22**, 581.

Waseda, Y., and T. Masumoto, 1975, *Z. Physik B* **22**, 121.

Yamamoto, R., H. Matsuoka, and M. Doyama, 1978, *Phys. Status Solidi A* **45**, 305.

Zachariasen, W. H., 1932, *J. Am. Chem. Soc.* **54**, 3841.

CHAPTER THREE

Chalcogenide Glasses and Organic Polymers

3.1 MOLECULAR SOLIDS AND NETWORK DIMENSIONALITY

For solid germanium, whether in crystalline or amorphous form, the covalent graph is *macroscopically* extended in *three dimensions*. Imagine a spider to be crawling along the "bonds" of the c-Ge model of Fig. 2.11. Given enough time, there is nothing to prevent an intrepid individual from visiting, in turn, every "atom" in the model. In other words, for the crystal, paths of covalent bonds connect every atom with every other atom in a macroscopic sample of the material. The same statement applies to the tetrahedral continuous random network (fragmentally represented in Fig. 2.15) of amorphous Ge. Covalently bonded solids like these possess covalent graphs that are, in the sense specified here, three dimensional in character. We now introduce solids characterized by covalent graphs of *low dimensionality*.

Many important amorphous solids are representable by *disconnected* covalent graphs, that is, they are *molecular solids*. Molecular solids are characterized by the coexistence of strong (primarily covalent) and weak (intermolecular, primarily "van der Waals") forces, with the attendant appearance of cohesive atomic groupings which mutually interact via the weak forces while being internally bound by the strong. Notable among molecular amorphous solids are the *chalcogenide glasses*, materials containing one or more of the chalcogen elements of the sixth column of the Periodic Table: sulfur, selenium, and tellurium.

The molecular unit in orthorhombic sulfur, a familiar elemental molecular crystal, is the puckered eight-atom ring shown in Fig. 3.1. That figure indicates the regular arrangement of these crown-shaped molecules in the crystal. The packing pattern of the molecules is rather intricate, with the

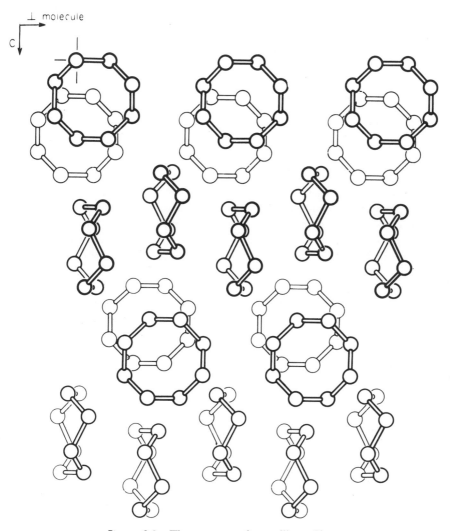

Figure 3.1 The structure of crystalline sulfur.

rings stacked in staggered array to form two sets of nearly perpendicular columns that interweave each other log-cabin style.

For rhombic sulfur, the covalent graph consists of a large set of disconnected eight-atom rings. Our hypothetical itinerant spider, finding itself on a model of this material and constrained (as usual) to move along the strong bonds, would no longer have the whole atomic assembly at its disposal but would instead be unhappily confined to a tiny island of eight atoms. *This fissioned feature, in which the solid-state covalent graph is fragmented into a large set of unconnected separate graphs, marks molecular solids as their defining characteristic.*

At this juncture, it is very important to emphasize that molecular solids are naturally classified into several distinct categories on the basis of the dimensionality of the component molecular graphs into which the solid-state covalent graph subdivides. This connectivity-derived notion of *molecular network dimensionality* is most easily expressed in terms of a few simple examples. All of the possibilities can be represented by materials constructed from elements contained in that small portion of the periodic table which is exhibited in Table 3.1. Atoms of these elements, from columns four, five, and six and from rows three and four of the periodic table, have a strong tendency to form covalent bonds with a small number of neighbors. Lighter elements (e.g., from row two, such as oxygen) have a greater tendency for ionic bonding; heavier elements (e.g., from row five, such as tin or tellurium) have a greater tendency for metallic bonding. The elements included in Table 3.1 are close to being *maximally covalent* in their bonding behavior. The number of *s* and *p* valence electrons *n*, and the most characteristic covalent coordination *z*, given in Table 3.1, satisfy the $z = 8 - n$ rule (discussed further in Section 3.4) expected on the basis of the number of shared electrons needed to attain a closed shell.

Examples of each of the four possibilities for molecular dimensionality are listed in Table 3.2 (Zallen, 1974). Included, for reasons of logical completeness, is the case of germanium, which is not a molecular solid at all since there is no molecular unit that can be "dissected out" of the covalent network (without breaking bonds, which is impermissible in this context). This example of a fully bonded covalent solid is included as the limiting case of a "molecule" that is macroscopically extended in all three dimensions. For concreteness let us consider samples measuring a few millimeters on a side, containing of the order of 10^{21} atoms. Using L to denote a characteristic linear dimension of the sample in units of the average interatomic separation, we note that L is of the order of 10^7 for this sample size (which is typical of many experimental situations). For a Ge sample of this size, the atoms that are connected together by covalent bonds (to make up the single all-encompassing "molecular unit") number, in order of magnitude, L^3. By contrast, in a

Table 3.1 A part of the Periodic Table with elements exemplifying covalent-bonding behavior

Number of Valence Electrons		
n = 4	*n = 5*	*n = 6*
Si	P	S
Ge	As	Se
z = 4	*z = 3*	*z = 2*
Normal Covalent Coordination		

Table 3.2 Examples of each type of molecular network dimensionality

Network Dimensionality	Crystal[a]	Coordination Number[b] z	Size of the Molecular Unit[c]
Zero	S	2	8
One	Se	2	10^7
Two	As_2S_3	3	10^{14}
Three	Ge	4	10^{21}

[a]Orthorhombic sulfur, trigonal selenium.
[b]For As_2S_3, $z = 3$ corresponds to the arsenic coordination, which dominates the network topology.
[c]Number of atoms connected by covalent bonds, for a macroscopic sample containing about 10^{21} atoms.

similar-size sample of rhombic sulfur the number of atoms comprising the molecular unit (again expressed as a power of L) is of order L^0. Of course for sulfur, the S_8 molecular unit is *independent* of sample size, while for Ge, the Ge_N molecular unit ($N \approx L^3$) is exactly the *same* as the sample size.

With these diametrically opposite examples in mind, the definition we seek follows quite naturally. Clearly, for these two limiting cases, the appropriate molecular dimensionality is three for germanium and *zero* for sulfur. *We define the network dimensionality as the number of dimensions in which the covalently bonded molecular unit is macroscopically extended.* Equivalently, it is the *number of dimensions in which the covalent graph belonging to a single molecule is indefinitely extendable*, that is, "infinite" in the normal parlance employed by the solid-state physicist (typically $\sim$ 1 mm). The three-dimensional-network character of Ge has already been emphasized. The corresponding network dimensionality of rhombic sulfur is zero because the molecular unit is macroscopically extended in *no* dimensions: It is *finite* on an atomic scale.

Zero-dimensional-network crystals form the largest and most familiar class of molecular solid. Our choice of S_8 for illustration of this class was adopted in order to use an example from column six, because the chalcogens underlie the important glasses to be discussed next. Examples from columns seven and eight form even simpler elemental molecular crystals. Iodine, for example, with $8 - n$ calling for a covalent coordination of unity, has little choice but to crystallize as a molecular crystal composed of I_2 molecules. For the solid rare gases of column eight, $z = 8 - n$ is zero and the resulting "molecular crystals" (if it is really appropriate to call them that) consist of weakly interacting closed-shell spherical "molecules" that are simply individual neutral atoms. From that point of view, the molecular unit in, say, solid argon would be written as Ar_1. Moving in the other direction in terms of complexity, a familiar broad class of zero-dimensional-network solids are organic crystals such as naphthalene ($C_{10}H_8$), anthracene ($C_{14}H_{10}$), pyrene ($C_{16}H_{10}$), etc.

3.2 ONE- AND TWO-DIMENSIONAL-NETWORK SOLIDS

Having discussed 0d and 3d networks, it is now time to fill in the two intermediate types of network dimensionality, which have obviously been thus far omitted. Important chalcogenide examples of 1d- and 2d-network *crystalline* solids have been incorporated in Table 3.2. Indeed it is *these* network dimensionalities, 1d and 2d, that best characterize the chalcogenide glasses. Like 0d-network solids, but in contradistinction to 3d-network solids (called by some authors simply "network solids," a simplified but ambiguous usage that we will avoid), 1d- and 2d-network solids are molecular solids because discrete molecular units are clearly identifiable. But unlike 0d-network solids, the molecular units are now macroscopically extended "macromolecules."

As usual, we begin with the corresponding crystals in order to have a secure point of departure for the discussion of structure. Figure 3.2 displays the molecular unit of the trigonal form of crystalline selenium (henceforth simply *c*-Se), which is the stablest solid form of this element. Like the sulfur atoms in S_8, each Se atom is twofold coordinated. However instead of closing back on itself to form a ring, the —Se—Se—Se— chain in *c*-Se keeps on going "indefinitely" (i.e., until it is terminated by encountering the surface or some other crystal "defect"). Each chain is a helical coil, with three atoms per turn of the helix. In the crystal the helical chains are arranged on a triangular lattice, as expected for the close packing of rods.

Trigonal Se is manifestly a 1d-network molecular solid. The molecular unit Se_N is macroscopically extended in one dimension. One-dimensional-network macromolecules are usually termed *polymers*, and trigonal Se is an elemental, crystalline, *polymeric* solid. In many ways it is the *simplest* known polymeric solid. Let us use L, as before, to denote the dimensionless measure of length for a macroscopic solid sample containing L^3 atoms. In the case of *c*-Se we may identify N with L (assuming an "ideal" crystal in which no defects interrupt the chains). The solid may be thought to contain roughly L^2 molecules, with each polymer molecule containing roughly $N \approx L^1$ atoms. N plays a role in this discussion that is precisely analogous to the *polymerization index* which occurs in discussions of organic polymers (and that appears, in that context, in subsequent sections). The notion of network dimensionality introduced earlier may be identified with the *exponent* of L appearing in the appropriate polymerization index $N \approx L^d$ for the molecular entity in question. For the helical-chain molecular unit of trigonal Se, that exponent is evidently 1. A crystalline chalcogenide for which the corresponding exponent is 2 is described next. Prototypical examples of $N \approx L^3$ (Ge) and $N \approx L^0$ (S_8) have already been described.

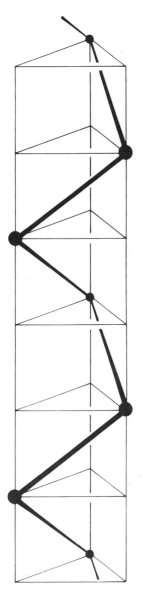

Figure 3.2 A single polymer chain in the trigonal form of crystalline selenium, a 1d-network solid.

Figure 3.3 shows the structure of crystalline As_2S_3. Topologically, the two-dimensionally-extended molecular unit in c-As_2S_3 is a physical realization of the covalently bonded decorated honeycomb lattice which Zachariasen (1932) invoked as the crystalline analog of his A_2B_3 crn glass of Fig. 2.13b. Although the bonding topology of its covalent graph is indeed identical to that of the decorated honeycomb of Fig. 2.12b, the actual geometric structure of the individual layer (single macromolecule) in c-As_2S_3 is atomically non-

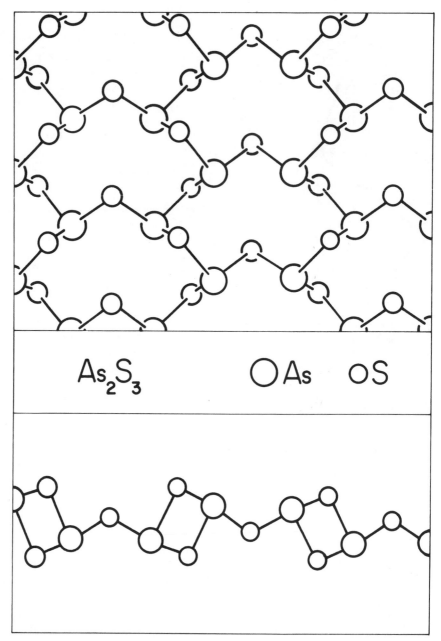

Figure 3.3 One of the extended layers which make up crystalline As_2S_3, a 2d-network solid. The upper panel views the layer broadside; the lower panel views it edge on.

coplanar and relatively nonsymmetric. The threefold coordination of the arsenic atoms dominates the topology of the network. Threefold coordination in the strong bonding is characteristic of two-dimensional-network solids, the simplest example being graphite, the archetype among layer-structure crystals. Crystalline As is also a three-coordinated elemental layer crystal, and in this instance the covalent graphs corresponding to the c-As and c-As$_2$S$_3$ layers *are* rigorously related by a decoration transformation (via the introduction of the bridging sulfurs).

The appropriate polymerization index in c-As$_2$S$_3$ (as in graphite and c-As) is $N \approx L^2$. Thus our habitual millimeter-sized ($L \approx 10^7$) sample contains roughly 10^7 molecules, with each extended-layer (As$_2$S$_3$)$_N$ macromolecule containing roughly 10^{14} atoms. This chalcogenide material is a 2d-network molecular solid. Table 3.2, with its exemplars of each type of network dimensionality, is now completed.

Having discussed their crystalline counterparts, we now turn to the structure of the amorphous chalcogenides. Again, despite the fact that these glasses are technologically important and have been widely studied, their structures are only partially understood. Nevertheless, the single most significant aspect can be stated with considerable confidence: *The short-range order in the glasses is essentially the same as in the corresponding crystals.* Thus Se in a-Se and S in a-As$_2$S$_3$ are twofold-coordinated, As in a-As$_2$S$_3$ is threefold-coordinated, and the bond lengths and bond angles are similar to those in the crystalline forms. Beyond that, however, the picture is less clear (Wright and Leadbetter, 1976).

Amorphous selenium is believed to be a 1d-network solid, composed primarily of Se$_N$ long chains with $N \approx 10^5$. Some small rings, such as Se$_8$, may also be present in this elemental glass, but there is no consensus on that. It is exceedingly difficult to resolve this chain/ring issue by means of diffraction-determined RDF curves. For a single molecule, the first peak expected in the RDF is the same for the Se$_8$ ring and the helical Se$_N$ chain, since the bond length and coordination are the same for both. But the second peak is *also* the same for ring and chain since they have the same bond angle (105°). One has to go out to *third*-nearest-neighbor intramolecular distances to find a difference between ring and chain. In other words, information about the dihedral angle distribution is needed to resolve this question. Such subtle structural information would require $I(k)$ diffraction data out to very high k values, in order to yield a sufficiently precise RDF.

The chains making up Se glass are not constrained to be straight and to maintain the perfectly helical conformation of trigonal Se; they are certainly flexible on a scale length larger than a few times the bond length. The structure of a-Se probably resembles the *random-coil* structure of polymeric organic glasses, which is discussed in detail in the second half of this chapter. (For a qualitative glimpse of the random-coil model of 1d-network amorphous solids, look ahead to Fig. 3.12.) Models based on such convoluted Se$_N$ chains, with random dihedral rotations, have yielded reasonable fits to the observed RDF data for a-Se.

Similar remarks as above apply to amorphous sulfur and tellurium. These elemental glasses are less stable than a-Se, which is easily quenched from the melt and is retained indefinitely at room temperature. Amorphous Te may be obtained by vapor deposition onto a cold substrate, and quickly crystallizes if raised to room temperature. The case of a-S is instructive. Near its melting point, sulfur forms a low-viscosity liquid composed entirely of S_8 rings. A glass *cannot* be readily produced by quenching the highly-mobile 0d-network molecular liquid; it crystallizes. However, when heated above its well-known polymerization temperature, liquid sulfur transforms into a very-high-viscosity liquid (i.e., with a long relaxation time for molecular motion) composed of S_N chains with $N \approx 10^5$. Quenching *this* liquid *does* produce a glass, although for this material the resulting 1d-network amorphous solid is not stable at room temperature and soon converts to rhombic sulfur. For selenium the liquid form contains chains at *all* temperatures, so that quenching the melt always gives rise to the glass.

Although chalcogenides primarily form 1d- and 2d-network glasses, three-dimensional networks can also be obtained by the introduction of column-four elements like Ge which enter the covalent graph in fourfold coordination. (Cross-linking via high coordination, in this context meaning $z > 2$, is discussed next in Section 3.4.) It is even possible to catch a fleeting glimpse of a zero-dimensional-network chalcogenide glass. As_2S_3 is such an excellent glass former that the "quenched" bulk glass may be quite easily prepared by even a very gentle cooling of the liquid form. However a-As_2S_3 may also be prepared by condensation of the vapor. Structural studies show that the vapor-deposited films adopt, upon annealing, essentially the same 2d-network crn structure of the melt-quenched material. Initially, however, it is known that the vapor-quenched a-As_2S_3 films are structurally somewhat different. Scattering experiments reveal the presence of small molecular units such as As_4S_4, a cagelike eight-atom molecule whose structure is sketched in Fig. 3.4. These molecules are present in the vapor and evidently some retain their identity in the "as-deposited" condensed film. Other molecular species, some rich in sulfur, are present as well (DeNeufville et al., 1974; Solin and Papatheodorou, 1977; Nemanich et al., 1978; Slade and Zallen, 1979; Takahashi and Harada, 1980).

It has been suggested that the initial structure of the vapor-condensed amorphous As_2S_3 film resembles the random close packing of nearly spherical chalcogenide molecules. Since the stable glass structure to which the film relaxes upon annealing is the 2,3-coordinated continuous random network, we then have an amorphous system that exhibits (at different stages) *both* canonical structures: rcp and crn. Actually an rcp $\rightarrow$ crn transformation is probably a bit too glib as a description of the structural relaxation occurring here. The crn network appears to be already partially formed upon deposition, and coexists with the molecular holdouts that eventually reluctantly link up with the extended network in a polymerization process that takes place as the sample anneals. Nevertheless, it is very nearly correct (and certainly quite

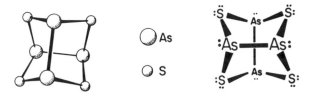

Figure 3.4   Structure of the As_4S_4 molecule.

pleasing) to look upon the annealing process in vapor-deposited a-As_2S_3 as having the nature, at least in part, of a *transition from a Bernal random close packing of ''hard-sphere'' molecules to a Zachariasen continuous random network*. This pretty picture brings together, in a *single material* but at *different times, two basic homogeneous models for the structure of amorphous solids*. Viewed a bit differently, in the context of our discussion of network dimensionality, it is a transition from a zero-dimensional-network to a two-dimensional-network molecular glass.

Figure 3.5 suggests one possible way in which ''bond switching'' may occur to effect the digestion of discrete molecules by a sulfur-rich As_2S_{3+} random-network matrix. (Such a schematic sketch, placing everything in one plane, suffers from unavoidable bond-length distortions but nevertheless conveys the main idea.) The four bonds represented by dashed lines break, while the four ''incipient'' bonds denoted by dotted lines form, in the annealing process which consumes the As_4S_4 molecule. Note that the net result is the

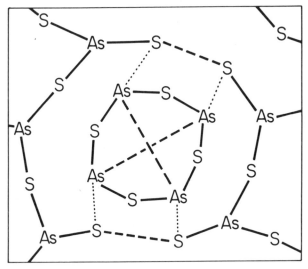

Figure 3.5   One possible way in which bond reconstruction can effect the incorporation of a small molecule into the random network of a chalcogenide glass. The bonds represented by the dotted lines replace the ones represented by the dashed lines.

replacement of like-atom bonds (As—As and S—S) by unlike-atom bonds (As—S). Energy is lowered primarily because the As—As bond is weaker (i.e., higher energy, less stable) than the other two bond types present.

One of the interesting properties of chalcogenide glasses, which underlies an exciting potential application as computer-memory materials, is the possibility of producing a change of structure by means of incident light. The simplest such photoinduced structural change is crystallization. Figure 3.6 displays the observation of photocrystallization in As_2Se_3 glass (Finkman et al., 1974). In this light-scattering experiment, a laser beam is focused on the

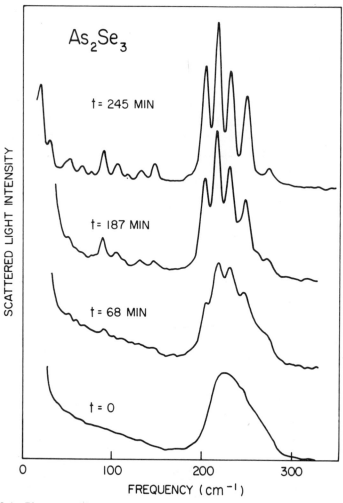

Figure 3.6   Photocrystallization of As_2Se_3, monitored via the appearance of the sharp-line Raman spectrum characteristic of the crystal (Finkman, DeFonzo, and Tauc, 1974).

sample and serves as both the cause of the glass→crystal transformation and the probe of that transformation. The conversion is monitored by means of the Raman scattering excited by the same laser beam responsible for the conversion. As discussed in Chapter Six, the Raman spectrum of an amorphous solid contains contributions from the full vibrational density of states, and is thus very broad. This is in marked contrast to the sharp-line Raman spectrum characteristic of a crystal. Figure 3.6 shows the evolution of the line spectrum of c-As_2Se_3 out of the broad continuum of the a-As_2Se_3 spectrum as the light-induced crystallization of the sample proceeds. These curves may be considered to present a light-scattering analog of the electron-scattering curves of Fig. 2.20. In both experiments, a discrete distribution substituting for a continuous distribution provides the evidence for the growth of crystalline regions within an initially amorphous sample.

3.3 COMPOSITIONAL FREEDOM IN CHALCOGENIDE GLASSES AND IN OXIDES

Amorphous Se and As_2Se_3 have been introduced as representative chalcogenide glasses embodying, respectively, 1d-network and 2d-network molecular glasses. In fact, melt-quenched amorphous solids are easily formed for *all* chemical compositions intermediate between these two, that is, for all compositions of the Se—As system in the range from 0 to 40 atomic percent arsenic: $Se_{1-x}As_x$ with $0 \leq x \leq 0.4$. For all of these "nonstoichiometric" glasses, the short-range order is well defined and similar to that of the parent glasses: Se is two-coordinated and As is three-coordinated. Chemical order is maintained as before for the arsenic atoms, each of which (as in As_2Se_3) is bonded to three Se atoms. The compositional variability is achieved by the accommodating selenium atoms, each of which is content to be bonded to two Se atoms (as in a-Se) or to two As atoms (as in a-As_2Se_3) or to one Se and one As. The energetically unfavorable As—As bonds are avoided in the glass.

An analogous situation applies to a system of bulk glasses that spans the composition range from Se to $GeSe_2$, i.e. $Se_{1-y}Ge_y$ with $0 \leq y \leq 0.33$. Here the cross is between a 1d-network glass (a-Se) and a 3d-network glass (a-$GeSe_2$ is assumed here to be isomorphic to a-SiO_2). Se atoms are two-coordinated, with neighbors of both types, while Ge atoms are four-coordinated to neighboring Se atoms.

The above two examples couple a chalcogen column-six element (Se) with one of its neighboring elements from either column five (As) or column four (Ge) of the same row of the Periodic Table. Each case comprises a family of variable-composition selenium-rich binary glasses. The corresponding Se-rich ternary system, $Se_{1-x-y}As_xGe_y$ with $x + y$ less than about 0.3, also forms a family of melt-quenchable bulk amorphous solids. Again the coordinations are $z = 2, 3, 4$ for Se, As, Ge. The chemical ordering is such that Se—Se, Se—As, and Se—Ge bonds abound in the structure; while As—As, As—Ge, and Ge—Ge bonds are essentially absent (i.e., occur with frequency very much less than that expected on the basis of nonpreferential random-mix statistics).

A highly schematic sketch, intended only to indicate aspects pertaining to the connectivity of the covalent graph, is presented in Fig. 3.7. Represented here is the bonding network of a $Se_{1-x-y}As_xGe_y$ glass with $x + y \ll 1$, i.e., amorphous selenium incorporating a small concentration of As and Ge atoms. For pure Se, the network dimensionality is one and the structural elements are extended chains. (The possible presence of some rings, interspersed between the chains, is an inessential complication, which we leave out of this discussion.) The introduction of three-coordinated As and four-coordinated Ge produces "branching" or "cross-linking" between Se chains. Without these cross-linking higher-coordinated ($z > 2$) elements, the glass network resembles a mass of entangled and intermeshed spaghetti (again, refer ahead to Fig. 3.12). With them, the topologically linear Se_N chains are sewn together at irregular intervals to form larger and more highly connected structures.

The connectivity changes in an essential way with the concentration of cross-linking constituents. To see this note that the number of disconnected individual molecular units drops sharply with increasing As or Ge concentration

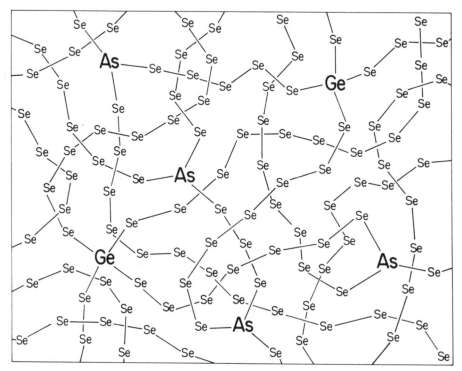

Figure 3.7   Schematic bonding topology of $Se_{1-x-y}As_xGe_y$ glass.

as the separate units link up via the $z > 2$ branch points. Amorphous Se starts out as roughly an assembly of N^2 Se_N linear chain molecules, and winds up (after adding Ge) at $Se_{0.67}Ge_{0.33}$ as a single completely-knit-together three-dimensionally-extended molecule. At this composition the number of molecules has been reduced from order N^2 to unity. (Strictly speaking, the initial "N^2" must be taken with a grain of salt because the polymerization index in amorphous selenium cannot in practice be increased indefinitely; in real samples it is typically about 10^5. Thus the number of chains in a millimeter-size sample with $N^3 \approx 10^{21}$ atoms is of order 10^{16} rather than 10^{14}.) Similarly, for the case of arsenic addition, the transformation from Se to $Se_{0.6}As_{0.4}$ also drastically reduces the number of molecules. This reduction in the number of disconnected parts of the covalent graph is the other side of the coin of increasing molecular size with increasing network connectivity.

The continuous variability of chemical composition (x, y) in these systems, taken together with the above discussion of the different network dimensionalities of the end members of each series, makes it tempting to consider the possibility of network dimensionalities that are *intermediate* in value between the integers, 1, 2, and 3. Indeed such noninteger or fractional dimensionalities will usefully appear, but in a quite different context (that of "similarity dimensionality"), in the next-to-last section of this chapter. However, in the present context of covalent networks, such a generalization to in-between dimensionalities is unwieldy and it will not be introduced here. One reason is the complication caused by percolation, a topic to be covered in the next chapter. For long chains of length N, cross-linking connections present in a concentration as low as of order $1/N$ are sufficient to produce a "percolating" three-dimensionally-extended network. This 1d → 3d vulcanization transformation is treated in Section 4.6. To characterize the intermediate network topology of glasses of intermediate chemical composition, a straightforward measure (which is simply the average coordination number) will come into play in our discussion of the structural aspects of the glass transition.

Three variable-composition glasses have been mentioned thus far: a "6-5" binary (two-component compound made up of elements of columns six and five), namely, Se—As; a 6-4 binary, Se—Ge; and a 6-5-4 ternary, Se—As—Ge. A great variety of other combinations are available to these chalcogenide systems. Atoms which may readily be woven into the fabric of a covalent network like the one sketched in Fig. 3.7 include S, Se, Te from group six; As, Sb from group five; and Si, Ge from group four. In addition to binaries and ternaries, more complex multicomponent systems (quaternaries, quinternaries, etc.) can be prepared. The compositional variability of this class of amorphous solids permits their properties to be tailored to specific applications; some of these will be spelled out in Chapter Six.

It probably has already occurred to the reader that, if the definition of chalcogen is a group-six element with the characteristic s^2p^4 outer-shell electron configuration, then there is an evident omission in the list which has been habitually recited: S, Se, Te. Why isn't oxygen usually included in this classification? The chalcogenide glass $GeSe_2$ is, after all, a (4,2)-coordinated solid as is the classic oxide glass SiO_2. In fact, a case *could* be made for considering fused silica as a chalcogenide glass (though this view would be highly nontraditional). Nevertheless, there are significant differences between oxide glasses and the family of glasses based on S, Se, and Te. These differences are mainly attributable to the high electronegativity of oxygen and its marked tendency to accept one or two electrons to form O^- or O^{2-} ions.

SiO_2 is itself predominantly covalently bonded. Ge can substitute for Si in the (4, 2) continuous random network to form a mixed $Si_{1-x}Ge_xO_2$ glass, in the same way that Ge can substitute for Si in the ternary chalcogenide $Si_{1-x}Ge_xSe_2$. In the convention used for oxide glasses, Si and Ge are termed "network formers." (All of the elements of groups four and five, which have been discussed in the context of chalcogenide glasses, are, analogously, network formers.) Thus far, the behavior is familiar. The ionic feature that arises because of oxygen's propensity to form polar bonds is in the manner of incorporation into SiO_2 of highly electropositive elements such as Li, Na, and K.

For concreteness we consider the case of sodium; the other alkalis behave similarly in multicomponent oxide glasses. The ternary compounds of the system Na—Si—O form glasses which may be written as $Na_{4x}Si_{2-x}O_4$, where x is continuously variable from $x = 0$ up to about $x = 0.5$. (The upper limit of the glass-forming range naturally depends on the speed of melt quenching; the faster the quench, the higher the value of x for which an amorphous solid is produced.) Unlike all of the cases discussed up to now, the sodium atoms do *not* form bonds with (and thereby become part of) the covalent network. Instead, each sodium atom gives up an electron and goes into the solid *interstitially* (i.e., into the interstices, the holes, which exist within the crn structure of the solid) as positively charged closed-shell spherically symmetric Na^+ *ions*. The electrons are taken up by oxygen atoms to form O^- ions.

While an O atom, with its six outer-shell electrons, is covalently divalent (the "8 – n rule" for covalent coordination is spelled out in the following section); an O^- ion, with its seven outer-shell electrons, is isoelectronic to fluorine and is monovalent. The silicon–oxygen continuous random network in $Na_{4x}Si_{2-x}O_4$ differs from the crn structure of SiO_2 in the presence of these singly bonded "nonbridging" oxygens, each of which is bonded to only a single silicon atom and is topologically "dead-ended." Each Si is bonded to four O's, as in silica, but now there are both $z = 2$ and $z = 1$ O atoms, and the average oxygen coordination is $2 - x$. Clearly the connectivity of the covalent graph decreases with increasing sodium content. Thus, although the sodium ions do not become a part of the crn (using a tailored metaphor invoked earlier, the Na^+ ions are not themselves directly "woven into the fabric of the covalent network," but instead sit in its pockets!), they significantly alter its topology via the physical requirement of charge compensation and the charge-accepting valence versatility of oxygen. Hence the alkalis are not network formers but are named "network modifiers" in the silicate glasses.

The stoichiometric variability in $Na_{4x}Si_{2-x}O_4$ is thus accomplished in a way that is quite different from the way that compositional variety is achieved in, say, $Se_{1-x-y}As_xGe_y$. Note that if the sodium content of the silicate glass could be increased to a concentration corresponding to $x = 1$, then *all* of the oxygens would be nonbridging and the covalent graph would be broken up into isolated SiO_4^{4-} ions. This situation corresponds, from a covalent viewpoint, to a 0d-network solid. However, it is an ionic rather than a molecular solid, held together by the net Coulomb attraction between the sodium and the silicate ions. Ionic forces are intermediate in strength between intermolecular forces and the strong covalent forces that constitute the Si—O bond. This 0d-network solid does indeed exist, but not in an amorphous form derivable from the melt; it occurs instead as crystalline sodium orthosilicate Na_4SiO_4. The reason that the melt of this composition cannot be quenched to form a glass is the high molecular mobility of the small ions that comprise the liquid; crystallization takes place too quickly to be bypassed at accessible cooling rates.

The situation described in the last sentence above, namely the difficulty of forming a glass by cooling a melt composed of agile small molecules (i.e., a 0d-network liquid) because of the low viscosity and consequent rapid crystallization rate characteristic of such a liquid, was encountered before in the case of sulfur mentioned in the preceding section. A third example of this will be mentioned here. In discussing the binary system $Se_{1-x}As_x$, we noted that melt-quenched glasses were easy to make for $x < 0.4$ and, by implication, were not easy to make for $x > 0.4$. Above $x = 0.4$, arsenic–arsenic bonds can no longer be avoided. Small Se—As molecules now compete favorably with the extended network, the most notable species being As_4Se_4. This cage-structure molecule, containing two As—As bonds, is isomorphic to the As_4S_4 molecule illustrated in Fig. 3.4. (All of this discussion for $Se_{1-x}As_x$ carries over as well to $S_{1-x}As_x$.) Mobile small molecules such as As_4Se_4 or related ones (such as As_4Se_3, which has three As—As bonds) appear in the melt, lower its viscosity, and make it difficult for glass formation to compete with crystallization under normal quenching conditions.

3.4 THE 8 – n RULE AND THE "IDEAL GLASS"

Look again at Fig. 3.7. For all of the geometrical and even topological disorder present in the structure implied by this schematic sketch, a very important *chemical* regularity persists in this drawing. In constructing it, the following rule was adopted as a central working assumption: The valence of *each* constituent atom is satisfied throughout the amorphous solid. Valence in this context is taken to mean the number of single (σ-type) covalent bonds necessary for the neutral atom to complete, via the shared electrons comprising the bonds, its outermost shell of s and p electrons. (The neutral atom is specified here because the ideal of valence satisfaction can be extended to cover chalcogens in ionized states, as discussed for charged defects in the next section.) Since the number needed to complete the shell is $8 - n$, where n is the number of s and p electrons of the atom in question (i.e., the column number locating the element in the Periodic Table, as at the top of Table 3.1), the stated assumption amounts to the following specification on the nearest-neighbor coordination number of the atom in the covalent graph of the glass:

$$z = 8 - n \tag{3.1}$$

That (3.1) is an equation and not an identity may be immediately seen by considering a familiar *crystalline* counterexample of crucial importance. A group-five atom such as arsenic, when present as an impurity in crystalline Ge or Si, enters the covalent network substitutionally in a tetrahedrally coordinated site. In this case, z is 4 while $8 - n$ is 3. The fifth electron of arsenic, not taken up in any bond, forms a loosely bound extra electron that is easily ionized into the conduction band of the host semiconductor crystal. What is described here of course is an example of doping by a donor impurity, representative of the techniques used for controlling electrical properties in the ubiquitous silicon technology that is central to the electronics industry. In a crystalline solid, an impurity atom that enters into the covalent network must take on a coordination that is characteristic of the host crystal, whether its own shell-filling requirements are met (as for Ge in Si, an isoelectronic case) or are *not* met (as for As in Si, as cited above).

The "eight-minus-n rule" of (3.1) was proposed by Mott (1969), in connection with the chalcogenide glasses, to account for the observation that the electrical properties of these solids do *not* exhibit great sensitivity to impurities, and, in fact, are often insensitive to composition changes of even a few percent. This behavior of this class of amorphous semiconductors, which is in such sharp contrast to experience with crystalline semiconductors, was explained by Mott by means of the $8 - n$ rule, which ensures that all of the valence electrons of the constituent atoms are taken up in bonds. This chemical condition is a reasonable one to invoke for melt-quenched glasses. During the relatively slow formation of such bulk amorphous solids, there is enough time to attain this local energy-minimizing chemical-bonding configuration. The validity of Eq. (3.1) is consistent with the bulk of the experimental evidence on these systems.

Although couched mainly in chemical terms, Mott's argument amounts to a significant statement about the structure of amorphous chalcogenides. This is the reason for including it here in a chapter on atomic-scale structure, although it will later reappear in Chapter Six on the *electronic* structure of glasses. To see how the $8 - n$ rule works in a simple case, consult the right-hand side of Fig. 3.4. Each of the heavy bars in this diagram represents a single covalent bond and accounts for a pair of opposite-spin atomically shared electrons occupying an orbital that contributes importantly to the internal chemical cohesion of the covalent network, which in this example is the 0d-network molecule As_4S_4. The pairs of dots denote pairs of opposite-spin electrons occupying nonbonding "lone-pair" atomic orbitals which provide negligible interatomic overlap and which are essentially uninvolved in the intramolecular cohesion. The As_4S_4 molecular structure of Fig. 3.4 provides the energetically optimum geometry for an As_xS_{1-x}, $x = 0.5$ stoichiometry, fulfilling the $8 - n$ topology condition by the incorporation of two As—As bonds in addition to the As—S bonds. [For the $x = 0.40$ stoichiometry, the lowest-energy $(8 - n)$-rule geometry is the 2d-network layer of Fig. 3.3.]

Satisfaction of the covalent-bonding needs of each constituent atom, via strict adherence to the $8 - n$ rule, minimizes the energy locally and guarantees that the glass has nearly as low an energy as the crystal. Thus Mott's suggestion for the chalcogenides echoes Zachariasen's earlier thought in constructing the continuous random network for silica. That such excellent short-range order can exist in the absence of long-range (crystalline) order has been abundantly demonstrated experimentally, and we have tried to evoke this structural situation in figures such as 2.13, 2.15, and 3.7.

The idea, in the sense discussed here, of perfect chemical short-range order, is sometimes expressed in terms of the conceptual construct of an *ideal glass*. This construct, conceptually well-defined but physically nonexistent, may be viewed to play a role in the physics of amorphous solids that is parallel to that played in the physics of crystalline solids by the conceptually useful (but similarly nonexistent "in real life") ideal crystal or perfect crystal. For covalent glasses, we define an ideal glass as a crn amorphous solid having no bonding defects and no impurities. By "no bonding defects" we mean that there are *no interior unsatisfied bonds* in the solid, the $8 - n$ rule being everywhere obeyed. (Like the translationally periodic "perfect crystal," the ideal glass is not allowed to have an exterior. Strictly speaking, it is infinite in extent and has no surface, since a surface clearly constitutes an enormous defect. *Interior* unsatisfied bonds are specified here to distinguish them from the unsatisfied bonds that obviously occur at the inevitable surface of any real solid.)

In view of the already discussed compositional freedom of chalcogenide glasses, the notion of an impurity or "foreign" atom becomes a bit ambiguous, in contrast to crystals in which it is easy to distinguish between an impurity and a "host" atom. Elements that are network formers, i.e., whose atoms enter into the covalent network as in Fig. 3.7, are more appropriately regarded as intrinsic, rather than as foreign, to the glass. Network modifiers such as the alkalis, which go into the glass interstitially and which sit there as ions forming no covalent bonds to the random network, are bona fide impurities and are point defects inadmissible in the ideal glass.

It should be quite evident to the reader that this idea of ideal-glass structure was met before in Chapter Two: It was embodied there in the continuous-random-network models introduced for amorphous Si and SiO_2. Explicit mention of the term "ideal glass" was deferred until this point in order to emphasize its close connection to the $8 - n$ rule and to generalize the idea somewhat so as to include variable-composition glasses, such as the case of $Se_{1-x-y}As_xGe_y$ illustrated by Fig. 3.7. Also, it is now time to mention a few of the ways in which real glasses differ from the ideal glass. (Just as there is not one but many idealized structures corresponding to perfect crystals, so there are many ideal-glass structures. "The" ideal glass, in any specific case, will depend on the amorphous solid under discussion; the appropriate ideal-glass model will be clear from the context.)

In the following section we will conclude our treatment of the structure of chalcogenide glasses with a discussion of defects, deviations from the ideal glass. But first, though it is a bit misplaced here among the chalcogenides, a remark will be inserted to describe the main aspect in which the structure of amorphous silicon and germanium differs from the *a*-Si continuous random network of Chapter Two. For these swiftly formed materials, the vapor-quenched thin-film solid is much more structurally defective (vis-à-vis the ideal glass) than in the case of the chalcogenides. Amorphous Si and Ge contain microvoids, holes up to several angstroms across, and the internal surfaces of the microvoids provide broken bonds within the bulk of these amorphous solids. These dangling bonds have profound consequences for the electrical properties of these amorphous semiconductors, and this story, which is related to the hydrogenation of amorphous silicon and to possible solar-cell applications, will be told in Chapter Six.

3.5 TOPOLOGICAL DEFECTS AND VALENCE ALTERNATION

We now wish to consider defects in chalcogenide glasses—ways in which the atomic-scale structure of these solids deviates from the ideal-glass model. Although Mott's picture of valence satisfaction is broadly confirmed and stands as an excellent approximation to the actual structures, the infrequent exceptions to the $8 - n$ rule are important for understanding the electrical properties of amorphous chalcogenides (Chapter Six). We are not concerned here with impurities—foreign atoms that are not network formers. Instead, we are interested in *native* defects in real glasses (analogous to, say, vacancies in real crystals). These take the form of *topological defects* (deviations from the normal $8 - n$ coordination) in the crn covalent graph of the glass.

The simplest topological defect is a broken or *dangling bond*. Consider a 1d-network glass such as amorphous selenium. The ideal-glass model calls for every Se atom to be bonded to two others. No chain ends are permitted in this idealization, since the atom terminating a chain would have coordination $z = 1$ instead of the required $z = 8 - n = 2$. A chain-end Se atom would have one of its electrons (which normally would be paired with a neighboring-atom electron in a bonding orbital) in a singly occupied lone-pair orbital, ready to bond with another such "dangling" electron if it could find one on a neighboring atom.

In actuality, chain ends (and, consequently, dangling bonds) are present in amorphous Se because its Se_N chain molecules, although long, are finite. The polymerization index N is of the order of 10^5, so that native defects are expected to be present in a concentration of at least 10 parts per million.

Dangling bonds were long thought to make up the dominant species of native defect in chalcogenide glasses. However, it is now believed that a large fraction of these convert, during the formation of the glass, into pairs of different defects described by the term *valence alternation* (Kastner et al., 1976). These new types of native defects will be illustrated here by means of a gedanken reconstruction process that is sketched in Fig. 3.8.

The top half of the figure suggests a topological transformation that takes us, in several easy stages, from the ideal glass to a glass with two conjugate defects, which together are termed a valence alternation pair. Selenium serves as the example, and in Fig. 3-8a we have represented a finite sample of the defect-free amorphous solid by a condensed 1d-network covalent graph. In order to avoid any broken bonds we have pretended that *all* of the atoms are twofold-coordinated and are linked in a *single* very long chain that closes back upon itself to form, in effect, a giant ring. The topology of the covalent graph is as shown in panel *b* of the figure, in which (as in the subsequent parts of Fig. 3.8) the graph has been schematically spread out as if on a tabletop in order to

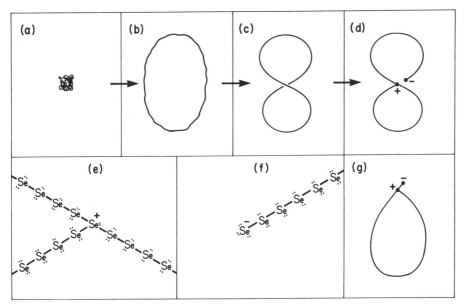

Figure 3.8 Schematic formation of a valence alternation pair in amorphous selenium. The ideal glass of *a* contains only two-coordinated atoms, as suggested in *b* and *c*. Bond switching and charge transfer at the crossover in *c* creates the valence alternation pair shown in *d* and, in close-up, in *e* and *f*. An "intimate" pair is shown in *g*.

simplify its configuration and expose its essential structure. Next, in *c*, we put back a small part of the conformational complexity which was taken out between *a* and *b* by twisting the ring into a "figure eight." At the crossover of the "eight" we now cut the lower arm (for concreteness, "upper" and "lower" refer to vertical position, at the crossover, with respect to the hypothetical tabletop, and can be permuted here) and connect one of its ends to the upper arm, leaving the other end dangling. Topologically, the original ring has been converted into a figure six or nine. There are now two discrete topological defects present in what had initially been a continuous structure. These two special points are the chain end and the threefold branch point of the reconstructed graph.

Physically, the above mathematical maneuver of cutting and (differently) reconnecting corresponds to a *switching of bonds accompanied by a transfer of electronic charge*. The bonding situations at the three-coordinated and one-coordinated defect sites are shown in panels *e* and *f*, respectively, in Fig. 3.8. The $z = 3$ role is played by a Se^+ ion, while the $z = 1$ chain-terminator role is played by a Se^- ion. Se^+, with five outer-shell electrons ($n = 5$), and Se^-, with seven such electrons ($n = 7$), are isoelectronic to As and Br, respectively, and their respective $8 - n$ covalent coordinations are 3 and 1. Viewed in this way, the $8 - n$ rule *for the ions* is obeyed! In this sense, these $z = 3$ and $z = 1$ *topological* deviants (aberrant with respect to the $z = 2$ norm in *a*-Se) are *not* bonding deviants, once the charge exchange that "prepares" them in ionic form is accepted as a *fait accompli*. They therefore differ in a fundamental respect from the $z = 1$ topological dead ends formed by neutral Se chain-ending atoms. Singly coordinated Se^- is directly analogous to the nonbridging O^- in Na—Si—O glass as discussed in Section 3.3, while triply coordinated Se^+ occurs in various crystalline solids such as GaSe and $MoSe_2$. Charge neutrality demands that these point defects occur in pairs. These are called *valence alternation pairs* ("VAPs", in the unfortunately proliferating jargon) to evoke the ± 1 alternation in bonding coordination about the normal value of *z*.

These types of defects were first proposed in chemically specific form in a pioneering paper by Kastner, Adler, and Fritzsche (1976). Those authors pointed out that such defect pairs are far less costly, in terms of excess energy relative to the ideal glass, than are the same number of pairs of dangling bonds. A simple argument for this is contained in Fig. 3.8. Suppose that the sample contains N atoms. The ideal-glass analog, represented as a ring in panel *b*, then has N bonds. Breaking the ring into a chain to produce a pair of chain-end dangling bonds, as in panel *e*, reduces the number of bonds to $N - 1$. The energy cost in creating two neutral chain ends is thus twice the energy needed to promote an electron from a bonding orbital to a lone-pair orbital. ("Promotion" of electrons is discussed in Chapter Six.) Now consider the configuration of panel *d*, with a pair of charged defects corresponding to valence alternation. The number of bonds present is N, the *same* as for the defect-free ideal glass! Thus the energy cost does not involve bonding to lone-pair promotion, but only the substantially smaller energy required for the transfer of an electron between two initially neutral Se atoms.

Valence alternation pairs provide a plausible picture of an important class of defects in chalcogenide glasses, although direct proof of their presence is, as usual for glasses, hard to come by. Their importance lies in connection with the continuing theoretical efforts to explain the unusual electrical and magnetic properties of the chalcogenides (as will be discussed in Section 6.4), and in fact the motivation for developing this structural model was to help solve that puzzle. Although they have been illustrated in Fig. 3.8 for the simple case of an elemental chalcogen glass, similar arguments also imply the existence of such charged topological defects in complex chalcogenides such as the Se—As—Ge system sketched (in defect-free form) in Fig. 3.7.

As a final graphical point in the present context, panel g of Fig. 3.8 illustrates a situation in which a $z = 3$ and a $z = 1$ defect are as close to each other as they can physically get—namely, one covalent bond length apart. Dubbed an *intimate* valence alternation pair ("IVAP," pronounced "eye-vap"), such a structural feature can be derived from the topologically equivalent situation of panel d by shrinking the along-the-chain separation between the two special sites down to a *single* bond. Note that for two defects remote from each other along a bond path (as in d), or unconnected by a bond path (i.e., on different chains), the closest possible spatial approach is an interchain van der Waals spacing. This is appreciably larger than the IVAP bond-length spacing indicated in Fig. 3.8g. IVAPs are energetically favored and may play an important role in the electronic properties of chalcogenide glasses.

3.6 THE RANDOM COIL MODEL OF ORGANIC GLASSES

Customarily, when seeking to identify the class of glass structure that is most characteristic of amorphous solids encountered in everyday experience, the candidate of choice in a historical context is the SiO_2-type continuous random network. That 3d crn structure is broadly representative of common "window glass," as well as of glass artifacts and utensils produced by mankind for more than three millenia. However, long attempts at repairing my children's "plastic" toys have provided me with ample personal evidence of the fact that a fair fraction of the materials familiar from use in current everyday life are molecular glasses, 1d-network amorphous solids composed of long-chain organic molecules, or polymers. Polystyrene provides a good prototype of such an *organic glass*. It is also among the most technologically important of the thermoplastic materials which, when heated, soften and flow controllably, enabling them to be processed at high speeds and on a large scale in the manufacture of molded products (such as: toys, tires, pens, innumerable household items, appliance bodies, building materials, synthetic fibers, automobile and airplane parts, etc.). Airplane windows, an application requiring transparency, light weight, and high strength, may be mentioned as a representative impressive use of the ubiquitous class of amorphous solids based on polymer chains.

Figure 3.9 shows a conventional graphlike symbolization of the structure of a single extended polystyrene chain. [A fine review of simple organic glasses has been given by Haward (1973).] Each carbon atom along the skeleton or backbone is in a fourfold-coordinated tetrahedrally bonded situation which is not dissimilar to the diamond-structure setting of Fig. 2.11. Two of its bonded neighbors are the carbons that precede and follow it along the polymer backbone, a third neighbor is a hydrogen atom, and the fourth is either another hydrogen or a carbon that is part of a sidegroup composed of several atoms. The sidegroups or "pendants" are attached to the backbone at regular intervals, strung along it like pearls on a necklace. In the case of polystyrene, the pendant is a benzene ring minus one hydrogen whose place is taken by the backbone carbon to which the ring is attached. Its chemical formula is C_6H_5 and it is symbolized by the hexagon in Fig. 3.9. One such "phenyl" sidegroup is attached to every second carbon along the skeleton of the chain; chemical regularity is maintained at a very high level.

The atomic-scale building block, or segment, or repeat unit, or *monomer* unit of the polymer chain is shown bracketed in Fig. 3.9. This, in the case of polystyrene, is the styrene monomer, and may be abbreviated (with the suppression of even more structural information) as $CH_2CHC_6H_5$. The polymer chain itself, which is one single large molecule ("macromolecule") and forms a covalently bonded 1d-network structural element of the solid, may be denoted as $(CH_2CHC_6H_5)_N$. The *polymerization index* N is typically of order 10^4. (In polystyrene, it is possible to reach N values which exceed 10^5.) Thus, if fully extended (which, in a polymeric glass, it is most definitely *not*, as we shall see), the molecule is roughly a long thin cylinder several microns long but only several angstroms thick. Many varieties of internally bonded atomic groupings may stand in for the sidegroup role that is played, in the case of polystyrene, by C_6H_5. Also, every polymer glass is characterized by a spread of lengths (N).

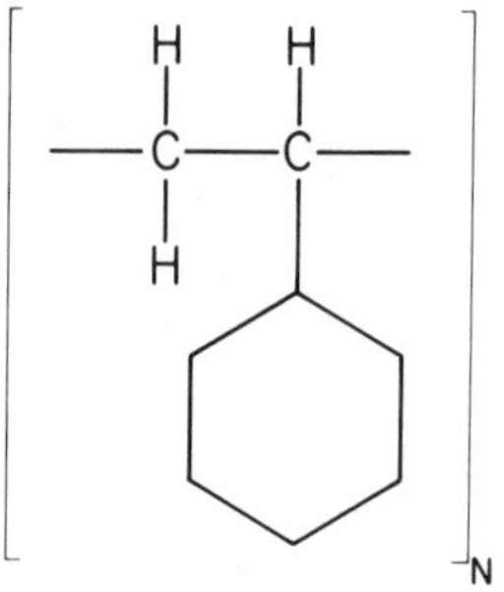

Figure 3.9 Graphlike representation of the monomer repeat unit within a polymer chain in polystyrene.

It is evident from Fig. 3.9 that even a relatively simple organic polymer such as polystyrene lacks the ultimate simplicity of polymeric selenium; for that elemental polymer the repeat unit contains but one bond and one atom. However, it is worth noting that the intrachain complexity in organic polymers mitigates against the presence of closed rings, a structural question mark that continues to plague amorphous selenium. *That* particular complication, at least, is absent in organic glasses because the bulky sidegroups prevent the closure needed to form small rings (large rings are very improbable in any polymer, and may be safely ignored).

The planar representation of Fig. 3.9 fails to convey a sense of the true atomic-scale geometry of an organic polymer chain. Figure 3.10 provides a perspective sketch of one possible conformation of a section of such a chain. The smaller spheres represent carbon atoms, the larger spheres represent the sidegroup pendants, and the hydrogens which were included in Fig. 3.9 are now omitted. Because of the clarity-necessitated omission of the hydrogens, the carbons in the drawing appear with either two or three bonds shown. They are actually tetrahedrally bonded, of course, with the missing bonds being the ones that connect to the not-shown hydrogens. Every pair of adjacent tetrahedra are mutually oriented, with respect to their shared bond, so as to be in the staggered configuration of Fig. 2.16. This is an energy-minimizing arrangement which allows nonbonded atoms to avoid each other, and keeps them from needlessly paying the price of overlap repulsion.

Let us focus on the course of the carbon skeleton in Fig. 3.10. This course is consistent with quite a few energy-minimizing constraints. First of all, all of the bond lengths are naturally equal to the proper crystalline value. Next, the bond angles also possess the correct (tetrahedral) value, so that second neighbors as well as first neighbors are properly placed. Furthermore, as mentioned above, even the dihedral angles have the right (staggered-bond configuration) values, so that even third neighbors along the carbon backbone reproduce separations characteristic of the diamond lattice. This time, however, two different diamond-structure distances crop up, depending on whether the third bond along the three-bond "walk" which takes us from the initial atom to its third neighbor along the chain is, or is not, parallel to the first bond of the "walk." In fact, the carbon-atom backbone of Fig. 3.10, zigging and zagging along its (topologically one-dimensional) path in three-dimensional space, can be viewed as being carved out of a diamond crystal.

In spite of its docile compliance with this rather stringent set of crystal-like requirements, which are much stricter than those discussed earlier for the *a*-Si structure in which the resemblance to the diamond lattice is relaxed for bond angles and very permissive indeed for dihedral angles, it is clear from Fig. 3.10 that the polymer chain manages to retain a great deal of configurational freedom. At each carbon, the bond to the next carbon along the chain can choose among three separate directions, each of which satisfies all of the above constraints. Any energetic preference that might exist among the three possibilities (and hence tend to impose long-range order) is in response to in-

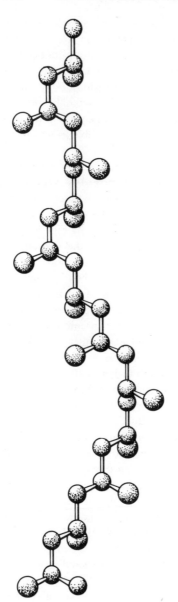

Figure 3.10 One possible conformation of an organic polymer chain.

teractions between atoms that are quite distant, so that the interactions are weak and the resulting preference is small. Consequently, polymeric glasses are among the easiest amorphous solids to form by cooling from the melt, while crystalline polymeric solids are actually quite difficult to prepare as bulk crystals.

In Fig. 3.10, the choice among the three possible bond directions for chain continuation was not actually randomly made. In order to stretch out the structure and clarify typical conformational relationships, the vertical bond direction was chosen whenever it presented itself as an option (on every alternate step). Thus, starting with the carbon atom at the top of the figure, the first bond "step" along the chain backbone is taken downward. Then the third bond, the fifth, seventh, etc., are all downward. On the even-numbered steps, the bonds adopt at random the three directions which point 19° below the horizontal plane. The projections onto a horizontal plane of the even-numbered bonds is shown in Fig. 3.11; it amounts to a two-dimensional triangular-lattice *random walk*. Without the artificial vertical-preference constraint introduced into the construction of Fig. 3.10, the chain would trace the path of a three-dimensional random walk. Random walks are discussed in some depth in the following two sections.

On a scale exceeding that of several bond lengths, the polymer chain is evidently extremely flexible. It is capable of changing direction drastically *many* times along its full length of N segments. In other words, its directional "memory" has a spatial extent much shorter than the chain length. Chain length, or more properly the *contour length* of the chain, will be used to denote the length of the chain in its fully extended configuration (i.e., Fig. 3.10). The contour length is of order of magnitude $N \cdot b$, where b is the bond length.

The most satisfactory model for the structure of amorphous solids based on organic polymers is the *random coil model*, a model most closely associated

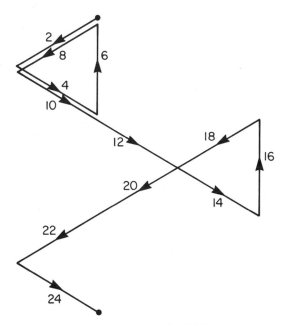

Figure 3.11 The two-dimensional random walk which corresponds to the horizontal projection of the backbone of the section of polymer chain shown in Fig. 3.10.

with the name of P. J. Flory (Flory, 1949, 1953, 1975). Each individual chain is regarded as adopting a random coil configuration, a configuration akin to that describable as a three-dimensional random walk. The organic glass then consists of *interpenetrating* random coils, coils which are very substantially intermeshed. A schematic illustration of the random coil model for the organic amorphous solid state is displayed in Fig. 3.12. One of the intermeshed chains has been "labeled" to single it out and (hopefully) aid in visualization. *Note that there is more than a two-order-of-magnitude difference between the length scales displayed in Figs. 3.10 and 3.12.* Figure 3.10 deals with distances on the scale of 1 Å (characteristic of atom–atom bond lengths), while Fig. 3.12 deals with distances of hundreds of angstroms (characteristic of the coil dimensions in polymer glasses).

The random coil model, like the continuous-random-network model and the random-close-packing model, is essentially a homogeneous single-phase model. Like the crn and rcp models, it has had its own historical conflict with competing heterogeneous models of a microcrystalline nature. And like crn and rcp, it has emerged successful in terms of compatibility with the preponderant weight of experimental evidence. The critical experiments in this case involve neutron scattering from samples in which a small fraction of the chains have been labeled (as in Fig. 3.12), say by the replacement of all of their hy-

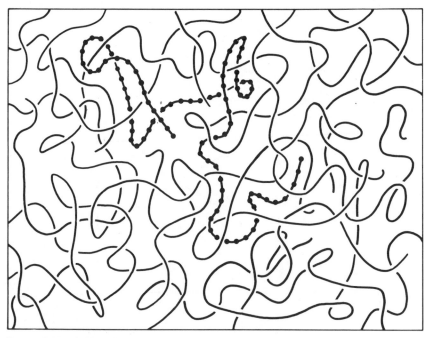

Figure 3.12 Schematic sketch of the random-coil model for the structure of organic glasses. One polymer chain is "labeled" for ease of visualization.

drogen atoms by atoms of deuterium, so that the scattering from the separated labeled chains can be observed to yield their individual structure. (Because the signal-to-noise ratio is better for this combination, the actual experiments are usually performed on a dilute solid solution of normal hydrogenated chains in a majority matrix of deuterated chains.) The data demonstrate that each coil encompasses or pervades a spherical space with a "radius" (to be defined below) typically about 300 Å. This result for the linear dimension of the three-dimensional region penetrated by a single polymer chain in the amorphous solid, which can be seen to be much smaller than the stretched-out contour length but much larger than the size of a tightly wound ball-of-string configuration (the latter fact contradicts a type of heterogeneous model based on "nodules"), is in good agreement with the random-coil picture. In particular, the observed coil radius is found to be proportional to the square root of the chain length (i.e., $N^{1/2}$) which is a random-walk hallmark and a central prediction of the Flory model (Benoit et al., 1973; Fischer et al., 1975; Kirste et al., 1975).

3.7 RANDOM WALKS, DRUNKEN BIRDS, AND CONFIGURATIONS OF FLEXIBLE CHAINS

The mathematical problem of random walks, already alluded to several times in the preceding section, will now be explicitly defined and discussed. We have already noted that this topic is relevant to the nature of the stochastic geometry of 1d-network glasses, notably the organic amorphous polymers. The Flory random coil model for the spatial organization of polymeric glasses is intimately connected to random-walk statistics, and that connection shall now be developed. In doing so it will also be necessary to discuss the closely related question of the conformation of polymer chains in dilute solution, as modeled by self-avoiding random walks (habitually abbreviated in the literature as SAWs).

The subject of random walks is valuable in many other contexts. For example, later on it will appear in connection with the topic of electronic transport in amorphous photoconductors (Section 6.3). In the present context it is convenient to illustrate by example, via their natural appearance in polymer applications of random walks, two theoretical approaches of broad applicability. One is the *mean-field technique*, often quite useful in dealing with random systems. Its application to the polymer-in-solution problem, described in the following section, provides a beautifully simple and successful example of the mean-field method. The other illustrated theoretical approach is the generalization of dimensionality d to noninteger values, as exemplified here by physical objects (termed *fractals*) that exhibit, in a certain scaling-law sense, fractional dimensionality. Both concepts are introduced in our analysis of Flory's classic treatment, in statistical-mechanical terms, of self-avoiding walks. Unexpected fractional exponents, which are dimensionality-dependent, appear in this phenomenological derivation in a way that provides a premonition of the similar surprises which occur in, say, Mott's theory of variable-range electron tunneling in amorphous semiconductors (Section 6.3).

The random-walk problem was posed by K. Pearson in a paper published in *Nature* in 1905:

> A man starts from a point $\mathcal{O}$ and walks ℓ yards in a straight line; he then turns through any angle whatever and walks another ℓ yards in a second straight line. He repeats this process n times. I require the probability that after these n stretches he is at a distance between r and $r + dr$ from his starting point $\mathcal{O}$.

This problem had actually been solved some years earlier, in the asymptotic limit $n \rightarrow \infty$, in a paper by Lord Rayleigh on the superposition of waves of equal amplitude and random phase (Chandrasekhar, 1943).

To ease the transition to the polymer-chain problem, we shall use N (rather than n) to denote the number of steps and b (rather than ℓ) to denote the length of each step. The central result is for the rms value of the magnitude R of the vector displacement $\mathbf{R}$ achieved in walks of N steps:

$$\langle R^2 \rangle^{1/2} = N^{1/2} \cdot b. \tag{3.2}$$

The angle brackets denote an average over all possible walks having exactly N steps, that is, a *configuration average.*

Equation (3.2) can be simply derived. Using $\mathbf{r}_i$ for the displacement at step i ($|\mathbf{r}_i| = r_i = b$), then $\mathbf{R} = \sum_{i=1}^{i=N} \mathbf{r}_i$ and

$$\langle R^2 \rangle = \sum_{i=1}^{N} \langle \mathbf{r}_i \cdot \mathbf{r}_i \rangle + \sum_{i \neq j} \langle \mathbf{r}_i \cdot \mathbf{r}_j \rangle \tag{3.3}$$

where the second summation is a double sum extending over all values of i and j except for those with $i = j$. All of the latter diagonal terms, corresponding to the appearance in $R^2 = \mathbf{R} \cdot \mathbf{R}$ of the self-products $\mathbf{r}_i \cdot \mathbf{r}_i$ which represent the square of the length of a given step of the walk, have been separately taken into account in the first sum of Eq. (3.3). Since there are N such self-product terms, and since each contributes b^2, that first sum is simply Nb^2. On the other hand, the second sum of Eq. (3.3), containing configuration averages over cross terms $\mathbf{r}_i \cdot \mathbf{r}_j$, necessarily *vanishes* because of the assured *randomness* of the walk. Since two different steps i and j are completely uncorrelated and all orientations of $\mathbf{r}_i$ and $\mathbf{r}_j$ occur with equal probability, the average $\mathbf{r}_i \cdot \mathbf{r}_j$ of their scalar product, taken over all possible configurations, must equal zero. Hence $\langle R^2 \rangle$ equals Nb^2, and we have the basic relation of Eq. (3.2).

It is time for a concrete illustration. Figure 3.13 presents a beautiful example of an *experimentally observed* random walk in two dimensions. The dots mark the positions of a colloidal particle suspended in water, as observed through a microscope by Perrin (1916), noted at 30-second intervals and connected in sequence by straight lines. (These lines are to order the observed points and do not represent actual interpoint trajectories. If the motion of the

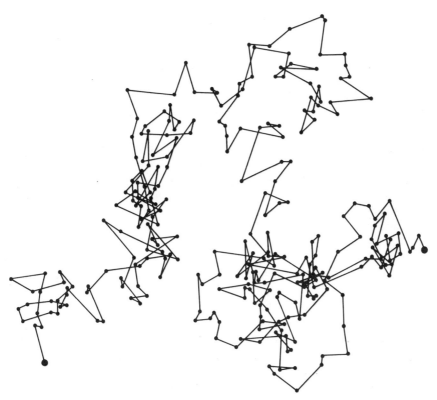

Figure 3.13 An experimentally observed random walk recorded by Perrin (1916) in his studies of Brownian motion.

particle were to be followed at more frequent intervals, each straight segment would be seen to be replaced by an entangled polygon similar in complexity to the whole walk shown in the figure. This is an example of *self-similarity*.) This of course is an observation of Brownian motion—irregular movement induced by the buffeting caused by collisions with the molecules of the fluid as they undergo thermal motion.

The *Brownian trail* of Fig. 3.13 illustrates a more general random walk than that described by Pearson and Lord Rayleigh, since here the steps vary in magnitude as well as direction. However, on a scale length much greater than that of any single step, the same statistical properties, such as the key proportionality between R^2 and N expressed in Eq. (3.2), apply to both types of walk. This was demonstrated in Einstein's famous first paper (Einstein, 1905) on the theory of Brownian movement in which he derived $\langle R^2 \rangle^{1/2} = (6Dt)^{1/2}$ for the rms displacement after time t of a particle diffusing with diffusion coefficient D. The time t plays the same role as does the number of steps N in Eq. (3.2). Perrin's experimental work (for which he earned the 1926 physics Nobel prize) was inspired by the fact that in the same 1905 paper, Einstein had provided an independent expression for D in terms of particle size, temperature, liquid viscosity, and Avogadro's number. Thus the tools were provided to permit the latter quantity, the fundamental connection between macroscopic and molecular dimensions, to be directly determined by means of Brownian-motion measurements.

Incidentally, the Brown of Brownian motion was an English botanist, the second to turn up in our own random walks in and around stochastic subjects. The first was Stephen Hales, in connection with his 1727 work on random close packing (Section 2.4.4). Robert Brown made his discovery, the "irregular swarming" of pollen particles dispersed in water, exactly one century later. This historical digression ends a brief flurry of name-dropping which has occurred in this section, not for its own sake, but because it happens to be true that random walks have cropped up in the work of men in many branches of science. (Among the names "dropped" have been those of five Nobelists: Rayleigh, Einstein, Perrin, Flory, Mott.)

In passing to the application of random walks to polymer structure, the walk is reinterpreted as the physical configuration (*static* configuration in a glass) of a flexible long-chain molecule, rather than the transitory path of a diffusing particle. Each step of length b (for a uniformly stepping walk) is interpreted as a chemical unit—a monomer segment of the chain. The net displacement magnitude R is now the *end-to-end length* separating the first and last monomers at the two ends of the chain. The situation is now a three-dimensional one, so that "random flight" is more appropriate nomenclature than "random walk."

Since a human drunkard cannot fly (though he might think he can), his random walk is at most two dimensional. For animate random *flights*, inebriated *flyers* are needed, hence the reference to "drunken birds" in the title of this section. It is hoped that the reader will forgive the use of this admittedly chimerical image for what will be, in the end, a chemical object: the form of a polymer molecule in an organic glass.

The configuration-averaged rms end-to-end length $\langle R^2 \rangle^{1/2}$ will henceforth usually be abbreviated as R_{rms}. For a random flight of fixed step length b this is known to equal $bN^{1/2}$. Nothing in the derivation of Eq. (3.2) depended on the dimensionality of the problem; $R_{rms} = bN^{1/2}$ holds for uniform random flights in one or two or three or more dimensions.

In addition to R_{rms}, the full distribution function $P(\mathbf{R})$ is also known for random flights. $P(\mathbf{R})$ is the probability of finding configurations with end-to-end *vector* $\mathbf{R}$. As no direction is favored over any other, it is isotropic and depends only on the scalar $R = |\mathbf{R}|$. Thus the frequency of occurrence of end-to-end lengths lying in the range from R to $R + dR$ in configuration space is, in three dimensions, $4\pi R^2 P(R)dR$. The distribution function $P(R)$ has the form of a Gaussian, with the single shape-determining parameter being R_{rms} ($= \langle R^2 \rangle^{1/2} = bN^{1/2}$):

$$P(R) = A \exp(-BR^2) \tag{3.4}$$

where

$$A = (2\pi/3)^{-3/2} R_{rms}^{-3} \tag{3.5}$$

and

$$B = (3/2) R_{rms}^{-2} \tag{3.6}$$

For large N, Eqs. (3.4)–(3.6) provide a superb approximation for the random-flight distribution of end-to-end lengths. That it is not exact may be seen by observing that the expression for $P(R)$ does not vanish for $R > R_{max}$, where R_{max} is the contour length Nb. The true $P(R)$ must be zero for $R > Nb$, since the end-to-end length cannot exceed the chain length of full extension. Note, however, that $P(R_{max})$ is smaller than $P(R_{rms})$ by a factor of order $\exp(-\frac{3}{2}N)$. For $N = 10^4$, this factor is about 10^{-26}; for $N = 10^5$, it is about 10^{-32}. Thus this failing is quite negligible, for realistic chain lengths.

The Gaussian expression $\exp(-x^2)$, a ubiquitous function very familiar as, *inter alia*, the form of the bell-shaped normal curve specifying the expected distribution of fluctuation-induced random errors, is an intuitively agreeable result. It was first obtained in the present context by Rayleigh for a random walk of discrete steps of equal length, and then later obtained by Einstein in a continuous formulation in which he derived and solved the diffusion equation $\partial \rho / \partial t = D \partial^2 \rho / \partial x^2$ for the probability density $\rho(x, t)$ of a kinetically bombarded Brownian particle. To appreciate this result (if not actually prove it) in a simple way, it should be sufficient to note the following connection. In one dimension, asking for the likelihood of an N-step walk arriving at location nb (when each step is either $+b$ or $-b$) is exactly the same as asking for the likelihood that a coin tossing sequence consisting of N independent tosses yields an excess $n = N_+ - N_-$ of heads over tails. Note that while the single most likely value of n is naturally zero, for sequences of very large N this almost never happens. In the limit $N \to \infty$, with $2x = n/N$ denoting the fractional excess, probability theory yields the well-known error function for the distribution

$$P(n) = (2\pi)^{-1/2} x_{rms}^{-1} \exp[-\frac{1}{2}(x/x_{rms})^2] \tag{3.7}$$

$P(n)$, where $nb = 2xNb$ is the net walk displacement, is the one-dimensional equivalent of the three-dimensional $P(R)$ of Eqs. (3.4)–(3.6).

The distribution in each of the three space components of R, taken without regard to the values of the other two orthogonal components, is individually of the form (3.7). Since the random walk is isotropic and the distribution is identical in each of the d independent directions of a d-dimensional walk, the dimensionality generalization of the Gaussian result is straightforward. It is easy to construct the correct general form by examining Eqs. (3.4)–(3.7). In d dimensions the distribution of end-to-end vectors retains the R dependence of Eq. (3.4), with A and B replaced by

$$A_d = c_d R_{\text{rms}}^{-d} \tag{3.8}$$

and

$$B_d = (d/2)R_{\text{rms}}^{-2} \tag{3.9}$$

In Eq. (3.8), c_d is a numerical constant determined by the normalization requirement that $P(\mathbf{R})$ integrate to unity. The prefactor A_d must have the physical dimensions of reciprocal volume in d-space, hence the term R_{rms}^{-d}. B_d, the dimensionality-dependent term within the argument of the exponential, is linear in d as it represents the sum-of-squares vector addition of d orthogonal components. Expressions (3.8) and (3.9), for the constants appearing in the generalized Gaussian

$$P_d(R) = A_d \exp(-B_d R^2) \tag{3.10}$$

for the endpoint distribution of (gedanken) random walks in d dimensions, will come in handy for the mean-field derivation for self-avoiding walks that is coming up soon in Section 3.8.

Returning now to the three-dimensional expressions (3.4)–(3.6) and their relationship to chain structure, we show in Fig. 3.14 the functions $P(R)$ and $4\pi R^2 P(R)$ for the end-to-end displacements of an assembly of chains assumed to have random-walk conformations and having an R_{rms} of 300 Å (a representative value for polymers with $N \approx 10^5$). $P(R)$ in Fig. 3.14 provides a linear

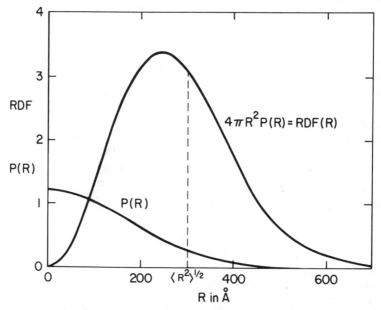

Figure 3.14 Distribution functions for the end-to-end distance R of an assembly of "ideal" (i.e., random-walk configuration) chains for which R_{rms} is 300 Å (after Yamakawa, 1971).

section, along any radial line, of the spherically symmetric three-dimensional endpoint distribution $P(\mathbf{R})$. The function $4\pi R^2 P(R)$ amounts to the pair correlation function (i.e., the RDF) for connected chain ends. Its peak occurs at $(2/3)^{1/2} R_{rms}$ and its second moment is, of course, R_{rms}^2. R_{rms} sets the scale. Although other linear measures of the region encompassed by a random coil chain might be adopted (such as the rms distance of the chain segments from its center of gravity, called its "radius of gyration"), they scale with each other and R_{rms} is the most tractable to treat.

Throughout this discussion, the end-to-end length R has been employed as a fantastically abridged one-parameter characterization of chain conformation. For each given R there exists an astronomical variety of possible configurations. (A conservative example of this multiplicity is given in the following section.) The complete configuration of any particular chain, that is, the full sequence of $\mathbf{r}_i$'s of Eq. (3.3), is virtually *unknowable*, as is the complete atom-by-atom structure of *any* amorphous system. Precisely because of the large numbers involved, the statistical approach becomes both necessary and valid.

By this time the reader may have become a bit uneasy about one aspect of the proferred connection between random walks and chain structure in amorphous polymers. (Or if not, probably *should* have become so.) There is, after all, something wrong about using a random trail like that of Fig. 3.13 to describe a random coil chain in the Flory picture.

The Brownian particle that traced the path shown in Fig. 3.13 was not inhibited by any memory of where it had been earlier in its two-dimensional wanderings. It was free to cross its own path, and indeed did so many times. Such self-intersections, although less evident, also occur for the trajectory of a particle diffusing in three dimensions. It is with respect to these self-intersections that the geometric analogy between Brownian motion and chain configuration breaks down. The flexible polymer chain, a palpable physical object occupying finite volume, cannot self-intersect. The spatial course of the chain structure, unlike that of the wandering particle, "cannot go home again." Different monomer segments must satisfy the obvious "steric" requirement that they do not superimpose on the same portion of space.

The above remarks would appear to suggest that the random walk model might better be replaced by one based on self-avoiding random walks, walks which are constrained to avoid self-intersections. Such self-avoiding walks (SAWs) have different statistical properties than those put forward above in Eqs. (3.2)–(3.10). SAWs are discussed in the next section. They are used to model polymers in dilute solution, the closest physical realization of a "gas" of isolated long-chain molecules. The statistical properties of self-avoiding walks are indeed found to be applicable to well-separated *nonoverlapping* single polymer chains. Nevertheless, in a nice ironic twist (which is rather in keeping with the convoluted nature of the structural subject matter at hand), it turns out that unconstrained ("ideal") random walks of the type we have been discussing, and *not* SAWs, provide the appropriate model for chain structure in the polymeric amorphous solid state. Thus neutron-scattering experiments

(Benoit et al., 1973; Fischer et al., 1975; Kirste et al., 1975) reveal that features expected for random-walk "Gaussian" chains, most notably the distinctive square-root dependence of coil radius on chain length, are exhibited by the chains in organic glasses. The reason for this will become clear in the next section. In brief it amounts to a *cancellation* (in a statistical/conformational sense), for any given chain, of the effect of self-avoidance by the effect of avoiding the *other* chains which interpenetrate its "turf." Like many another initially surprising result, it becomes "obvious" in retrospect.

3.8   SAWs, MEAN FIELDS, AND SWOLLEN COILS IN SOLUTION

A self-avoiding random walk (SAW) differs from a simple random walk in that it remembers where it has been and is forbidden to go back there. The effect of this inhibiting memory is to force the path to spread out and cover more ground. As usual, the effect is most spectacular in one dimension. In a one-dimensional SAW the walk *must keep going* in the same direction as that taken on the first step; it can never turn around. An N-step 1d SAW of equal steps of unit length is necessarily of length N; it is *fully extended*. This is drastically different than the result $N^{1/2}$ for the rms length of N-step 1d random walks. A random walk in one dimension revisits its starting point (or any other point, for that matter) *over and over again*; but after *its* start, a 1d SAW *never* sees the origin again.

The spreading out of the path, which results from the effect of self-avoidance or *excluded volume*, also occurs in higher dimensions. However, the effect decreases dramatically with dimensionality; the self-avoidance restriction rapidly loses its severity as the options for path continuation increase. Figures 3.15 and 3.16 show some two-dimensional examples. Figure 3.15 exhibits several 18-step SAWs. At the top, to provide a comparison, is the one-dimensional case: the straight, fully extended chain with end-to-end length $R = 18$. Below it are three two-dimensional SAWs. For purposes of enumeration, the problem has now been digitized by considering the case of walks inscribed on a lattice. Every step now connects a site of a two-dimensional square lattice to an available nearest-neighbor site. Each of the three 2d walks in Fig. 3.15 has $R = (34)^{1/2} = 5.8$. There are many distinct 18-step self-avoiding walks with this end-to-end length. In fact, it is known that there are *exactly* 782,030 18-step SAWs that start at (0,0) and end at (5,3), i.e., that have the same end-to-end *vector* **R** as those in the figure! (Thus there are eight times this number of SAWs with the same *value* of R, that is, which start out at the origin and which arrive, 18 steps later, at a site a distance R away.)

This information on the number of possibilities for relatively short (in comparison to $N \sim 10^4$–10^5 for high polymers) self-avoiding lattice walks, which is known from computer-assisted enumerations (Domb et al., 1965), provides some idea of the configurational freedom of long-chain polymer

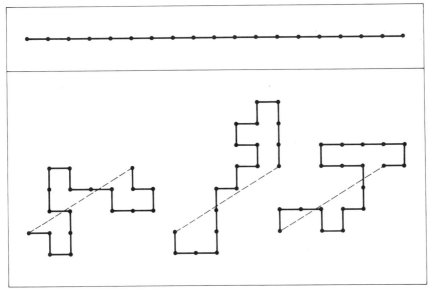

Figure 3.15 Self-avoiding lattice walks of 18 steps. The upper panel shows the one-dimensional case; the lower panel shows three two-dimensional SAWs on the square lattice, all of which have the same end-to-end distance.

molecules in their random coil form. Clearly a significant gain in entropy accompanies the transition from an extended to a random coil configuration (exemplified, respectively, by the upper and lower parts of Fig. 3.15). The enormous variety revealed by even such short and restricted (two-dimensional, tied-to-a-lattice) walks as those in Fig. 3.15 is relevant to an even more fundamental issue: the very validity of the random coil picture of polymer glasses. Perhaps the chief motive for the appearance of rivals of the random coil model is a perceptual one, namely the human difficulty in visualizing a three-dimensional intermeshed-coil arrangement (like that of Fig. 3.12, but with finite-thickness fleshed-out chains) that succeeds in densely filling space. But it turns out that this packing problem presents no real difficulty in three (or more) dimensions, because of the protean diversity of polymer form. The situation is portrayed by Flory (1975), in characteristic insightful fashion:

Whereas dense packing of polymer chains may appear to be a distressing task, a thorough examination of the problem leads to the firm conclusion that macromolecular chains whose structures offer sufficient flexibility are capable of meeting the challenge without departure or deviation from their intrinsic proclivities. In brief, the number of configurations that the chains may assume is sufficiently great to guarantee numerous combinations of arrangements in which the condition of mutual exclusion of space is met throughout the system as a whole. Moreover, the task of packing chain molecules is not made easier by partial ordering of

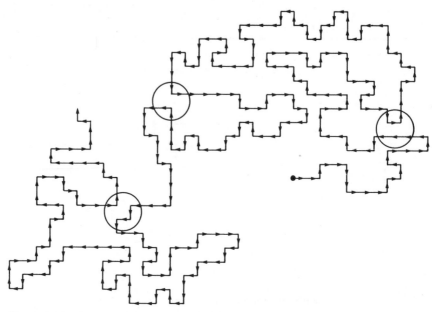

Figure 3.16   A two-dimensional square-lattice self-avoiding walk of about 250 steps. The circles mark three places where the walk is deflected when it encounters itself.

> the chains or by segregating them. Any state of organization short of complete abandonment of disorder in favor of creation of a crystalline phase offers no advantage, in a statistical-thermodynamic sense.

The last two sentences of the quoted paragraph have a wider generality than amorphous polymers; they correctly imply the inappropriateness of heterogeneous (e.g., microcrystalline) models for amorphous solids. The proper models for the structure of amorphous solids are the homogeneous ones: random close packing, continuous random network, and the random coil model. Also, take note of the comment about the ability of the chains to follow their "intrinsic proclivities" in the bulk glass. This refers to the fact that random walks, rather than self-avoiding walks, describe the statistics of the random coils. Flory predicted this in 1949, but it was only recently confirmed because of the by-now familiar difficulty of experimentally determining structural information for glasses.

Figure 3.16 displays a two-dimensional square-lattice self-avoiding walk of about 250 steps (or chain segments). The circled locations mark three places where self-avoiding prohibited intersections occur between monomer segments which, although *spatially close* to each other as a result of an accidental near-return-to-a-previous-spot of the meandering chain, are *remote in sequence along the chain*. For this reason these avoided intersections are sometimes termed, rather misleadingly, "long-range interactions." Obviously their physical origin is *short-range overlap repulsion between impenetrable atoms*. In any case, their effect is to deflect the chain generally outward, causing it to swell. This swelling of random-coil-configured chains, to dimensions larger than those expected from the simple (non-self-avoiding) random-walk model, applies primarily to isolated polymer chains suspended in a good solvent (defined below). The coil-enlargement phenomenon is known by several names: the *excluded-volume effect, self-repulsion effect*, long-range interaction effect, and intramolecular interference effect. Remarkably, this swelling does *not* apply to the random coil polymers in an organic glass, for reasons which will be clearer after the following derivation and which will be explicitly put forward in the following section.

The expanded dimensions of self-avoiding random walks vis-à-vis unconstrained random walks is best seen in their respective asymptotic ($N \to \infty$) forms for the contour-length dependence of the rms configuration-average end-to-end length:

$$R_{rms} \text{ (random walk)} = \text{const. } N^{1/2} \tag{3.11}$$

$$R_{rms} \text{ (self-avoiding walk)} = \text{const. } N^{\nu} \tag{3.12}$$

Equation (3.11) is a restatement of Eq. (3.2). Equation (3.12) is the corresponding statement for self-avoiding walks. Unlike Eq. (3.11), the basis for Eq. (3.12) is largely empirical. The mathematical problem of self-avoiding walks is a very hard one. No analytic solutions, analogous to the $N \to \infty$ asymptotic Gaussian form (3.4) for random walks, have been obtained. Nevertheless, there is available a highly instructive, albeit approximate, derivation of Eq. (3.12).

The swollen size of a typical SAW relative to that of a typical random walk can be seen from the fact that, in all practical dimensions ($d = 1, 2, 3$), the exponent ν in Eq. (3.12) is greater than the random-walk exponent ½ (the same in all dimensions). We already know that, in one dimension, $\nu = 1$. In three dimensions, computer calculations (Domb, 1963; Domb et al., 1965) show that ν is very close to 0.60:

$$R_{rms} \text{ (3d SAW)} = \text{const. } N^{3/5} \tag{3.13}$$

The molecular-weight (N) dependence of the random coil radius (R_{rms}) that is indicated by Eq. (3.13) is indeed experimentally borne out for polymers in liquid solution. In a good solvent, the relationships between the different types of intermolecular interactions present are such that the monomer segments of a polymer chain much prefer (energetically) the company of solvent molecules to the company of other monomers. The self-repulsion or excluded-volume effect holds sway, so that the self-avoiding walk provides a good model for the random coil chain configuration of isolated polymers in dilute solution in good solvents.

Equation (3.13), with its rather surprising fractional exponent, was first proposed by Flory (1949). Flory's derivation was approximate and phenomenological, and it has occasionally been criticized as nonrigorous (or worse). Nevertheless, his result, notably the unexpected exponent, is certainly correct; it has been decisively supported by numerical results of computer experiments on self-avoiding walks, as well as by neutron- and light-scattering experiments on polymer solutions. Convincing evidence affirms that the intuitive arguments Flory used, in arriving at Eq. (3.13), include the essential physics of the excluded-volume problem.

The Flory derivation has been recast into a transparent mean-field form by Fisher (1969), who has also generalized the statistical-mechanical analysis to display the dimensionality dependence of the exponent ν in Eq. (3.12). In d dimensions, for $d = 1, 2, 3,$ or 4, $\nu(d)$ is given by the beautifully simple relation

$$\nu = \frac{3}{d + 2} \tag{3.14}$$

Equation (3.14) neatly describes the decreasing influence, with increasing dimensionality, of the self-avoidance constraint. For $d = 1$, $\nu = 1$ corresponds to the severest effect possible: the fully extended chain configuration. For $d = 2$, $\nu = {}^3\!/_4$ is consistent with computer experiments on 2d SAWs, while for $d = 3$, Eq. (3.13) is recovered with its applicability to actual polymer chains in solution. For $d = 4$, the exponent ν adopts the random-walk value of ${}^1\!/_2$ and this remains true in all higher dimensions as well.

The interpretation of this convergence between self-avoiding and unconstrained random walks is that, in high dimensionalities, the self-avoidance constraint is no constraint at all! From the Brownian trail of Fig. 3.13, it is plain to see that a two-dimensional random walk intersects itself enormously often. Hence the great expansion of 2d SAWs relative to 2d random walks. In three dimensions the self-intersections of a random walk are much rarer, but nevertheless recur with finite frequency. The excluded-volume-induced expansion of SAWs vis-à-vis random walks is thus appreciable in 3d, though not as great as in 2d. But for $d \geq 4$, Brownian-motion random walks self-intersect with probability zero, so that the distinction between self-avoiding and unconstrained walks disappears completely. This is an example of a *marginal dimensionality* d^*, which marks the transition from a complex behavior (for $d < d^*$) to a simpler and theoretically tractable behavior (for $d \geq d^*$). For the self-avoiding walk problem, $d^* = 4$. Also, ν is an example of a dimensionality-dependent *critical exponent*. Both of these concepts are prominent in the theory of phase transitions, and will be encountered again in Chapter Four in connection with percolation theory.

The Flory–Fisher derivation of Eqs. (3.12)–(3.14) proceeds as follows. The first step is the construction of a partition function for an N-monomer chain in solution using, as the sole configuration-specifying variable, the end-to-end length R:

$$Z(R) = \text{RDF}(R) \cdot \langle \exp(-U/kT) \rangle_R \tag{3.15}$$

The first term is the density of (configurational) states in the form of the radial distribution function RDF(R), and this configuration-counting term is taken to be the known random-walk result. In three dimensions this is $4\pi R^2 P(R)$, with $P(R)$ given by Eqs. (3.4)–(3.6). Since we are interested in d dimensions, we replace $P(R)$ by the generalized Gaussian of Eqs. (3.8)–(3.10) and $4\pi R^2$ by $S_d R^{d-1}$. S_d is the surface area of a d-dimensional sphere of radius unity ($S_1 = 2$, $S_2 = 2\pi$, $S_3 = 4\pi$, $S_4 = 2\pi^2$, etc.). Combining Eqs. (3.2), (3.8), (3.9), and (3.10) yields the R dependence of the configurational density-of-states function RDF$_d(R)$ for the d-dimensional random-walk problem, with the polymerization index N entering parametrically:

$$\mathrm{RDF}_d(R) = C_d R^{d-1} \exp(-dR^2/2Nb^2). \qquad (3.16)$$

The constant prefactor C_d will drop out in the subsequent differentiation with respect to R. For completeness we note that it is the product of S_d and c_d [the dimensionless normalization constant of Eq. (3.8)] and b^{-d} (b is the bond length or step length) and $N^{-d/2}$. Dimensionality enters into both exponents on the right-hand side of Eq. (3.16).

The brackets $\langle\ \rangle_R$ enclosing the second term of Eq. (3.15) denotes an average over all configurations having end-to-end length R. While the first term on the right-hand side of Eq. (3.15) describes the statistics of an ensemble of "ideal" or "Gaussian" chains having random-walk configurations, the second term introduces excluded-volume effects via a monomer–monomer repulsion energy U. This repulsive intramolecular-interaction term introduces an energy-driven bias favoring looser, expanded coils (R large) over tighter, contracted ones. Without the second term in Eq. (3.15), the following analysis would simply recover the random-walk result $R \sim N^{1/2}$ for the characteristic coil diameter. With it, we intend to show that for self-avoiding walks, $R \sim N^{\nu(d)}$ with $\nu(d)$ obeying Eq. (3.14).

Now comes the remarkably simple, and successful, simplifying assumption of the Flory–Fisher argument. The question is how to handle the inter-monomer interactions in the face of the frightfully unruly geometry. The answer lies in Flory's image of a random coil polymer molecule as a *continuous cloud of monomer segments* that is spatially distributed about the molecular mass center according to a smooth (Gaussian) probability function, a gently varying probability density epitomized by $P(R)$ of Fig. 3.14. In this "smoothed-density model," the chance of any given monomer encountering (interacting with) any of the other monomers is proportional to the density at the position of the monomer in question.

The crudest picture of the cloud is simply to treat it as a *uniform* distribution of N segments over a sphere of radius R (or R multiplied by some factor of order unity). The smooth function is now a constant within the sphere, zero outside. In this model every monomer sees the same density, that density being given, apart from a numerical factor, by N/R^d. (Remember that we are in d dimensions.) Since the N monomers each have the same probability of experiencing the repulsive interaction, and that probability is proportional to N/R^d, the total repulsion energy is proportional to N^2/R^d:

$$U = EN^2/R^d. \tag{3.17}$$

E is a measure of the strength of the interaction between a pair of monomers. It has dimensions of energy $\times$ volume, the latter entering to reflect the excluded zone surrounding each monomer (within that zone it "feels" the presence of another, costing repulsion energy). Strictly speaking, N^2 in Eq. (3.17) should be replaced by $\frac{1}{2}N(N - 1)$, where $N(N - 1)$ is the number of possible interacting pairs and the factor $\frac{1}{2}$ takes care of double counting. But the $\frac{1}{2}$ has simply been lumped together with other numerical constants in E, while $N(N - 1)$ differs negligibly from N^2 because $N \approx 10^4$.

Substituting into Eq. (3.15) the density of states in configuration space (i.e., R-space) of Eq. (3.16) and the configuration-dependent (i.e., R-dependent) energy of Eq. (3.17) yields for the chain partition function

$$Z(R) = C_d R^{d-1} \exp(-dR^2/2Nb^2) \exp(-N^2E/R^dkT). \tag{3.18}$$

The most probable configuration, that is, the optimum R, is the value that maximizes Z and minimizes the free energy $F = -kT \ln Z$. Note that the effect of the energy-driven third term of Eq. (3.18) is to push the maximum of $Z(R)$ out to larger R than would be the case in the absence of this term. The repulsion energy would be minimized for $R \to \infty$, but R is of course limited at the upper end by the Gaussian fall-off in available configurations. The position of the maximum is a compromise between these two effects.

Differentiating $Z(R)$ with respect to R and equating the result to zero yields, with some rearrangement of terms,

$$\frac{R^2}{Nb^2} = \frac{E}{kT} \frac{N^2}{R^d} + \frac{d-1}{d} \ . \tag{3.19}$$

The simplest procedure now is to try a power-law solution $R \sim N^\nu$ with $\nu > \frac{1}{2}$. If this guess is correct, it then follows that the left-hand side of Eq. (3.19) diverges at large N. In that case the constant term $(d - 1)/d$ on the right-hand side must be negligible relative to the N-dependent term, and

$$\frac{R^2}{Nb^2} = \frac{E}{kT} \frac{N^2}{R^d} \ , \tag{3.20}$$

which yields:

$$R = (b^2E/kT)^{1/(d+2)} N^{3/(d+2)} \ . \tag{3.21}$$

For $d = 1$, 2, and 3, this verifies the trial function and confirms Eq. (3.12)–(3.14), which was one objective of this exercise.

For $d \geq 4$, $R \sim N^{1/2}$ satisfies Eq. (3.19). It is the second term that dominates for $d \geq 4$, so that the term containing the repulsion energy E has no effect and the random-walk result is obtained. For $d = 4$, the borderline or marginal dimensionality, the term in E does contribute although the random-walk exponent of $\frac{1}{2}$ applies. Of course, if we set E equal to zero in Eq. (3.19), then we obtain, in *all* dimensionalities, the random-walk result $R = [(d - 1)/d]^{1/2} \times bN^{1/2}$ corresponding to the peak position of the Gaussian RDF of Eq. (3.16).

It is worthwhile to reiterate the central assumption of this successful Flory–Fisher treatment, namely the model of the polymer coil as a *cloud of N segments distributed with constant density over a sphere of radius R*. This is a classic example of a *mean-field theory*, in which wild spatial fluctuations are replaced by a smoothed gently varying environment, subject to self-consistency. In a certain sense the example considered here is the *meanest* of mean fields because, within the coil, the actual widely varying density is supplanted by the ultimate in smoothing: a completely uniform density with no variation remaining whatsoever!

3.9 WHY OVERLAPPING COILS ARE "IDEAL"

In the preceding two sections, both random and self-avoiding walks have been introduced as mathematical models for the spatial layout of polymer coils. Now it might be supposed that only self-avoiding walks are suitable for the description of actual chain molecules, since it is always true that each monomer segment is excluded from occupying the space already occupied by another. While SAWs do describe the structure of nonoverlapping polymer coils in good solvents, the closing paragraph of Section 3.7 contained the seemingly nonsensical assertion that random walks, and *not* SAWs, provide the proper model for the important case that interests us: bulk polymers in the amorphous solid state. This is an important and subtle point, worthy of adequate explanation. Neutron-scattering experiments have established it, confirming that polymer coils in the bulk glass (Fig. 3.12) are not swollen as they are in dilute solution, and that their linear dimensions scale as $N^{1/2}$ [Eq. (3.11)] rather than as $N^{3/5}$ [Eq. (3.13)]. It was anticipated quite early by Flory, but the idea has been "digested by the scientific community only rather slowly" (quoting deGennes, 1977).

Consider first the discussion preceding Eq. (3.18), for the isolated chain suspended in a liquid in which the interaction between the chain monomers and the solvent molecules is energetically favored over monomer–monomer and solvent–solvent interactions. The net effect of intrachain repulsion results in an energy term that decreases monotonically with coil diameter [Eq. (3.17)]. This is the source of the excluded-volume swelling effect; the chain can lower its self-repulsion energy by spreading out.

Consider now the situation of a random coil chain in a bulk polymer, overlapping and intertwined with very many other similar chains as sketched in Fig. 3.12. Monomer–monomer repulsive interactions are again very much present. But now there is nothing to be gained for the chain by spreading out beyond its intrinsic topologically disordered (random-walk) dimensions, because there are monomers *everywhere*. The total monomer density, from *all* chains, is spatially homogeneous, so that a given chain cannot evade monomer-monomer interactions by spreading. If it expands, all that happens is that the number of (intrachain) interactions *between its own monomers* decreases while the number of (interchain) interactions *with monomers of other chains* increases. The total number (intrachain + interchain) of overlap-repulsion interactions experienced by the chain does *not* change with its conformation, because the total number of chain segments per unit volume is unaffected by the posture of any particular polymer. Whereas an isolated coil in solution can control the total monomer density with which it itself interacts, because it is *solely responsible* for that density, a single coil in a polymer melt or bulk glass is helpless in that respect. Whatever it does to alter its shape, it cannot change the number of monomer–monomer contacts.

The argument can be phrased in mean-field terms. Each monomer feels a repulsive potential proportional to the smoothed density. For a single coil in solution, the density profile is peaked in the midst of the coil and falls off with distance from its center. The potential gradient is directed radially outward, constituting an outward force that swells the chain until its influence is balanced by the radially decreasing availability of configurations (i.e., the entropy term). In a molten or amorphous dense polymer system, the density profile is flat. So is the repulsive potential; therefore, no potential gradient acts on the chains, and their distribution of configurations is "unperturbed" from the distribution determined by disorder alone. It is worth noting that the repulsive potential in the concentrated medium, although spatially featureless and thus *without distorting effect on polymer form*, is much larger in *magnitude* than the contoured repulsive potential that acts on (and *does* distort) a polymer coil in dilute solution. The latter sees a monomer density (its *own*) that is roughly 1% of that in the bulk-polymer phase, so that its self-repulsive potential is also reduced by a like factor relative to the repulsive potential arising from *all* chains in the dense melt or glass.

The upshot of these considerations is the definite (and correct, i.e., experimentally proven) conclusion that, in a bulk amorphous polymer, there is no inducement for the random coil chain molecules to adopt configurations that are expanded beyond the dimensions set by disorder. But in the preceding section it was just such expansion beyond random-walk dimensions that was attributed, for an isolated coil in solution, to the *effect* of self-avoidance or excluded volume. How is it possible for the excluded-volume *effect* to "go away" in the bulk glass?

The *basic principle* underlying the *effect* in solution is obviously just as valid in the glass: One chain segment is excluded from the space occupied by another chain segment because the superposition of one monomer on another is utterly prohibitive in energy cost. This stubborn resistance of molecules to superposition naturally does *not* "go away" in the glass. But the point to bear in mind is that only for the single-coil-in-solution case is the rule of monomer–monomer exclusion synonymous with *intrachain* nonintersection, i.e., self-avoidance. For the polymer-in-a-glass case, *self-avoidance* is a negligible consideration in comparison with *other-avoidance*, because nearly all monomer–monomer nonbonded contacts (nonintersecting near misses) are between segments of *different* chains. *Interchain* evasion is the order of the day, and since chains are distributed everywhere, there is no directional preference analogous to the outward bias of the polymer coil in solution. Hence the swelling of the latter which occurs, in that instance, as a consequence or effect of excluded volume, is absent for coils in a bulk polymer. Nevertheless, the excluded-volume *principle* of atomic nonsuperposition remains inescapably operative in the amorphous solid. No contradiction or paradox is at work here.

It is for these reasons that random flights, *not* self-avoiding ones, provide the correct mathematical model for describing the characteristic dimensions and statistical distributions of the random coils in the random coil structure of polymeric organic glasses. By this time, it should be "obvious." In the amorphous condensed phase, the effect of intramolecular interference is simply cancelled out by the compensating effect of intermolecular (interchain) interference. The cancellation is in the sense that the *net* result is to recover the *statistical* properties of the "ideal" random-flight configuration.

There is another interesting twist here. The disorder that distinguishes the condensed phase in this case is *simpler* than that of the corresponding "gas" (dilute solution). Random walks are simpler than self-avoiding random walks because the former represent "free" or unconstrained chance, while the latter represent a clear case of constrained chance. The relationship here between gas/glass structure and constrained/unconstrained chance is exactly backward from the usual situation which was illustrated earlier in Fig. 1.6. A gas of point particles is the epitome of unconstrained chance, while rcp and crn glasses, with their well-defined short-range order, exemplify constrained chance. Partly because the theory of random walks developed earlier than that of SAWs, and partly because they represent the simpler case of unconstrained chance, it has become conventional in the polymer literature to refer to random-flight configurations as "free" or "unperturbed" or "ideal." Thus in the gas/glass comparison for flexible polymer chains, it is the *condensed* phase (with its overlapping, intermeshed, entangled coils) in which the chains adopt their "free" configurations. It is in *this* sense that the chains in the glass are, ironically, "ideal."

3.10 SCALING EXPONENTS AND FRACTAL DIMENSIONS

The exponent ½ (or ν) in the relation $R \sim N^{1/2}$ (or $R \sim N^\nu$) is an example of a *scaling exponent*. There is analogy between these asymptotic ($N \to \infty$) polymer-physics relations and similar power-law relations which occur in the theory of phase transitions or critical phenomena. In phase-transition theory, the power laws apply close to the transition temperature T_0 in the limit $\epsilon \to 0$, where ϵ is the parameter of smallness $(T - T_0)/T_0$. The analogous parameter in the polymer case is $1/N$, that is, the correspondence is $\epsilon \longleftrightarrow N^{-1}$. In the case of critical phenomena, scaling exponents are also called critical exponents, and are extensively used to characterize the critical behavior (notably the divergence of various physical properties) in the vicinity of the transition. Such exponents are discussed further in the next chapter.

The simplest and most familiar of all scaling exponents is nothing more than the *Euclidean dimensionality* d. If we take a regular d-dimensional object and, while preserving its shape (i.e., its internal proportions), scale up its linear dimensions by a factor of L, then the *content* (generalized volume) of the object increases by a factor of L^d. This volume-versus-length scaling relation is probably the mathematical expression that is most closely related to the intuitive notion of dimensionality, and it provides a painlessly straightforward basis for the generalization to dimensions higher than three.

Thus far, two distinct dimensionalities coexist in our discussion of polymer glasses: the topological dimensionality of the covalent graph of the structural units comprising the glass (this *network dimensionality* is one for long-chain polymer molecules) and the ever-present Euclidean dimensionality of the three-dimensional physical space of the solid, in which the topologically one-dimensional covalent graphs of the polymer chains are embedded. We will now introduce a third type of dimensionality, one that appropriately characterizes the *stochastic-geometry aspect* of polymer-chain configuration. Unlike Euclidean and topological dimensionalities, this new dimensionality can take on noninteger values!

The new dimensionality is a generalization of a *similarity* definition of dimension. Let us take a concrete two-dimensional example. Suppose that we take a rectangle and dissect it into 16 identical smaller rectangles, each similar (same shape) to the original rectangle. Each new rectangle is a rescaled version of the old one, with each linear dimension reduced relative to that of the original by the similarity ratio $r = 1/4 = 16^{-1/2}$. More generally, if the initial rectangle can be divided into N equal rectangles similar to itself, the similarity ratio is $r(N) = N^{-1/2}$. Likewise, in three dimensions, if a rectangular parallelepiped is divided into N such equal similar-to-the-whole parts, the similarity ratio is $r(N) = N^{-1/3}$. In d dimensions,

$$r(N) = N^{-1/d}. \tag{3.22}$$

Equation (3.22), specifying the scaling ratio r that accompanies a division into N equal parts similar to the original, provides a similarity or scaling approach to dimensionality d. For regular shapes such as rectangles and parallelpipeds, Eq. (3.22) recovers for d the Euclidean dimensionality of the embedding space. But for highly irregular *stochastic* shapes, a *different* dimensionality is arrived at, one that yields a very valuable *measure of the irregularity itself*. In mathematics, it is called the Hausdorff–Besicovitch dimension after the two mathematicians who developed it as an abstract concept for dealing with wild "pathological" geometric sets not envisaged in the old "classical" mathematics. We shall use instead the term *fractal dimension*, coined in the book by Mandelbrot (1977). Mandelbrot's book (which is filled with beautiful graphics) makes an eloquent case for the widespread applicability of this concept to natural phenomena. Two examples he mentions are the realizations of random-walk geometries which have been discussed in this chapter: Brownian motion and polymer structure.

In Section 3.7, it was noted that the Brownian trail of Fig. 3.13 was displayed for points (in a planar projection of an actual motion) recorded by Perrin at intervals of 30 seconds. To quote Perrin's own discussion of this figure:

> ... diagrams of this sort, ... in which a large number of displacements are traced on an arbitrary scale, give only a very meager idea of the extraordinary discontinuity of the actual trajectory. For if the positions were to be marked at intervals of time 100 times shorter, each segment would be replaced by a polygonal contour relatively just as complicated as the whole figure, and so on.

What we have here is an excellent example of *self-similarity*: When we examine an apparently featureless piece of the original "shape" with finer resolution, we find a complex spatial fine structure that is a miniature replica of the whole. The "straight-line" piece of the original turns out to be a very crooked line, every bit as crooked as the full original itself!

Naturally, for a random-walk geometry the "similarity in shape" between the initial large polygon and the final small polygon (built on one edge of the large polygon) is in a statistical sense. Suppose we now ask for the similarity ratio $r(N)$ for this problem in statistical geometry. The end-to-end length of the reduced-scale final walk of N steps is equal to a single step length (which we may set equal to unity) of the large walk. We then need to know the step length b corresponding to the reduced walk. Using the knowledge that the rms end-to-end length of an N-step random walk of step length b is $bN^{1/2}$, and setting this equal to unity, yields $N^{-1/2}$ for b and hence for $r(N)$. Comparing this result to Eq. (3.22) reveals that the random-walk Brownian trail of Fig. 3.13 has a fractal dimensionality of 2.

The topological dimensionality of a Brownian trail is 1, because the trail is of course a line. The fractal dimensionality of 2 reflects the enormous irregularity of that line. To appreciate the irregularity-induced two-dimensional aspect, note that a Brownian trail is a "plane-filling" curve. It completely fills a finite domain of the plane, meaning that it approaches arbitrarily close to any point within that domain. If we interpolate 100 intermediate steps between each point of the walk recorded in Fig. 3.13, and then do so again for the new, more complicated (and much longer in contour length) walk, and continue this process, then before long the drawing becomes a solid blotch, no matter how fine a pen is used to draw it.

Although such a wild curve as that in Fig. 3.13 is quite different from the smooth curves (which are topologically *and* fractally of dimension 1) most familiar to physicists, it certainly *cannot* be viewed as pathological or *unnatural*. Remember that Fig. 3.13 is a record of experimental observations! A Brownian trail is (in Mandelbrot's words) a curve "drawn by Nature itself." The amount by which the fractal dimension exceeds the topological one is a mathematically defined measure of the *wildness* of the curve, that is, it provides a well-defined statement of the curve's degree of irregularity or complexity or convolutedness or crookedness.

The notion of self-similarity that is so well epitomized by Brownian motion is closely related to *scaling invariance*, a concept crucial to phase-transition theory. No matter how much we may "zoom in" on the structure, it still "looks the same." Of course this change-of-scale process is ultimately limited at the low end (despite the provocative words "and so on" which end the quote from Perrin's work) by atomic dimensions, and at the high end by macroscopic practicality, but it nevertheless has a scope of many orders of magnitude. In the theory of phase transitions, the gedanken process goes in the other direction. The minds-eye camera retreats and the scale increases as the correlation length grows from interatomic to macroscopic dimensions with the approach of the transition. The technique of the *renormalization group* mathematically formalizes the scaling-invariance concept, and will be discussed in Chapter Four in connection with percolation theory and the sol ⟷ gel transition, and in Chapter Five in connection with the scaling theory of electron localization.

Since we now know that random-walk geometry characterizes the configuration of polymer chains in organic glasses, it follows that a like *threefold multiplicity of coexisting dimensionalities applies to different facets of a polymeric amorphous solid*:

1. A *network dimensionality* of 1, corresponding to the topological dimensionality of the covalent-bond backbone of the individual chain molecules.

2. A *fractal dimensionality* of 2, corresponding to the convoluted and irregular shapes and forms adopted by the intermeshed random coils in the solid.

3. A *Euclidean dimensionality* of 3 for the bulk glass in real space, in which the lower-dimensional aspects are embedded.

The amorphous polymer is an even better example of this (literal) multidimensionality than the Brownian-motion case. A Brownian trail is, after all, ephemeral. On the other hand, chain configurations in a polymeric glass are (metastability of the amorphous solid state notwithstanding, as we have seen in Chapter One), for all intents and purposes, permanent structures.

Fractal dimensionality for polymers in glasses is 2, but at the outset of this section it was noted that this measure of shape complexity could take on *non-integer* values. This, in fact, is the case of polymers in dilute solution. We could show this by replacing $N^{1/2}$ by N^{ν} in the self-similarity argument given previously. Instead we shall use another approach, which is simpler once we accept the interpretation of the number of monomer segments as a measure of the content or volume or bulk of a chain, and R_{rms} as a measure of its linear extent. Then, for a chain in a self-avoiding-walk configuration, we simply invert the relation $R_{rms} \sim N^\nu$ and compare the resulting $N \sim R_{rms}^{1/\nu}$ with the usual volume-versus-length scaling relation $V \sim L^f$. We have used f rather than d in the latter relation because this procedure leads to the fractal dimensionality of the chain shape rather than the Euclidean dimensionality of the space in which the chain is embedded. The fractal dimension is then $1/\nu$, and since $\nu = {}^3/_5$ (for SAWs in three dimensions) it follows that $f = {}^5/_3$ for polymers in solution. This is a case, as promised, for which the *fractal* dimensionality is indeed *fractional*. In some sense, the chain shape is "in between" a line and a plane. Note that the fractal dimensionality for chains in solution ($1\,{}^2/_3$) is smaller than that for chains in the glass (2). This is a reasonable result, reflecting as it does the fact that constrained self-avoiding walks are tamer, less irregular in form, than are unconstrained random walks. The extension to noninteger values allows dimensionality to be considered as a *continuous* variable, an analytical viewpoint which is widely adopted in the theory of critical phenomena and which will enter into the discussion of percolation in the following chapter.

REFERENCES

Benoit, H., D. Decker, J. S. Higgins, C. Picot, J. P. Cotton, B. Farnoux, G. Jannink, and R. Ober, 1973, *Nature (Physical Sciences)* **245**, 13.

Chandrasekhar, S., 1943, *Rev. Mod. Phys.* **15**, 1.

de Gennes, P. G., 1977, *Riv. Nuovo Cimento* **7**, 363.

DeNeufville, J. P., S. C. Moss, and S. R. Ovshinsky, 1974, *J. Non-Crystalline Solids* **13**, 191.

Domb, C., 1963, *J. Chem. Phys.* **38**, 2957.

Domb, C., J. Gillis, and G. Wilmers, 1965, *Proc. Phys. Soc. (London)* **85**, 625.

Einstein, A., 1905, *Ann. Phys.* **17**, 549.

Finkman, E., A. P. DeFonzo, and J. Tauc, 1974, in *Proceedings of the Twelfth International Conference on the Physics of Semiconductors*, edited by M. H. Pilkuhn, Teubner, Stuttgart, p. 1022.

Fischer, E. W., G. Leiser, and K. Ibel, 1975, *Polymer Letters* **13**, 39.

Fisher, M. E., 1969, *J. Phys. Soc. Japan* **26**, supplement, p. 44.

Flory, P. J., 1949, *J. Chem. Phys.* **17**, 303.

Flory, P. J., 1953, *Principles of Polymer Chemistry*, Cornell University Press, Ithaca.

Flory, P. J., 1975, *Science* **188**, 1268.

Haward, R. N., 1973, in *The Physics of Glassy Polymers*, edited by R. N. Haward, Wiley, New York, p. 1.

Kastner, M., D. Adler, and H. Fritzsche, 1976, *Phys. Rev. Letters* **37**, 1504.

Kirste, R. G., W. A. Kruse, and K. Ibel, 1975, *Polymer* **16**, 120.

Mandelbrot, B. B., 1977, *Fractals: Form, Chance, and Dimension,* Freeman, San Francisco.

Mott, N. F., 1969, *Phil. Mag.* **19**, 835.

Nemanich, R. J., G. A. N. Connell, T. M. Hayes, and R. A. Street, 1978, *Phys. Rev. B* **18**, 6900.

Pearson, K., 1905, *Nature* **77**, 294.

Perrin, J., 1916, *Atoms,* English translation by D. L. Hammick, Constable and Company, London.

Slade, M. L., and R. Zallen, 1979, *Solid State Commun.* **30**, 357.

Solin, S. A., and G. N. Papatheodorou, 1977, *Phys. Rev. B* **15**, 2084.

Takahashi, T., and Y. Harada, 1980, *Solid State Commun.* **35**, 191.

Wright, A. C., and A. J. Leadbetter, 1976, *Phys. Chem. Glasses* **17**, 122.

Yamakawa, H., 1971, *Modern Theory of Polymer Solutions,* Harper and Row, New York.

Zachariasen, W. H., 1932, *J. Am. Chem. Soc.* **54**, 3841.

Zallen, R., 1974, in *Proceedings of the Twelfth International Conference on the Physics of Semiconductors,* edited by M. H. Pilkuhn, Teubner, Stuttgart, p. 621.

CHAPTER FOUR

The Percolation
Model

4.1 INTRODUCTION

Few theoretical techniques are available for dealing with severely disordered systems and stochastic-geometry situations. One of the nicest of these techniques, percolation theory, is the subject of this chapter. The intellectual appeal of the percolation model resides in its almost gamelike mathematical aspects and the fact that it provides a well-defined, transparent, and intuitively satisfying model for spatially random processes. The practical importance of percolation theory resides in its applicability to a broad (and growing) range of physical phenomena. (The reader should look ahead to Table 4.1 for a glimpse of this range, which extends to fields outside of physics.) Amorphous solids, in which the role of topological disorder is of vital importance, provide a natural setting in which unruly geometries abound. In this setting, the ideas of percolation theory can be fruitfully applied.

Percolation theory deals with the effects of varying, in a random system, the *richness of interconnections present.* From the perspective of condensed-matter physicists (who have been the main ones to adopt this mathematical subject for use in their own discipline), the single most seductive aspect of the percolation model is the presence of a sharp phase transition at which *long-range* connectivity suddenly appears. It is this *percolation transition*, which occurs with increasing connectedness or density or occupation or concentration, that makes percolation a natural model for a diversity of phenomena. The percolation transition provides a splendid prototype for second-order phase transitions in general, so that the percolation model is a fine pedagogic instrument for illustrating concepts central to the physics of phase transitions and critical phenomena. Many of these same concepts are useful for amorphous solids.

Two of the most prominent examples of the uses of percolation concepts in amorphous solid-state physics, the glass transition (for atomic motion) and the Anderson transition (for electronic motion), are discussed in the following chapter. Another application ("variable-range hopping") to electronic transport is also contained in that chapter. Among the illustrative examples introduced in the present chapter is one (the sol ↔ gel transition, see Fig. 4.11) that is a special type of glass transition. For a quick preview of the percolation transition itself, peek ahead to Figs. 4.4 and 4.5.

4.2 AN EXAMPLE: THE VANDALIZED GRID

As an introductory example of a percolation process and the qualitative event called the *percolation threshold*, consider the gedanken experiment rather fancifully illustrated in Fig. 4.1. A communication network, represented by a very large square-lattice network of interconnections, is attacked by a crazed saboteur who, armed with wire cutters, proceeds to cut the connecting links *at random* (not a realistic scenario, obviously, but necessary for our purpose here). His aim is to break contact between two well-separated but well-connected communication centers or command posts, represented by the heavy bars which, in Fig. 4.1, form the left and right boundaries of the network. Question: What *fraction* of the links (or bonds) must be cut in order to isolate the command posts from each other?

This question, which can be given a definite answer by percolation theory, illustrates the central issue at the heart of the percolation model: the existence of a sharp transition at which the *long-range connectivity* of the system disappears (or, going the other way, appears). This basic transition, which occurs abruptly as the composition of the system—or some generalized density—is varied, constitutes the percolation threshold. At the percolation threshold, significant properties may change qualitatively in a go/no-go manner. In the context of our present example, certainly the question of whether the joint command posts can communicate or not is a yes-or-no matter of some importance.

Other, more simple, physical situations can be discussed in the context of Fig. 4.1. If the square-lattice graph is interpreted, not as a communication grid but instead as an electrical network with the (intact) bonds as unit conductors, the percolation threshold corresponds to the onset/disappearance of electrical conduction between the two bounding electrodes (formerly, the "command posts"). Starting with the wholly occupied network (all conducting bonds present) and then turning the random wire-cutter loose, the current drops as the curve indicated in the lower part of Fig. 4.1 is traversed from right to left. The position of the arrow roughly corresponds to the stage depicted in the upper part of the figure: about 21% of the bonds in the network have been cut, 79% remain uncut. At this stage, current still flows between the electrodes (which are connected externally to a voltage source and an ammeter), although less than initially. Denoting the fraction of uncut bonds remaining as p, $I(p)$ continues to decrease as p decreases until a critical bond concentration, denoted as p_c, is reached at which the current I vanishes. For p less than p_c, I is zero (not merely small, but *zero*!). For $p < p_c$, there ex-

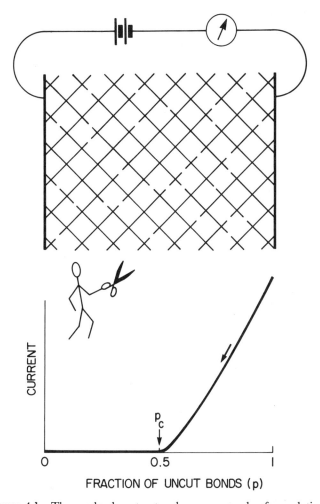

Figure 4.1 The randomly cut network as an example of percolation.

ists *no connected path* of conducting bonds that traverses the network from one electrode to the other.

The second, electrical-circuit, interpretation of Fig. 4.1 belongs to a class of problems that are generally labeled by the descriptor of *random resistor network*. This type of model is useful for the analysis of several varieties of transport phenomena occurring in amorphous solids, and will appear again. A third, mechanical, interpretation can be attached to Fig. 4.1, which will also prove relevant. Viewed as a screen, a two-dimensional structural unit, at $p = 1$ the structure has its greatest mechanical strength. As structural elements are snipped away and p decreases, this strength declines until, at p_c, the remainder of the badly thinned-out screen simply falls apart completely into discrete (finite) bits and pieces.

The existence of a perfectly sharp percolation threshold, a well-defined p_c at which, with decreasing p, the joint command posts lose touch with each other or the random resistor network becomes an open circuit or the screen structure disintegrates, has involved a tacit assumption which we now make explicit. When we stated that the lattice of Fig. 4.1 should be understood to be "very large," the quantitative meaning is that the scale length of interest, for example, the distance L between the command posts, greatly exceeds the lattice constant or bond length a of the square lattice: $(L/a) \gg 1$. The connectivity threshold is only mathematically sharp in the limit $(L/a) \to \infty$, an infinite system. For a finite system, repeated experiments such as that of Fig. 4.1 will yield a spread of observed thresholds which bracket p_c. (Obviously, for a very small system, our hypothetical randomly snipping saboteur could have significantly good or bad luck, succeeding at a value of p appreciably above or below p_c.) Henceforth we will generally assume a very large system; the limit $(L/a) \to \infty$ (the "thermodynamic limit") provides a very good approximation to the cases of interset. Typically, a will be an atomic-scale length and L a macroscopic length.

For the question ("What fraction of the bonds...?") posed at the beginning of this section in connection with Fig. 4.1, the answer provided by percolation theory is $p_c = 0.5$. It is known that $\frac{1}{2}$ is the value of the "percolation threshold for bond percolation on the square lattice." This is a rare case in which an "exact" analytically derived value is available for p_c. The percolation threshold is known exactly for a few other two-dimensional lattices, but not for any lattices in three (or higher) dimensions. Note immediately that in one dimension, $p_c = 1$; any nonzero value for the fraction of cut bonds breaks the network into finite sections and destroys long-range connectivity. In $d = 1$ there is no way to "get around" a blockage (as can be done in $d \geq 2$), so no blockages can be tolerated. In effect, there is no percolation in one dimension.

Note also the use of the term *bond percolation* in the above specification of the percolation threshold for the scenarios envisaged via Fig. 4.1. A lattice (or graph) is composed of sites (vertices, intersections between bonds) and bonds (edges, links, pairwise connections between sites). There are two basic types of percolation processes on lattices: bond percolation and site percolation. In both cases we start (as in Fig. 4.1) with a regular geometric object, a periodic lattice. Now a *nongeometric* two-state property, which carries the statistical character of the problem (and converts it into a stochastic-geometry situation) is randomly assigned to each site or bond. In bond percolation, such as we have been considering thus far, each bond is either *connected* (which occurs with probability p) or *disconnected* (which occurs with probability $1 - p$). In place of connected/disconnected, the corresponding terms unblocked/blocked are often used in order to evoke a fluid-flow image which (as mentioned in the next section) provided the original motivation for the use of the term percolation for the connectivity threshold. The assumption of a completely random system means that the probability p, for every bond, is *independent* of the state of the neighboring bonds.

In *site percolation*, each bond is considered to be connected, and it is now the *sites* which carry the random-connectivity character of the structure. Each site is either connected (unblocked) or disconnected (blocked), with probabilities p and $1 - p$, respectively. Again, p is the same for each site, and is not influenced by the state of the neighboring sites. To convey the concentration- or density-dependence that is an important aspect of most phenomena which can be modeled as site-percolation processes, unblocked and blocked sites are usually referred to as *filled* and *empty* sites, respectively. This usage will be adopted.

Just as, in bond percolation, adjacent unblocked bonds are connected to each other, so, in site percolation, adjacent filled sites are connected to each other. Adjacent has an intuitive, nearest-neighbor, meaning here. (For mathematically oriented readers requiring more "rigor": Two bonds are adjacent if they are incident with the same site; two sites are adjacent if they are incident with the same bond.) A set of connected bonds or sites is called a *cluster*. Thus, in site percolation, two filled sites belong to the same cluster if they are linked by a *path* of nearest-neighbor connections joining a string of filled sites. Likewise, in bond percolation, two unblocked bonds belong to the same cluster if they are linked by at least one unbroken path of unblocked bonds.

4.3 THE PERCOLATION PATH

The term percolation for this class of statistical-geometry model was coined in 1957 by the mathematician J. M. Hammersley, who had in mind the passage of a fluid through a network of channels, with some of the channels (at random) being blocked. A sketch of such a bond-percolation situation, with the channel network idealized as a 2d honeycomb lattice weaving its way through hexagonal "coffee grounds," is shown in Fig. 4.2. A map of the network is shown in the lower part of the figure, with the unblocked bonds represented by heavy lines and several clusters labeled. One cluster has been labeled as a possible *percolation path*, as explained below. Bond-percolation processes can thus be viewed in terms of the flow of some generalized "fluid" through a medium represented by interconnecting pipes, some of which are valved shut as shown in Fig. 4.3b. Such a "plumbing analogy" can similarly be constructed for site percolation, as indicated in Fig. 4.3a. Now the valves are placed at the joints (intersections), rather than in the pipes, of the plumbing network. It is clear from this that a combined percolation process can be considered which corresponds to valves placed *both* in the pipes *and* at the joints. This is called, not surprisingly, site-bond percolation, and is an example of an interesting and useful generalization of "conventional" percolation theory. Several such generalizations will be touched upon in this chapter, usually (as here) quite briefly. One in particular,

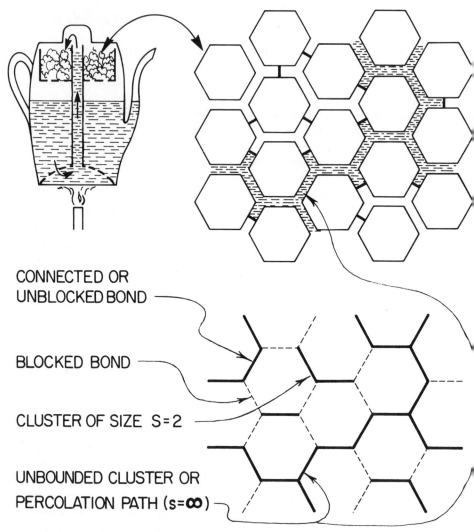

CONNECTED OR
UNBLOCKED BOND

BLOCKED BOND

CLUSTER OF SIZE S = 2

UNBOUNDED CLUSTER OR
PERCOLATION PATH (s=∞)

Figure 4.2 The percolation of a fluid through a porous medium that is modeled as a network of interconnected channels (some of which are blocked, at random) constitutes a bond-percolation process. The unblocked channels in this example play a role analogous to that played by the uncut wires in Fig. 4.1. The connectivity map shown in the lower part of the figure corresponds to the array of blockages shown in the upper diagram. In the latter, liquid is shown in the channels which belong to the percolation path.

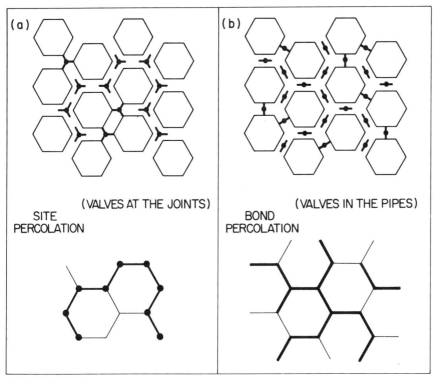

(a) SITE PERCOLATION (VALVES AT THE JOINTS)

(b) BOND PERCOLATION (VALVES IN THE PIPES)

Figure 4.3 Plumbing analogy for the distinction between site percolation and bond percolation.

continuum percolation, is especially important in its application to amorphous solids, and will be developed more fully.

The most striking aspect of a percolation phenomenon is the dramatic change that occurs in the connectivity of the system at the percolation threshold. Our introduction of this sharp transition, given in the previous section in the context of Fig. 4.1, was slightly unorthodox. There we started with an extended connected network and proceeded to progressively *dilute* it, in a random manner, until it fell apart (at p_c) into finite fragments. In other words, we traversed Fig. 4.1b from right to left, in the direction of *increasing dilution* or *decreasing concentration*. The more usual visualization of the percolation threshold views the process as "proceeding" in the other direction, that of *increasing concentration*, i.e., from left to right in Fig. 4.1b: we increase the probability p that a bond, chosen at random, is unblocked, from $p = 0$ through $p = p_c$ to $p = 1$. Since there is no *a priori* time bias built into percolation, "directionality" in this sense is simply a matter of custom and both viewpoints are obviously equally good. It is useful to compare the two way of approaching p_c, and this can easily be done with the aid of an illustrated case such as that shown in Fig. 4.4.

Figure 4.4 portrays a portion of a square lattice upon which a site-percolation process has been superimposed. The three frames represent the same region of the lattice at three different concentrations: The fraction of filled sites (p) grows from 0.25 to 0.50 to 0.75 in going from Fig. 4.4a to 4.4b to 4.4c. Filled sites are shown as heavy dots, nearest-neighbor filled sites (which are connected, i.e., belong to the same linked cluster) are shown joined by heavy lines, and several cluster sizes are indicated.

The three panels of Fig. 4.4 can be thought of as the middle three frames of a five-frame sequence spaced at equal Δp = 0.25 intervals. The frame preceding (a) is the completely empty lattice, the low-density limit corresponding to p = 0, while that following (c) is the high-density p = 1 limit in which every site is filled and connected with every other site in one giant cluster which completely fills the lattice. The latter situation is the site-percolation analog of the fully connected bond-percolation situation which was the starting point for the treatment of Fig. 4.1. As indicated there, beginning at p = 1 and then progressively diluting the giant cluster (cutting bonds in bond percolation, removing filled sites in site percolation) first introduces small holes, then thins out the giant cluster and renders it more and more ragged, and eventually destroys its long-range connectivity by breaking it up into finite pieces. Now let us, as promised, view the process the other way, from top to bottom in Fig. 4.4.

For $p \ll 1$, the low-concentration regime, nearly all of the filled sites occur as isolated singlets, clusters of size one. Choosing a site at random, the chance of it being a filled site is p. Suppose that we have landed on a filled site; what is the likelihood that it belongs to a cluster of size 2? Since there are four nearest neighbors in the square lattice, and since $p \ll 1$, the chance of this being the case is $4p$, which is negligible. By similarly counting the possibilities, the probability of a given filled site belonging to a cluster of size 3 is $18p^2$ for the square lattice, which is even more negligible in the small-p limit. Put another way, at low concentrations the probability of encountering a cluster of size s is of the order of p^s. Thus the *cluster-size distribution*, in the $p \rightarrow 0$ low-density limit, is sharply peaked at s = 1 and falls off exponentially with increasing s. Henceforth, s will be used to denote cluster size.

The cluster-size distribution is typically expressed as a discrete function $n(s)$, defined at s = 1, 2, 3, 4, ..., where $n(s)$ is normalized per site, that is, $n(s)$ is the number of clusters of size s divided by the total number of sites in the system (for a large system, of course). Away from $p \approx 0$, it is not so easy to specify $n(s)$ analytically. [For site percolation on the square lattice, the previous paragraph showed that, for $p \approx 0$, $n(s)$ = p, $4p^2$, $18p^3$, for s = 1, 2, 3. These are approximations expressed "to lowest order in p," quite accurate for p less than 0.1, but "exact" only in the $p \rightarrow 0$ limit.] Nevertheless, for many lattices of interest, $n(s)$ is reasonably well known over the full concentration range as a result of computer simulations. For the moment, we are not concerned with quantitative aspects of the evolution of $n(s)$ with increasing p, but are interested instead in the qualitative change that takes place at the per-

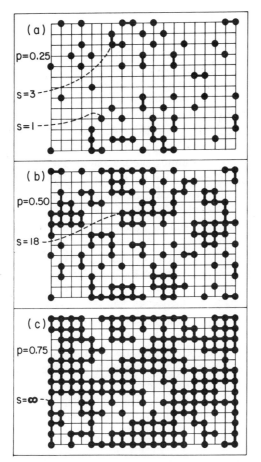

Figure 4.4 Site percolation on the square lattice, illustrating various cluster sizes (s) for three values of p, the fraction of filled sites. For $p = 0.75$, an unbounded cluster is present.

colation threshold. Thus we may discuss what happens, as we move away from $p \approx 0$, qualitatively.

As p increases, the proportion of filled sites belonging to $s \geq 2$ clusters increases, since the probability of continuing a cluster—by finding an adjacent filled site—becomes larger. The mean cluster size, which we will denote as s_{av}, grows. Since the number of sites occurring in clusters of size s is proportional to $sn(s)$, then the site-weighted average cluster size is given by

$$s_{av} = \frac{\sum_{s=1}^{s=\infty} s^2 n(s)}{\sum_{s=1}^{s=\infty} sn(s)}. \tag{4.1}$$

The sum in the denominator of this expression is proportional to the total number of filled sites. [In fact, since the proportionality constant contained in $n(s)$ is the reciprocal of the total number of sites in the system, this sum is simply equal to p.] The numerator is a corresponding sum in which each filled site is weighted by the size of the cluster to which it belongs. In other words, if we choose a filled site at random and note down its s value, the average value characterizing the resulting list of numbers is s_{av}.

s_{av} starts out (for small p) at unity, reflecting the total dominance of singlet clusters at low concentrations. As p increases, so does s_{av}. Consulting Fig. 4.4, our illustrative example examining site percolation on the square lattice, panel *a* represents a typical configuration for $p = 0.25$. (It was constructed with the aid of a random number generator, the computer equivalent of a coin-tossing sequence.) With 25% of the sites filled, the mean cluster size, as defined above, is about 3.5. Singlets still account for nearly a third of the filled sites, but appreciable numbers of larger clusters now appear.

Increasing p to 0.5 has the effect indicated in Fig. 4.4*b*, which is one possible evolution developed out of Fig. 4.4*a* by the (random) filling in of one-third of the remaining empty sites. Clusters have grown larger, and some have joined together to form quite large clusters. s_{av} has now reached a value well in excess of 20, and this quantity is now increasing, as p increases, very rapidly indeed.

Although the situation depicted in Fig. 4.4*b* has come a long way from the situations obtaining at small concentrations, in one extremely crucial respect the situation *has not changed* from that of Fig. 4.4*a* or even from the singlets-only low-concentration limit: *All clusters are finite in size.* Before examining this statement further, let us look at Fig. 4.4*c*, in which the concentration has been increased again (by randomly occupying half of the empty sites remaining in Fig. 4.4*b*) to $p = 0.75$.

In Fig. 4.4*c* we observe a large cluster which extends across the entire sample (from the top edge to the bottom, from left to right). This extended cluster is called a *spanning cluster* for the finite sample at hand. For $p = 0.75$, such a spanning cluster persists as the lattice sample grows indefinitely large. This infinitely extended or *unbounded* cluster is called the percolation cluster or *percolation path*. Note that although the percolation cluster is infinite in size ($s \rightarrow \infty$), it does not fill the entire lattice (except, of course, at $p = 1$, the high-density limit). Instead it coexists with finite clusters and islands of empty sites.

The critical event in the connectivity has occurred between 4.4*b* and 4.4*c*, at a composition between $p = 0.5$ and $p = 0.75$. If a sequence of frames such as that shown in the figure were executed using a very large lattice size and very small density increments (equivalent to what is done in some Monte Carlo computer simulations of percolation), the percolation path would first be observed at $p = 0.59$, which is the critical concentration p_c for site percolation on the square lattice. With increasing concentration, the percolation threshold p_c marks the appearance of the unbounded, lattice-spanning, percolation cluster. It signals the point at which the system of interconnections has thickened sufficiently for a connected network to propagate indefinitely.

The above statement is the standard one used for defining the threshold. Clearly, it is equivalent to the approach taken earlier (Fig. 4.1) in which we watched the infinite network dissolve and disappear as the system of interconnections was thinned out. The discussion of Fig. 4.1 was for a bond-percolation case, while that given for Fig. 4.4 is for site percolation, but it is easy to see how the analogous quantities (cluster size, percolation path, percolation threshold p_c) carry over between the two cases.

Above p_c, the percolation path exists; below p_c, it does not. From $p = p_c$ to

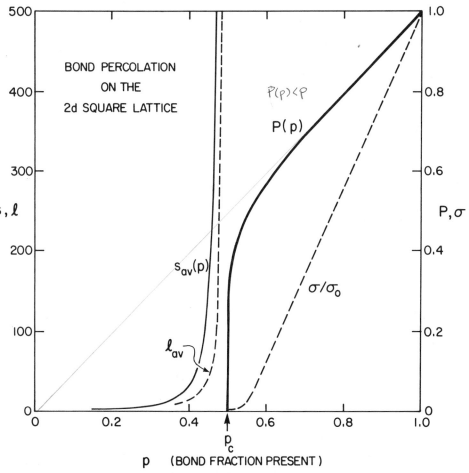

Figure 4.5 The behavior, as a function of the fraction (p) of filled bonds, of key properties that characterize bond percolation on the square lattice in two dimensions. The percolation probability $P(p)$ and the average cluster size $s_{av}(p)$ are results of computer studies carried out by Nakanishi (1980); the random-network conductivity σ is from the work of Kirkpatrick (1973); and the curve shown for the mean spanning length is schematic.

$p = 1$, the percolation path fleshes out until it fills the entire lattice. The growth in volume of the infinite cluster, for bond percolation on the square lattice (the situation corresponding to Fig. 4.1), is shown as the heavy curve in Fig. 4.5. This curve describes a function $P(p)$, which has the following significance: When the fraction of connected bonds is p, the chance of a randomly chosen bond being a connected bond *that belongs to the infinite cluster* is P. Obviously P is less than p (except for the filled system, when $P = p = 1$). Put slightly differently, P is the fraction of the entire system that is taken up by the percolation path. It is called the *percolation probability*.

The percolation probability vanishes (i.e., P is not merely small, but is identically zero) until the concentration of connections reaches p_c, then it rises very steeply. Eventually $P(p)$ approaches p as p approaches 1; the infinite cluster consumes the finite ones. Several other curves, representing functions to be described below, are displayed in Fig. 4.5. All of these do something special ("exhibit singularities") at the percolation threshold. But the percolation probability $P(p)$ is really the key function characterizing a percolation process. It marks the qualitative change at p_c as the opportunity for long-range connectivity goes from nothing to something, and it provides the principal measure of the growth in volume of the extended network as the concentration increases beyond p.

4.4 APPLICATIONS TO PHASE TRANSITIONS

The qualitative shape of the $P(p)$ curve of Fig. 4.5 evokes, for condensed-matter physicists, the image of a phase transition. It has the look of the behavior of the "order parameter" in a second-order thermodynamic transition, which goes to zero rapidly, but continuously, as the transition temperature is approached. In fact, the percolation model serves as a splendid paradigm for the theory of critical phenomena, which is concerned with the properties of a system near its transition point. Analogies between percolation and, for example, the traditional models for magnetic transitions will be drawn later in this section and the next.

The curve labeled $\sigma(p)$ in Fig. 4.5 relates directly to our original example of Fig. 4.1. It sketches the behavior of the macroscopic conductance of the randomly diluted resistor network, and simply mirrors the current-versus-concentration curve of Fig. 4.1. Like $P(p)$, $\sigma(p)$ is zero for $p < p_c$ and increases monotonically above p_c. However, one is immediately struck by the dramatic difference in the behavior of these two functions near the percolation threshold. Just above p_c, P rises very steeply. Right at threshold, in fact, it rises with *infinite slope (i.e., dP/dp becomes arbitrarily large as $p - p_c$ is chosen to be arbitrarily small)*. The conductance, on the other hand, shows a very *soft* rise: its initial slope is zero at threshold ($d\sigma/dP$ approaches zero as $p - p_c$ goes to zero).

The striking difference between the way that the percolation probability and the resistor-network conductivity "take off" above p_c illustrates an aspect belonging to the realm of *critical phenomena*, the topic which focuses on the region very close ($|p - p_c| \ll 1$) to the critical point. Behavior in this regime is controlled by the ubiquitous quantities known as critical exponents, such as those we have met earlier in connection with polymers and random walks. Critical exponents for percolation processes, and a comparison to those which are standard in magnetic phase transitions, will be discussed in the following section. In the meantime, it is not hard to understand the physical reason for the contrast between the near-threshold behavior of P and σ. (With the benefit of hindsight, it is easy to say this. Historically, however, the point mentioned in the next paragraph suffered a substantial time delay prior to its appreciation.)

The explosive growth in P reflects the rapidity with which finite clusters link up with the infinite one as the concentration exceeds p_c. Consider a finite cluster which, with the addition of one unblocked bond (or filled site, the argument applies to site percolation as well as bond percolation), joins up with the previously formed percolation path. Since it now is connected to, and forms part of, the infinite cluster, it contributes to the percolation probability $P(p)$. But from the point of view of macroscopic current flow, the new bonds add nothing in the way of a new parallel path for current to pass through because *they form a cul-de-sac on the original extended network.* They do not lead anywhere ("no outlet") and thus they *do not contribute* to $\sigma(p)$. Just above threshold, such dead ends dominate the percolation path, and only a negligible fraction of it (which constitutes its skeleton or "backbone") contributes to the conductivity. This is the explanation for the slow start in $\sigma(p)$ just above p_c. As p increases, so does the portion of the percolation path which participates in conductivity until, at $p \to 1$, *all* of it contributes.

Another way of saying what is said in the previous paragraph is that near threshold, the impedance of the extended network is enormous. When it first appears, the percolation path is terrifically tortuous. It is so lacy and stringy in character that distinct parallel paths available for supporting current are extremely scarce. It turns out that this tenuous nature of the infinite cluster, when it appears at p_c, can be characterized by a fractal dimensionality which is somewhat smaller than the space dimensionality of the percolation process being considered. This fractal dimensionality is related to the critical exponents controlling the properties of a percolating system near its critical point, and its discussion is included in the next section.

The percolation transition, which takes place as the occupation or concentration or density is varied in a disordered system, makes percolation a natural model for describing a diversity of phenomena. Table 4.1 lists some 15 different physical (also chemical and biological) situations to which percolation ideas have been applied. Nearly half of these refer to macroscopic phenomena, the rest to microscopic phenomena. The macroscopic/microscopic demarcation is easy to spot in the center of the table, since the two extreme cases have deliberately been placed in juxtaposition in order to dramatize the diversity. The astrophysical application concerns the propagation of star formation by supernovae, and the percolation transition occurs as a function of interstellar gas density. The particle physics application concerns the confinement (or lack

Table 4.1   Applications of percolation theory

Phenomenon or System	*Transition*
Flow of liquid in a porous medium	Local/extended wetting
Spread of disease in a population	Containment/epidemic
Communication or resistor networks	Disconnected/connected
Conductor–insulator composite materials	Insulator/metal
Composite superconductor-metal materials	Normal/superconducting
Discontinuous metal films	Insulator/metal
Stochastic star formation in spiral galaxies	Nonpropagation/propagation
⏜Quarks in nuclear matter	Confinement/nonconfinement
Thin helium films on surfaces	Normal/superfluid
Metal-atom dispersions in insulators	Insulator/metal
Dilute magnets	Para/ferromagnetic
Polymer gelation, vulcanization	Liquid/gel
The glass transition	Liquid/glass
Mobility edge in amorphous semiconductors	Localized/extended states
Variable-range hopping in amorphous semiconductors	Resistor-network analog

thereof) of quarks in nucleons, and the relevant density is that of nuclear matter in the cores of neutron stars. Note that the characteristic length scales in these two examples differ by a factor of 10^{35}(!): the characteristic galactic dimension is of the order of 10^{22} cm, while the nucleon dimension is of the order 10^{-13} cm.

The applications of percolation to amorphous solids appear in the lower part of the table. Notable among these are the connection of percolation to the localization of electrons (the mobility edge, or Anderson transition, in amorphous semiconductors) and of atoms (the glass transition itself). The characteristic length scales in these condensed-matter contexts range from atomic dimensions upward, typically 10^{-8}–10^{-2} cm. The Anderson and glass transitions are individually discussed in the following chapter; some of the other entries in Table 4.1 are more briefly treated here.

Figure 4.6 illustrates two of the situations mentioned in the table. One of these is a macroscopic example, the other, microscopic; one involves bond percolation, the other, site percolation. Although one phenomenon is essentially two dimensional in character and the other is three dimensional, for clarity, both have been represented as occurring in two dimensions (using the easy-to-draw square lattice).

Having mentioned length scales appropriate to galaxies and quarks, it may be reassuring to first consider the human-scale enterprise sketched in Fig. 4.6a. This figure is meant to illustrate the second entry in Table 4.1: the propagation of a disease (or a rumor, or an idea, etc.) through a susceptible population. Imagine a uniformly planted orchard which is composed of a species of fruit tree that is subject to a certain highly contagious blight. Assume that we know the function, call it $p(r)$, which gives the probability of a diseased tree infecting a neighboring healthy tree located a distance r away. Given this func-

(a) SPREAD OF BLIGHT IN AN ORCHARD

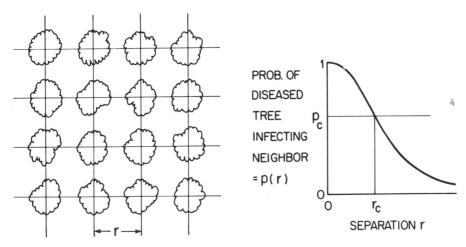

(b) DILUTE FERROMAGNETS

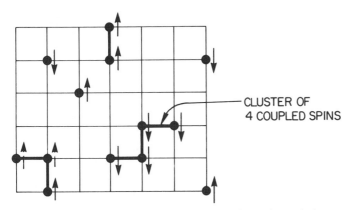

Figure 4.6 Two applications of percolation.

tion, and the natural wish of the farmer to maximize his yield by planting the largest practical number of trees on the available acreage, we can present the following problem: What is the highest density permissible for the planting which avoids the risk of blight-induced wipeout? We assume that a few widely scattered individuals will inevitably develop the disease, and define a wipeout as an event in which a single sick tree produces an epidemic that spreads throughout the orchard, destroying a finite fraction of the entire tree population.

It is evident that the percolation model answers the problem posed here in the following way: The orchard lattice constant a must be chosen large enough to ensure that $p(a) < p_c$. The spacing must exceed the critical distance r_c for which, as sketched in Fig. 4.6a, $p(r)$ falls below p_c. Thus the percolation solution is to take $a \gtrsim r_c$. Contamination is now confined to a finite cluster surrounding the initially infected tree.

At first glance it perhaps might seem that the situation described is one corresponding to site percolation, but this is not so. This is a bond-percolation process. The "fluid" is the disease, and its "flow" occurs along the bonds between neighboring trees. Although the simplest percolation model considers only nearest-neighbor bonds, it is relatively straightforward to construct the generalization which includes more distant neighbors. Such a long-range-interaction model would make more realistic use of the interaction function $p(r)$ of Fig. 4.6b, invoking its value not only at $r = a$ but also at $r = a\sqrt{2}$, $r = 2a$, $r = a\sqrt{5}$, $r = a\sqrt{8}$, etc. By not neglecting the small possibility for the disease to leap-frog over nearest neighbors to infect a more distant tree, a better estimate for the epidemic-avoiding lattice constant would be obtained (it would be slightly larger than the first estimate).

In a related example, experience with forest fires (in hot, dry weather) shows that the nearest-neighbor spacing in most woods is appreciably smaller than the r_c corresponding to the $p(r)$ appropriate for describing the spread of the fire from a burning tree to a fresh one. Thus forest fires, if not caught very early, are almost certain to propagate. One technique for fighting such fires is to attempt to surround it with a ring of burnt-out land (cleared by the use of backfires) of width w which is large enough so that $p(w) \ll 1$. Note that the connection between forest fires and phase transitions is by no means as fanciful as might seem at first sight. The phenomenon involves the nucleation and growth of a new phase (burnt-out land) within an old one (virgin forest), occurring in a process which is, rather manifestly, exothermic.

In these examples, the objects (trees) which define the sites of the lattice, although alive, are immobile. Nearest neighbors remain nearest neighbors, second neighbors remain second neighbors; *the topology of the system is time-independent*. This static character is characteristic of "pure" percolation; the medium is spatially stochastic, but does not fluctuate with time. Suppose we attempt to analyze the spread of disease among a population composed of individuals that are not sedentary (e.g., trees) but are highly mobile (e.g., people). Now we must contend with the fact that people change partners: The set of other individuals who are nearest neighbors of any particular individual *changes with time*. Thus the study, for example, of the spread of Asian flu in Manhattan involves, not only percolation, but also *diffusion*. There is no static lattice, but only one which is constantly *reconstructing* (old bonds are broken, new ones formed) with the passage of time. The distinction here is one which should by now be familiar; it is the same as that which distinguishes a liquid from an amorphous solid. This aspect is one of those which complicate the analysis of the glass transition, as discussed in Chapter Five.

In moving on to Fig. 4.6b, we reenter more familiar territory in solid-state physics in order to establish an analogy between percolation and magnetism. Figure 4.6b schematically represents the case labeled "dilute magnets" in Table 4.1. Consider a ferromagnetic crystal which is diluted by the random substitution of nonmagnetic atoms for magnetic ones. Assuming nearest-neighbor exchange interactions only and spin-½ magnetic ions (the "Lenz-Ising model"), the spin Hamiltonian is then $-J\Sigma_{i,j}S_z(i)S_z(j)$. $S_z(i)$ is the z-component of the spin at site i occupied by a magnetic ion, and can take on values $+½$ and $-½$. J is the nearest-neighbor exchange integral, and the sum extends over all pairs (i, j) of sites which are nearest neighbors and for which both sites are occupied by magnetic ions. The fraction of magnetic ions is p, of nonmagnetic ions, $1 - p$.

Let us look at the magnetic character of the system at zero temperature, as the dilution $(1 - p)$ is varied. At $T = 0$, all adjacent spins must be aligned parallel, so that the undiluted $(p = 1)$ pure system is a ferromagnet with all spins parallel and all sites contributing to the macroscopic magnetization M. As p decreases with the addition of nonmagnetic ions, M also decreases. When p falls below p_c, the site-percolation threshold for the lattice, M vanishes. The reason for this is clear, as illustrated in Fig. 4.6b. Although plenty of spins are still present, for $p < p_c$ there can be no macroscopically extended cluster of exchange-coupled spins. All of the magnetic ions now occur in finite clusters, and although *within* each cluster all spins are coupled and they must either all be up $(S_z = +½)$ or all down $(S_z = -½)$, the *separate* clusters can independently be either up or down so that the net magnetization cancels out on a macroscopic scale. There is no longer a spontaneous magnetization; the heavily diluted system is paramagnetic, not ferromagnetic.

Percolation theory thus provides the critical concentration of magnetic ions necessary for the appearance (at $T = 0$) of magnetic order. The critical composition corresponds to the site-percolation threshold. This is a site process rather than a bond one because the coupling, although it obviously "travels" along nearest-neighbor bonds, only exists for adjacent sites occupied by

Table 4.2 The analogy between bond percolation and the thermodynamic transition in a ferromagnet

Percolation	*Magnetism*
Unblocked-bond probability p	Exchange-interaction coupling (J/kT)
Percolation threshold p_c	Critical temperature T_c
Percolation probability P	Spontaneous magnetization M
Mean cluster-spanning length l_{av}	Correlation length ξ
Mean cluster size s_{av}	Magnetic susceptibility χ
Resistor-network conductivity σ	Exchange-stiffness elastic constant A
Bethe-lattice approximation	Mean-field approximation

magnetic ions. If two spins occupy nearest-neighbor sites, their coupling, at zero temperature, is certain. From this discussion, the composition-dependent magnetization $M(p)$ would be expected to mimic $P(p)$. Evidence for this general behavior has been found in some ferromagnetic alloys, although usually not simple $A_p B_{1-p}$ binaries as assumed here. Note that the critical concentration depends only on the lattice and on the assumption of nearest-neighbor-only coupling. The same result obtains in the Heisenberg model as in the Ising model for ferromagnetism $[-J\mathbf{S}(i)\cdot\mathbf{S}(j)$ for the coupling, instead of $-JS_z(i)S_z(j)]$, for spin greater than ½, and even for dilute antiferromagnets ($J < 0$).

The above connection between models for dilute magnets at $T = 0$ and the percolation model is quite close. There exists another connection, not quite as close but much more significant, between percolation and the thermodynamic paramagnetic/ferromagnetic transition at finite temperature. This analogy is important because it connects percolation theory to a vast body of theory that has been developed for thermal phase transitions and critical phenomena. The magnetic case is the prototype of a second-order thermodynamic transition, and some key elements of the analogy to percolation are indicated in Table 4.2.

In the dilute-magnet case, we were concerned with the behavior of the system as we varied the composition (p) at fixed temperature ($T = 0$). Now we are interested in varying the temperature for a system of fixed composition ($p = 1$, the pure ferromagnet). At zero temperature, the parallel alignment of adjacent spins is mandated by the nearest-neighbor exchange coupling. When we "turn on" temperature to a finite value, parallel alignment of adjacent spins is no longer fully obligatory because the nonvanishing Boltzmann factor $\exp(-J/kT)$ permits the occurrence of low-energy configurations in which a few spins are flipped. Eventually, at very high temperature ($kT \gg J$), adjacent spins decouple completely as the Boltzmann factor approaches unity and the exchange-induced bias in favor of parallel spins disappears. In between, at a critical temperature T_c, which is of order J/k, the solid ceases to be a ferromagnet.

The situation described in the previous paragraph resembles a bond-percolation process, with temperature playing the part (as the agent active in bond scission) of the "little man with the shears" in Fig. 4.1. The "fluid" flowing along the bonds is the control or influence exerted by the alignment of one spin on that of its neighbor. At $T = 0$ this control is complete; at $T \to \infty$, it is nonexistent. In order for any spin to have a chance of propagating information about its alignment over a macroscopic distance (i.e., belong to the infinite cluster, or macroscopic magnetization), the "influence" must, in some sense, exceed the bond-percolation threshold. As indicated in Table 4.2, the dimensionless ratio which plays a role analogous to that of p is (J/kT), which is a measure of the exchange-interaction coupling between neighboring spins. The bond-probability/coupling-constant analogy is made closer by considering, instead of (J/kT), a function of this ratio such as the complement of the Boltzmann factor: $1 - \exp(-J/kT)$. Such a form for the coupling strength goes from unity at $T = 0$ to zero at $T \to \infty$, making the correspondence to p closer.

While the connection between site percolation and the variable-composition magnetic alloy case of Fig. 4.6b is really quite close, the analogy between bond percolation and the thermodynamic ferromagnetic/paramagnetic transition is looser. This will be evident when the critical exponents for percolation and magnetism are compared in the next section; they are not the same. Nevertheless, the correspondences indicated in Table 4.2 provide an invaluable mapping of the percolation model onto *the* classic model of a continuous (second-order) phase transition. Many readers already familiar with the magnetic case can use Table 4.2 to ease this introduction to percolation theory. For others, a left-to-right reading of the table may be of interest, because magnetism is more subtle than the inherently geometric percolation model. As stated at the outset of this chapter, percolation provides a reasonably painless introduction to the field of phase transitions and critical phenomena.

The first three correspondences listed in Table 4.2 have already been discussed, and most of the others will be discussed in the following section. Because it allows us to hark back to a structural aspect of the percolation transition which was noted earlier in connection with Fig. 4.1, we close the present section with an explanation of the next-to-last entry in Table 4.2. The analogy pointed out here also connects the paramagnet/ferromagnet transition to the insulator/metal transition that occurs so frequently, in different guises, in Table 4.1 as an intuitively natural application of percolation. In discussing Fig. 4.1, we observed that the same dilution at which the mesh screen ceased to function *electrically* as a macroscopically conducting network, it also ceased to function *elastically* (it fell apart). Indeed, this indicated connection between electrical conductance and elastic stiffness is demonstrated in Table 4.2 for a magnetic counterpart of mechanical rigidity, the "exchange stiffness" A. A is essentially a *magnetic elastic constant*; it is a measure of the stiffness of the direction of the magnetization M against tilt away from its stable direction. This magnetic stiffness coefficient is the analog, in the magnetic case, of the resistor-network conductivity in percolation. The analogy between conductance and spin-tilt stiffness, stated here simply as a plausible pronouncement in view of the simultaneous disappearance of electricity and rigidity in the gedanken scenario of Fig. 4.1, can be justified mathematically by a one-to-one correspondence in form between Kirchhoff's laws for the voltage drops in the resistor-network case and the equations governing spin directions in the magnetic case. An analogy to an actual *elastic* stiffness will appear in Section 4.6 in connection with polymer gels.

4.5 CLOSE TO THRESHOLD: CRITICAL EXPONENTS, SCALING, AND FRACTALS

Of the four percolation functions that have been exhibited in Fig. 4.5 for the case of bond percolation on the two-dimensional square lattice, the two which have been discussed thus far, the percolation probability $P(p)$ and the resistor-network conductivity $\sigma(p)$, are functions which take off from zero at the percolation threshold p_c and which then grow monotonically to a finite maximum value at $p = 1$. Both P and σ vanish throughout the low-density regime $0 < p < p_c$. The other two functions shown in the figure behave in an entirely different way. For them the region of primary interest is *below* p_c, where they provide advance warning of the impending event. Both are finite at $p = 0$, grow monotonically and with increasing slope as p increases, and increase without bound as p approaches p_c.

The two quantities which diverge as the density draws near to p_c are the mean cluster size $s_{av}(p)$ and the mean spanning length $l_{av}(p)$. s_{av} was defined earlier, via Eq. (4.1), in Section 4.3. It starts out at $s_{av}(p \approx 0) = 1$ in the singlet-dominated low-density limit, grows as p increases, becomes very large as $p \to p_c$ as very large clusters form, and finally becomes infinite at $p = p_c$ with the emergence of the infinite cluster.

Whereas s, the number of connected sites or bonds, provides the natural measure of the content or volume or "mass" of a given cluster, l (as yet undefined) provides a measure of the cluster's characteristic *linear* dimension. l and s are separate and distinct attributes. Most notably, because of the subtlety and complexity of the question of cluster *shape* for the large clusters that occur close to threshold, we shall find that the conventionally expected relationship $s \sim l^d$ (d is the spatial dimensionality) does *not* hold as l, $s \to \infty$.

There are several possible choices for the characteristic length, such as the average or the rms distance from the center of gravity of the cluster. Since they are all essentially equivalent (same order of magnitude, same scaling behavior), it is simplest to take l to be the spanning diameter or *spanning length* of the cluster. The spanning length is defined as the maximum separation of two sites (or bond centers, for bond percolation) in the cluster:

$$l \equiv \max \{|\mathbf{r}_i - \mathbf{r}_j|\}_{i,j \text{ in cluster}}. \tag{4.2}$$

For a given p, the characteristic length obtained by averaging l over all clusters is the mean spanning length $l_{av}(p)$. This quantity plays a part, in percolation, analogous to that played by the correlation length ξ in phase transitions. Both provide length scales which correspond to the *graininess* of the system. This graininess, very fine far away from the percolation threshold or phase transition, coarsens dramatically as the transition is approached.

In a magnetic phase transition, the function whose characteristic length is specified by ξ is the spin–spin correlation function. The corresponding function in percolation theory, for which l_{av} sets the scale, is the *pair connectedness* $g(r)$. $g(r)$ is the probability that two sites i and j, separated by the distance $r = |\mathbf{r}_i - \mathbf{r}_j|$, belong to the same cluster. From what we know thus far, we may immediately deduce the asymptotic limit for $g(r)$ as $r \to \infty$. If the concentration of filled sites is less than the percolation threshold, then the asymptotic value $g(\infty)$ is zero since all clusters are finite, and there is no chance whatever of two infinitely separated sites being connected. But if the concentration exceeds p_c, a pair of widely separated sites will be connected to each other if both of them belong to the infinite cluster. Since *both* must belong, the $r \to \infty$ limit of $g(r)$ must be the square of the percolation probability P:

$$\lim_{r \to \infty} g(r) = [P(p)]^2. \tag{4.3}$$

While the above relation follows directly (i.e., is "derived") from the definitions of $g(r)$ and $P(p)$, the following statement appeals to higher authority. Like many other statements presented in this chapter about the behavior of various functions which describe percolation, it is asserted *as an experimental fact!* (The experiments referred to are, for the most part, Monte Carlo simulations of site or bond percolation on various lattices in two and three dimensions. In addition to such computer experiments, some more traditional laboratory experiments have been carried out for both site and bond versions of random resistor networks.) The *approach* of $g(r)$ to its limiting value at large r occurs *exponentially*, with a characteristic decay length l_0, which differs from l_{av} only by a numerical factor of order unity:

$$r \to \infty: \qquad g(r) - [P(p)]^2 \sim \exp(-r/l_0). \tag{4.4}$$

Although the asymptotic behavior expressed in Eq. (4.4) should be regarded, along with many of the other mathematical relations discussed for percolation, as empirically established, and although it is really *extremely* dangerous to attempt to generalize from behavior in one dimension, it is nevertheless instructive to reproduce a simple argument for 1d which was noted by Kirkpatrick (1979). For a 1d chain in which unblocked bonds occur with probability p, the probability $g(r)$ that two sites separated by distance r (with $r = na$, where a is the lattice constant and n is the number of bonds) are connected is just p^n. Hence $g(r)$ can be written as $p^{(r/a)}$, which in turn is $\exp[(r/a) \ln p]$, which in turn is $\exp(-r/l_0)$ with the decay length l_0 given by $a/(-\ln p)$. In 1d, $g(r)$ decays exponentially, not only for large r (as in $d > 1$), but for all r. It also asymptotes to zero for all p, since $p_c = 1$. Note that $l_0 \to \infty$ as $p \to p_c$. This important aspect of the 1d behavior is also true in higher dimensions; the correlation length diverges as the percolation threshold is approached [as shown for $l_{av}(p)$ in Fig. 4.5].

The below-threshold regime of the asymptotic form of the pair connectedness, as contained in Eq. (4.4), is important enough to warrant a mild redundancy in the form of a separate, explicit equation:

$$p < p_c, r \to \infty: \qquad g(r) \sim \exp(-r/l_0). \tag{4.5}$$

As discussed in the previous paragraph and displayed in Fig. 4.5, the decay length l_0 ($\approx$ the mean cluster-spanning length l_{av}) depends sensitively on density. The significance of Eq. (4.5) is the association between *localization* ($p < p_c$, all clusters finite) and a spatial dependence showing an *exponential fall off* at large distances. These two concepts will prove to be essentially synonymous, as will be seen in the discussion of disorder-induced localization of electron wave functions.

The preceding discussion of l_{av} and $g(r)$ presents some parallelisms to features that appeared previously in connection with glass structure. The pair connectedness in percolation processes resembles the pair-correlation function [which we have usually called the radial distribution function RDF(r)] that is used to characterize the structure of amorphous solids. Specifically, $g(r)$ is analogous to the reduced radial distribution function $G(r) = 4\pi r[n(r) - n_0]$, which was shown in Fig. 2.22 for the case of a metallic glass. Both $g(r)$ and $G(r)$ damp out beyond a characteristic distance. (For $g(r)$, the characteristic radius is a measure of the typical linear dimension of a cluster ($\approx l_{av}$). For $G(r)$, it is a measure of the range of the short-range order.

A connection with polymer structure also presents itself. The mean spanning length l_{av} plays the part for percolation clusters which R_{rms}, the configuration-averaged end-to-end length, plays for polymers and self-avoiding walks. Later on in this section, the percolation/polymer comparison will be analyzed in terms of the asymptotic scaling behavior and fractal dimensionalities of percolation clusters and polymer coils.

Although other characteristic functions for percolation can be introduced, the four displayed in Fig. 4.5 are probably the most basic. They suffice to illustrate the main features. Two of the functions, $s_{av}(p)$ and $l_{av}(p)$, describe geometric aspects of the growth of the clusters below the percolation threshold. These quantities are finite below p_c and infinite above p_c. They may be regarded as being complementary to the quantities $P(p)$ and $\sigma(p)$, which vanish below p_c and are finite above p_c. P and σ describe different measures of the filling out of the percolation path above threshold.

It should be noted that, although s_{av} and l_{av} as defined here (averaged over *all* clusters) remain infinite above p_c because of the presence of the infinite cluster, they are sometimes defined to exclude the latter. Above p_c they then characterize the finite clusters which coexist with the infinite one, falling off sharply from $p = p_c$ to $p = 1$ in a way which is roughly a mirror image of their behavior between 0 and p_c. Near $p = 1$, this definition of s_{av} and l_{av} describes the geometry of the holes or ''pores'' in the ''sponge'' formed by the extended cluster. This version of s_{av} and l_{av} (which lacks the nice ''complementarity'' aspect, noted above, vis-à-vis P and σ) will not be considered further. Our attitude is that the main value of s_{av} and l_{av} resides in what they tell us about the localization of the clusters below p_c. Once the percolation path appears, it absorbs our interest. Our attention then shifts to P and σ, which describe the macroscopically extended (delocalized) cluster.

We now focus down on the important region very close to the percolation threshold, the *critical region* in which $|p - p_c| \ll 1$. In this regime, the percolation functions are observed to obey power-law dependences on the distance-from-threshold $p - p_c$. For the divergence of the average size and linear dimension of the clusters as p_c is approached from below,

$$(p_c - p) \to 0: \qquad s_{av} \sim \frac{1}{(p_c - p)^\gamma}, \qquad (4.6)$$

$$l_{av} \sim \frac{1}{(p_c - p)^\nu}. \qquad (4.7)$$

For the initial increase in the content and conductivity of the percolation path,

$$(p - p_c) \to 0: \qquad P \sim (p - p_c)^\beta, \qquad (4.8)$$

$$\sigma \sim (p - p_c)^t \qquad (4.9)$$

In Eqs. (4.6)-(4.9), the exponents γ, ν, β, and t are found to be, for lattices in two and three dimensions, positive numbers that are *not* integers. The experimentally observed exponents are presented in Table 4.3.

Now as we will find in the following section, the percolation threshold p_c is a quantity which varies greatly from lattice to lattice. (To quote a three-dimensional demonstration of p_c's variability, the critical concentration for site percolation on the diamond lattice is more than double that for the fcc lattice.) The remarkable feature of the power-law dependences of Eqs (4.6)-(4.9) is that *these exponents do not depend on the details of lattice geometry, they are the same for all lattices of the same dimensionality.*

Called critical exponents because they govern the scaling behavior (power-law dependences) in the critical region, these quantities are examples of *dimensional invariants*. Each exponent has a fixed value for a given dimensionality d, without regard to the specific nature of the short-range structure. The values for lattices in two and in three dimensions are shown in Table 4.3 in the columns headed $d = 2$ and $d = 3$, respectively. The column headed $d \geq 6$ has a special significance which is mentioned at the end of this section.

Table 4.3 Critical exponents controlling the near-threshold scaling behavior of key quantities in percolation[a]

Functional Form close to $p = p_c$	Exponent	Value of Exponent in d Dimensions		
		$d = 2$	$d = 3$	$d \geq 6$
$P(p) \sim (p - p_c)^\beta$	β	0.14	0.40	1
$\sigma(p) \sim (p - p_c)^t$	t	1.1	1.65	3
$s_{av}(p) \sim (p_c - p)^{-\gamma}$	γ	2.4	1.7	1
$l_{av}(p) \sim (p_c - p)^{-\nu}$	ν	1.35	0.85	½
$s \to \infty: n(s) \sim s^{-\tau}$	τ	2.06	2.2	2½
$s \to \infty: l(s) \sim s^{(1/f)}$	f	1.9	2.6	4
Magnetism				$(d \geq 4):$
$M(T) \sim (T_c - T)^\beta$	β	0.125	0.32	½
$\partial M/\partial H \sim (T - T_c)^{-\gamma}$	γ	1.75	1.24	1
$\xi(T) \sim (T - T_c)^{-\nu}$	ν	1.0	0.63	½

fractal dimension (handwritten annotation pointing to f row)

[a]For three exponents, the analogies to the corresponding quantities in magnetism are shown in the lower part of the table.

The impressive generality of the critical exponents has been given an impressive name in the field of phase transitions, in which such exponents have long been known and appreciated. It is called "universality." The Greek symbols used for the percolation exponents have been chosen to correspond to those which have become traditional for phase transitions. For example, the exponent ν, which describes the near-threshold divergence of the mean cluster diameter, is named in correspondence with the exponent that describes the divergence of the correlation length near a second-order transition. Magnetic analogies for β, γ, and ν are indicated in the last three rows of Table 4.3. The analogies for the correlation-length exponent ν and the order-parameter exponent β are straightforward, but the correspondence of the cluster-size exponent γ to the exponent which describes the response function (susceptibility $\partial M/\partial H$) of the magnetic system is more subtle and will not be discussed here.

No distinction between site percolation and bond percolation has been made in the table. This is a further tribute to the generality of the critical exponents: For the same dimensionality, they are observed to be the same for bond processes as for site processes. In the jargon of phase transitions, site percolation and bond percolation are said to belong to the same "universality class."

In Eqs. (4.6)–(4.9), the exponents have been defined, for convenience as well as for concordance with tradition, so that their values are all positive numbers. Of course, exponents which describe properties that diverge as a function of the distance-to-threshold $|p - p_c|$, such as s_{av} and l_{av}, are inherently negative, while those which describe properties that grow from zero upon passing threshold (e.g., P, σ) are inherently positive. With respect to the latter two, the numerical magnitude of the exponent has an important bearing on the qualitatitive behavior near threshold. It is quite noticeable, in Fig. 4.5, that $P(p)$ and $\sigma(p)$ begin their growth in completely different ways. While the ascent of $P(p)$ is extremely steep, that of $\sigma(p)$ is extremely gentle. More precisely, at p_c, $P(p)$ starts out with *infinite* slope while $\sigma(p)$ starts out with *zero* slope. Both features follow from the form of Eqs. (4.8) and (4.9) and from the fact (Table 4.3) that $\beta < 1$ while $t > 1$. Note that the behavior of the percolation probability is that expected for an order parameter in a second-order phase transition: at p_c, $P(p)$ is continuous while dP/dp is discontinuous.

The exponent τ, defined and tabulated in the fifth row of Table 4.3, is a bit different from those which have been discussed thus far. Instead of describing the asymptotic behavior of some function as $|p - p_c| \rightarrow 0$, τ describes the behavior, *at $p = p_c$*, of the cluster-size distribution in the asymptotic limit of large cluster size,

$$p = p_c, \quad s \rightarrow \infty: \quad n(s) \sim s^{-\tau}. \tag{4.10}$$

We introduce this exponent because it will allow us to make contact with the concept of fractal dimensions in the context of the percolation model, as we have done earlier in the context of random walks and polymer structure.

Before going ahead with the discussion of Eq. (4.10), we must take care to emphasize that this relatively gradual power-law falloff in $n(s)$ appears *only* at

p_c. Throughout the entire range below threshold, for each value of p the large-s falloff in $n(s)$ is much more rapid:

$$p < p_c, \quad s \to \infty: \qquad n(s) \sim \exp[-\text{const.} \times s]. \tag{4.11}$$

This exponential decay of the frequency of occurrence as a function of cluster size is the asymptotic behavior appropriate to a regime in which localization (all clusters finite) applies, and was met previously in Eq. (4.5) for the pair connectedness. The constant in Eq. (4.11) is fixed for fixed p, but clearly changes when p changes. Incidentally, the pair connectedness $g(r)$ also exhibits a property analogous to Eq. (4.10): Right at p_c, with the correlation length l_0 of Eq. (4.5) having gone out of sight, $g(r)$ has a power-law decay at large r with its own characteristic exponent (which will go, mercifully, unmentioned; it is all too easy to let these exponents proliferate!).

The exponent τ must be a number lying between 2 and 3. To see this, look at the sums in the denominator and numerator of Eq. (4.1). The contribution of large clusters to these sums may be replaced by an integral from s_0 to infinity, where s_0 is a cluster size large enough for the asymptotic form given by Eq. (4.10) to be accurately obeyed for $s > s_0$. The large-cluster contribution to $\Sigma sn(s)$ is then $\int_{s_0}^{\infty} sn(s)ds$, which is proportional to $\int_{s_0}^{\infty} s^{-\tau+1} ds$. This integral must be finite, because $\Sigma sn(s)$ is equal to p (here, actually, p_c, since we are discussing the situation at the critical concentration). For this integral to be finite, it is necessary that $\tau > 2$. In similar fashion, the condition $\tau < 3$ can be derived from the form of the integral that measures the large-cluster contribution to the numerator of Eq. (4.1), taken together with the knowledge that s_{av} diverges at p_c.

Enter now the sixth, and last, of the critical exponents which we will discuss for the percolation model. This one is especially interesting because it says something about the wild and peculiar geometry of the infinite cluster when it makes its debut at p_c. To appreciate the need for such a measure of the special geometric character of the percolation path, we may pause to look at the large cluster shown in Fig. 4.7.

This figure displays a typical cluster observed in a careful computer study (Leath and Reich, 1978) of site percolation of the 2d triangular lattice. The concentration of filled sites is $p = 0.48$, just below threshold ($p_c = \frac{1}{2}$), and this particular cluster contains $s = 4741$ filled sites (shown as solid dots). Also shown in the figure (as open circles) are the surrounding empty sites which block continuation of the cluster. The blocking sites are said to form the boundary, and the number of empty sites comprising it is denoted by t. (This usage will appear only in this paragraph and the next, so there should be no confusion with the conductivity exponent t.) For this cluster, $t = 5102$.

Now we normally ("normality" being defined, geometrically, by Euclid) think of a d-dimensional object as being bounded by a $(d - 1)$-dimensional surface, and we normally expect the surface/volume ratio to go to zero as the size of the object goes to infinity. This does *not* happen for percolation clusters near threshold. Let us look at the ratio t/s which, for want of a better alternative, we

Figure 4.7   A large cluster observed just below the percolation threshold by Leath and Reich (1978) in a computer study of site percolation on the triangular lattice. The dots are filled sites belonging to the cluster, while the open circles are the blocking open sites at the boundary.

take to be the surface/volume ratio for a cluster. For the sizable cluster of Fig. 4-7, this ratio is not small; t is actually *larger* than s. As s increases, t/s is found to approach a limiting value *of order unity*, and indeed it can be rigorously shown that $\lim_{s \to \infty}(t/s)$ is exactly given by $(1 - p_c)/p_c$. (This ratio happens to be exactly unity for site percolation on the triangular lattice, which explains why $t \approx s$ for the example shown in Fig. 4.7.) Thus the surface-to-volume relationship for large clusters is an unexpected one. To quote Mandelbrot (1977), these quantitative observations bear out the qualitative appearance of the large clusters as being "all skin and no flesh."

Although symptomatic of something unusual about the dimensionality of the large clusters, the property described above is, by itself, inconclusive. After all, Swiss cheese is a familiar example of an entity for which the surface area (of the internal voids) is proportional to the volume of the sample. Another three-dimensional example, more apt in relation to near-threshold percolation, is that of a sponge. More meaningful is the demonstration of a fractal dimensionality, different from the space dimensionality, for the very large clusters. This demonstration follows.

It is borne out by experiment that for large clusters, the average relationship between the cluster size s and the spanning length l is well described by a power law,

$$p = p_c, \quad s \to \infty: \quad s \sim l^f,$$

or

$$l(s) \sim s^{(1/f)}. \tag{4.12}$$

Relation (4.12) links a length dimension (l) with a measure of content or volume (s), and the exponent that normally appears in such a relation [in the place where f appears in Eq. (4.12)] is the space dimensionality d. But f (the sixth and final critical exponent to be identified here) does *not* coincide with d for large percolation clusters. Instead, it is found to be *smaller* than d: f is approximately 1.9 in two dimensions, 2.6 in three dimensions.

From the definition of f in Eq. (4.12), and from the discussion of fractal dimensions given in Section 3.2.5, it is quite natural to interpret f as a fractal dimensionality which characterizes the very large clusters (and, thereby, the percolation path itself very close to p_c) which occur close to threshold. f provides a quantitative measure of a very interesting quality, evident in a glance at Fig. 4.7, which is possessed by the nearly percolating large clusters and which has already been briefly alluded to ("all skin and no flesh"). It is the feature that underlies the disparity between the content and the conductivity of the percolation path just above threshold (i.e., the P/σ dichotomy of Fig. 4.5, with $\beta < 1$ and $t > 1$).

The key quality in question, a basic attribute of the large clusters, is their wispy, tenuous, *insubstantial* character. This quality evokes many nice adjectives, including (in addition to the three already mentioned): lacy, stringy, sparse, ragged, airy, rarefied, ramified, and ethereal. Many of these terms have been applied to percolation clusters in the physics literature. Our favorites for this purpose are diaphonous and, especially, *gossamery*. Alas, physicists must eventually abandon verbal niceties for mathematical ones, and the fractal dimensionality f is the quantity which captures the quality of flimsiness inherent in the large clusters near threshold. More aptly, it is the difference $d - f$, the amount by which the cluster fractal dimensionality falls short of the space dimensionality (and thereby fails to fill d-space), that provides the best measure of "gossameriness." It gives an answer to the question, How wispy is it? The dimensionality deficit $d - f$ is called by mathematicians the "Hausdorff codimension." For an ordinary, substantial, d-dimensional object (such as the percolation path *away* from the critical region, say $p - p_c > 0.05$), f equals d and the deficit vanishes.

There exists, and it is instructive to deduce it, a connection between f, the exponent governing the linear dimension $l(s)$ of large clusters of size s, and τ, the exponent governing the frequency of occurrence $n(s)$ of such clusters. The following argument, much simpler than others used for this purpose, was first proposed by Harrison, Bishop, and Quinn (1978). The argument is phrased in a way to apply, in general fashion, to dimensionality d, an approach frequently adopted in this book.

Consider a large d-dimensional box of side L. For a cluster to span this box (reach from one face to the opposite one), it needs to be of size $s(L) \sim L^f$ or larger. The probability of finding such a cluster is proportional to

$$L^d \int_{s(L)}^{\infty} n(s)ds \sim L^d [L^f]^{-\tau+1}$$
$$\sim L^{d-f(\tau-1)}. \tag{4.13}$$

The factor of L^d in Eq. (4.13) is proportional to the total number of sites in the box. [Recall that $n(s)$ is a normalized quantity specifying the frequency of occurrence per site.] The integral is proportional to that portion of the cluster-size distribution corresponding to clusters larger than $s(L)$, the box-spanning size.

Now let L go to infinity. Since Eq. (4.13) represents a probability and therefore is finite, the exponent of L cannot be positive. Since we are at the percolation threshold (let us cheat slightly and assume that we are *very slightly* above p_c), we may expect the probability of occurrence of a spanning cluster to remain nonzero as $L \to \infty$. Hence the exponent is not negative. It must, therefore, be zero, and setting it to zero yields

$$f = \frac{d}{\tau - 1} \tag{4.14}$$

This is the desired connection between f and τ; it is seen to include the space dimensionality d. Since we have seen that τ is larger than 2, it follows that f is smaller than d.

A relation such as Eq. (4.14), which connects two critical exponents (here, f and τ), is known as a *scaling law*. Similar scaling laws appear in percolation theory as in the theory of phase transitions (Stanley, 1980). One very important example of such an interdependence among exponents, the Josephson scaling law, will appear below [as Eq. (4.17)] in connection with the critical dimensionality ($d = 6$) for percolation processes. Other scaling laws will not be explicitly mentioned, except to state, without proof, that they can be used to recast the result for f [Eq. (4.14)] in terms of the other exponents introduced earlier,

$$f = d - (\beta/\nu) \tag{4.15}$$

$$f = \frac{d + (\gamma/\nu)}{2} \tag{4.16}$$

It is evident from Eqs. (4.14)–(4.16) that there are scaling laws in sufficient supply to ensure that only *two* of the critical exponents can be independently set.

Equation (4.15) is nice because it exhibits a remarkably simple result (β/ν) for the dimensionality deficit $d - f$, our measure of the large-cluster wispiness at p_c. Even more interesting is Eq. (4.16). First note that, from Table 4.3, the ratio (γ/ν) is rather insensitive to dimensionality—it stays close to 2. So Eq. (4.16) may be approximated by $f = (d + 2)/2$. This reminds us of something. Recall Eq. (3.14) for the exponent governing the characteristic radius ($R_{rms} \sim N^\nu$) of d-dimensional self-avoiding walks: $(1/\nu_{SAW}) = (d + 2)/3$. The similarity suggests a connection, and there is one, since in Section 3.2.5 we saw that the fractal dimension for large SAWs is given by $(1/\nu_{SAW})$. Having noted this, it is then straightforward to proceed to write down the correspondences which form an attractive analogy between the properties of percolation clusters and of polymer configurations:

	Percolation	Polymers
Characteristic length scale	l	R_{rms}
Characteristic size or "volume"	s	N
Asymptotic (scaling) regime	$s \to \infty$	$N \to \infty$

Having arrived, via the unusual route of fractals, at the analogy between percolation and random flights, and having a partiality toward Cole Porter (1935, "Just One of Those Things"), the author cannot resist the temptation to compare the "gossamer wing" of Fig. 4.7 with the "fabulous flight" of Fig. 3.13. For a large near-threshold cluster in a percolation process occurring in two dimensions, the fractal dimension ($f = 1.9$) is intermediate between that of a Brownian-motion random flight ($f = 2$, Fig. 3.13) and that of a self-avoiding random flight on a 2d lattice ($f = 4/3$, Fig. 3.16).

We close this section by finally coming to grips with the mysterious fifth column of Table 4.3. This column, labeled $d \geq 6$, provides a bridge to a topic covered in the next section. The exponents listed here correspond to the critical exponents that appear in a mean-field theory for percolation, a theory which is the percolation counterpart of the Weiss molecular-field theory for ferromagnetism or the Flory–Fisher theory for polymer configurations. For percolation, mean-field theory corresponds to the theory of a cascade process on a Bethe lattice (these terms are defined in the next section). Now for phase transitions, it is known that there exists a *marginal dimensionality* d^* such that the critical exponents take on their "classical" (mean field) values for $d \geq d^*$. Though the justification for this is beyond the scope of this book, a rough argument for its plausibility is contained in the observation that the higher the dimensionality, the larger is the number of neighbors with which each site is in touch, and thus the closer is its environment to an average (i.e., mean field) environment. Another way to say this: If d is sufficiently large, then no essential mistake is made by assuming that *every site is an average site.* Conditions are such that spatial fluctuations are unimportant. [It is interesting to note that the coordination number $z_{cp}(d)$ for a close-packed lattice in d dimensions is known up to $d = 8$. The number of nearest neighbors goes up much more rapidly than d: z_{cp} = 2, 6, 12, 24, 40, 72, 126, 240 for d = 1 through d = 8, respectively.] For magnetism, d^* = 4; in four or more dimensions the exponents are the mean-field ones given in the lower right of Table 4.3. The question arises: What is d^* for percolation?

The answer to this was given by Toulouse in 1974, who reasoned that the Bethe-lattice values of the percolation exponents (see the next section) were indeed mean-field values, and to this added the observation that d^* is the value of d for which the Josephson scaling law is satisfied when β, γ, ν have their mean-field values (1, 1, ½, respectively). The Josephson is a particularly important scaling law because it connects the key exponents and [like Eq. (4.14)] contains the dimensionality:

$$2\beta + \gamma = \nu d. \tag{4.17}$$

Equation (4.17) is satisfied for $d \leq d^*$; it fails for $d > d^*$ where β, γ, ν retain their mean-field values, independent of dimensionality. Plugging these values into Eq. (4.17) yields d^* = 6 for percolation.

Toulouse quietly proposed d^* = 6 as a "nonimplausible conjecture" for the marginal dimensionality for percolation, and this nice idea was neatly confirmed by Kirkpatrick (1976) in a classic computer study of percolation in higher dimensions. The important part played in percolation theory by Monte Carlo computer calculations has already been emphasized several times; indeed, it can hardly be overemphasized. Dimensionalities higher than three present no special problems for such "computer experiments," that is to say, no problems other than the usual ones of machine speed, memory, and, of course, cost. (Note that Kirkpatrick did this work at IBM, where these problems are less severe than elsewhere). By analyzing the observed near-threshold behavior for site percolation on simple cubic lattices in d = 2 through d = 6,

Kirkpatrick was able to show that the critical exponents do indeed attain their mean-field values at $d = 6$. This established the validity of a marginal dimensionality for percolation, $d^* = 6$, further strengthening the analogy to the theory of second-order phase transitions.

By this time, noninteger dimensionalities should be rather familiar objects; such generalizations, strange at first, are simply a matter of getting accustomed to. (Then, going perhaps too far in the other direction, they are taken for granted!) One of the most important uses of the concept of dimensionality as a continuous variable, in condensed matter physics, is as an expansion parameter in the theory of phase transitions. The basic idea, which invokes the marginal dimensionality discussed above, is a form of perturbation theory. A pioneering 1972 paper, by Wilson and Fisher, was called "Critical Exponents in 3.99 Dimensions." For magnetic phase transitions, the question of critical exponents is a solved problem in $d^* = 4$ dimensions, where mean-field theory applies. If we are interested in d dimensions (say, $d = 3$), the program is to obtain each quantity as a power-series expansion about the known $(d = 4)$ regime, using $(4 - d)$ as the expansion parameter. The technique used to calculate the expansion is called renormalization group theory (see Section 4.8), and the approximation works when $(4 - d)$ is small.

This approach is sometimes referred to as the "epsilon expansion," where epsilon refers to the parameter of smallness $\epsilon = d^* - d$. It has been applied to percolation, for which $\epsilon = 6 - d$. For percolation processes, physically realizable dimensionalities ($d = 3$ and $d = 2$) are much further away from the "solved" dimensionality ($d^* = 6$) than is the case for magnetism ($d^* = 4$). This is the reason that the critical exponents for percolation in three and in two dimensions deviate much more markedly from their mean-field values (see Table 4.3) than do the corresponding exponents in magnetic phase transitions.

The unifying concept of universality was already mentioned. For the percolation critical exponents which have been introduced and collected in Table 4.3, namely β, t, γ, ν, τ, and f, universality means the applicability of the value of a given exponent to the near-threshold behavior of *all* site or bond percolation processes on *all* lattices of a given dimensionality. The scaling laws, the relationships *between* exponents, of which Eq. (4.17) is a beautiful example, are even *more* universal than the exponents themselves because they cut across different dimensionalities. The high degree of universality exhibited by the scaling laws is well exemplified by the validity (within experimental error) of Eq. (4.17) to all six sets of (β, γ, ν, d) values contained in Table 4.3, both those for percolation *and* those for magnetism. This is so even though percolation and magnetism belong to different universality classes, have different values for the exponents.

From the height of generality epitomized by the Josephson scaling law of Eq. (4.17), we are about to proceed to the other extreme in a treatment of the *position* (p_c) of the critical point. The percolation threshold is a thoroughly nonuniversal, but enormously important, parameter of percolation phenomena,

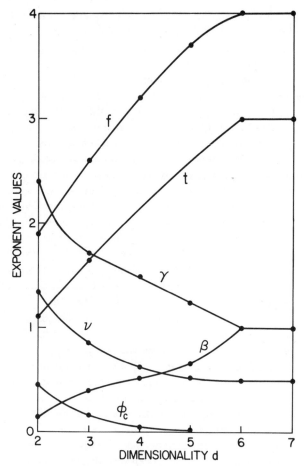

Figure 4.8   The dimensionality dependence of the critical exponents for percolation. The meaning of the exponents β, γ, ν, f, and t are given in Table 4.3; while ϕ_c is the critical volume fraction of Section 4.7.

and its dependence upon local structure is described in the next section. First, however, we show in Fig. 4.8 a diagram which is a fitting concluder for the present section.

The *x* axis in Fig. 4.8 is dimensionality *d*. Plotted against it are the values of critical exponents for percolation, as introduced in this section. (The quantity ϕ_c included in the figure is not an exponent, but a different type of dimensional invariant to be discussed in Section 4.7.) There are three main messages:

1.   Each of the quantities plotted has a well-defined value for a given dimensionality (dimensional invariance, the universality aspect).

2.   They show a quite systematic variation from *d* = 2 to *d* = 6.

3. For $d \geq 6$, mean-field theory is "exact" for the values of the percolation exponents.

The smooth curves that have been "eyeballed" through the experimental points are given in the spirit of the notion of dimensionality as a continuous variable.

4.6 TREES, GELS, AND MEAN FIELDS

Figure 4.9 displays four of the most basic two-dimensional lattices, the triangular, square, kagomé, and honeycomb lattices, along with their respective percolation thresholds. (The fifth structure illustrated, part of a Bethe lattice, will be discussed shortly.) All of these are "regular" lattices, every site has the same coordination z as every other site. They are also very symmetric structures, every site (or bond) is equivalent to every other side (or bond). Because of the high symmetry and simplicity of these lattices, it has been possible to analytically derive exact values for thresholds in several cases. Those p_c values that are known exactly are the ones given in Fig. 4.9 to four significant figures, namely, the critical concentrations for bond percolation on the triangular, square, and honeycomb lattices, and for site percolation on the triangular and kagomé lattices. Two of the five numerical values so quoted correspond simply to $p_c = \frac{1}{2}$, a third corresponds to the solution of $3p_c - p_c^3 = 1$, and the other two correspond to the complement ($p_c' = 1 - p_c$) of the latter solution.

The detailed arguments that are used to deduce the above exact values will not be given here, since they cannot be extended to less symmetric lattices. Also, no exact results for p_c are known for lattices in three (or more) dimensions. However, the general type of reasoning involved in the derivations of p_c for the four 2d lattices can be nicely illustrated by a pretty application to an electrical-network setting (Thouless, 1979). The basic idea exploited for the two-dimensional derivations is the notion of special relationships between pairs of lattices. The duality relationship between the triangular and the honeycomb lattices was illustrated earlier in Fig. 2–10; two lattices are duals of each other if (sites, bonds, cells) of one can be mapped in one-to-one correspondence to (cells, bonds, sites) of the other. The dual of a square lattice is another square lattice, and this self-duality property is the key to the argument now sketched.

Consider a resistor network arranged as a square lattice with the place of each bond being taken by a conductor with conductance equal to either zero (perfect insulator) or to one (normal metal) or to infinity (superconductor). (As discussed in Section 4.8, this three-species situation corresponds to a problem in *polychromatic* percolation.) For such a planar resistor network, a dual resistor network may be constructed by replacing each bond conductance by the intersecting bond of the dual lattice, with the new conductance having a value given by the reciprocal of the old. This transformation converts the original lattice into its dual and transforms electrical elements so that insulators be-

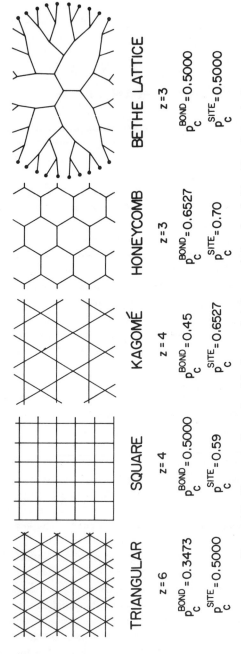

TRIANGULAR

z = 6

$p_c^{BOND} = 0.3473$

$p_c^{SITE} = 0.5000$

SQUARE

z = 4

$p_c^{BOND} = 0.5000$

$p_c^{SITE} = 0.59$

KAGOMÉ

z = 4

$p_c^{BOND} = 0.45$

$p_c^{SITE} = 0.6527$

HONEYCOMB

z = 3

$p_c^{BOND} = 0.6527$

$p_c^{SITE} = 0.70$

BETHE LATTICE

z = 3

$p_c^{BOND} = 0.5000$

$p_c^{SITE} = 0.5000$

Figure 4.9 Four two-dimensional lattices and their percolation thresholds. Also shown is the threefold-coordinated Bethe lattice, whose dimensionality is effectively infinite.

come superconductors, superconductors become insulators, and unit conductors become unit conductors. It can be shown (Straley, 1977) that the conductivity of the dual network is the reciprocal of that of the original network. Suppose a square-lattice random-resistor network has unit conductors present with probability p and both insulators and superconductors present with equal probability $\frac{1}{2}(1 - p)$. Since the square lattice is self-dual and since the composition in terms of electrical elements is unchanged, the transformed network is macroscopically indistinguishable from the original one. Since its conductivity equals the reciprocal of that of the original network, it therefore follows that the network described has unit conductance.

Turn back briefly to glance at Fig. 4.1. It happens that the square lattice was used to illustrate the resistor network under siege by the little man with the shears. That little man is, of course, in the process of effectively replacing unit conductors by insulators, thereby reducing the conductivity of the network. What has just been demonstrated is that the influence of the shearer (on the conductivity of the system) may be exactly neutralized by a second little man who randomly replaces unit conductors by superconductors at the same rate as the shearer replaces other conductors by insulators.

Having broached the idea of a random network containing superconducting links, we should briefly note that the conductivity-versus-composition curve of a network composed of normal conductors and superconductors would resemble, *not* the $\sigma(p)$ curve of Fig. 4.5, but instead the $s_{av}(p)$ and $l_{av}(p)$ curves. Starting out with finite conductivity when all links are normal ($p = 0$), the conductivity increases when some links are made superconducting ($p > 0$), and becomes infinite when the superconducting fraction exceeds p_c and the super-conducting links percolate. Moreover, the reciprocity theorem mentioned above for dual pairs of electrical networks provides an immediate result for the critical exponent describing the divergence of the conductivity of such a normal/superconductor resistor network as p_c is approached: At least in two dimensions, that exponent must be $-t$.

An inventory of lattices for which p_c values are known is given in Table 4.4. The threshold concentrations for bond percolation and for site percolation are shown in the third and fourth columns of the table. The main message evident from a glance at these data is the large variation from lattice to lattice, the sensitivity of p_c to local structure. For the two-dimensional lattices listed, the p_c values extend from 0.35 to 0.70; for the three-dimensional lattices, p_c spans the range from 0.12 to 0.43. As emphasized earlier, the percolation threshold is by no means a dimensional invariant.

Systematic behavior is quite discernible in the table. The coordination number z is included in the table as a measure of lattice connectivity, and it is easy to see the correlation between increasing connectivity and decreasing p_c. The more highly connected a lattice is, the lower is the concentration of filled sites (or unblocked bonds) needed for the formation of the infinite cluster. Similarly, because lattices in higher dimensions are more highly connected than those in lower dimensions, the trend is to decreasing p_c with increasing d.

Table 4.4 Critical concentrations for bond (p_c^{bond}) and site (p_c^{site}) percolation on a variety of lattices.[a]

Dimension-ality d	Lattice or Structure	p_c^{bond}	p_c^{site}	Coordination z	Filling Factor v	$z p_c^{bond}$	$v p_c^{site} \equiv \phi_c$
1	Chain	1	1	2	1	2	1
2	Triangular	0.3473	0.5000	6	0.9069	2.08	0.45
2	Square	0.5000	0.593	4	0.7854	2.00	0.47
2	Kagomé	0.45	0.6527	4	0.6802	1.80	0.44
2	Honeycomb	0.6527	0.698	3	0.6046	1.96	0.42
						2.0 ± 0.2	0.45 ± 0.03
3	fcc	0.119	0.198	12	0.7405	1.43	0.147
3	bcc	0.179	0.245	8	0.6802	1.43	0.167
3	sc	0.247	0.311	6	0.5236	1.48	0.163
3	Diamond	0.388	0.428	4	0.3401	1.55	0.146
3	rcp		[0.27][b]	[2.35]	0.637 [0.6][b]	[1.5]	[0.16][b]
						1.5 ± 0.1	0.16 ± 0.02
4	sc	0.160	0.197	8	0.3084		0.061
4	fcc		0.098	24	0.6169	1.3	0.060
5	sc	0.118	0.141	10	0.1645		0.023
5	fcc		0.054	40	0.4653	1.2	0.025
6	sc	0.094	0.107	12	0.0807	1.1	0.009

[a]The last two columns indicate a pair of approximate dimensional invariants that are useful in unifying threshold data for different lattices.
[b]Results of experiments of the type shown in Fig. 4.13.

Another point made by Table 4.4 is that bond percolation occurs more readily than site percolation; for a given lattice $p_c^{bond} < p_c^{site}$. One way to view the reason for this may be illustrated in terms of the square lattice shown in Fig. 4.9. While a site on this lattice is adjacent to *four* other sites, a bond is seen to touch *six* other bonds. In general, while a site has z nearest-neighbor sites, a bond has $2(z - 1)$ nearest-neighbor bonds. In this way a given lattice presents a richer connectivity to a bond process than it does to a site process, which is why bond percolation is ''easier'' than site percolation.

The statement $p_c^{bond} < p_c^{site}$ is violated by the equality of the two thresholds given in the first row of Table 4.4, but this exception is not very interesting since the one-dimensional lattice (included in the table for completeness) lacks percolation anyway because $p_c = 1$. More significant is the $p_c^{bond} = p_c^{site}$ situation shown on the right of Fig. 4.9. This will now be discussed, but first a few comments about the other material contained in the table. The last row among the 3d entries refers to results obtained for the random close-packed lattice, and will be discussed in the next section (as will the rows for $d > 3$). The last three columns relate to efforts to find *approximate* dimensional invariants (''almost-invariants'') related to the position of the percolation threshold, and are also discussed in Section 4.7.

The structure shown at the right of Fig. 4.9 represents a part of an infinite network of a type known as a *Bethe lattice*. The finite portion shown in the figure is an example of a *Cayley tree*, a graph containing *no closed loops* (= no cycles = no circuits). A Bethe lattice is defined as an infinite, regular, Cayley tree.

A Bethe lattice is not a lattice in a conventional sense. Obviously it is not periodic, but this lack is unimportant. The continuous random networks of Chapter Three are not periodic, but they are statistically well-defined lattices possessing the crucial property of being *uniform in the large*—macroscopically homogeneous. This physically essential property of macroscopic homogeneity is dispensed with by a Bethe lattice in spectacular fashion. The prohibition against closed loops enforces an endlessly increasing density as the structure is extended. The small portion of a $z = 3$ Bethe lattice that is shown in Fig. 4.9 in a two-dimensional setting already exhibits serious crowding at the boundary. (Among the criteria set down in Chapter Two for continuous random networks was the requirement that the surface density of dangling bonds remain constant as the lattice is extended. For a Bethe lattice that surface density is not merely inconstant, it grows without limit!)

The crowding problem can be alleviated, but it *cannot be removed*, by setting the Bethe lattice in a space of higher dimensions. A Bethe lattice in three dimensions *also* suffers from a density that diverges with size, as is nicely demonstrated by the drawing reproduced as Fig. 4.10. (The wintry plant represented in Marianne Lehmann's graceful ink drawing actually corresponds fairly well in structure to a $z = 3$ Cayley tree.) A given severity of crowding is reached at a larger sample size in three dimensions than in two, but the problem is only postponed, not eliminated. Similarly, in still higher dimensions the regions of unbearable density are pushed further out (delaying the Malthusian

Figure 4.10   "Wells Family Tree," by Marianne Lehmann.

catastrophe), but eventually the branching process overwhelms the available space. This property of unbounded density persists in a space of *any* finite dimensionality d, because the number of branches in a tree of radius r grows exponentially with r, while the available space grows only as r^d. The mathematical statement is that a Bethe lattice "cannot be embedded in any finite-dimensional space."

The upshot of these considerations is that a Bethe lattice may legitimately be regarded as the equivalent of a conventional lattice in a space of *infinite dimensionality*. If this is so then it follows, using reasoning taken over from the theory of phase transitions and discussed in the previous section, that the percolation properties of such a lattice constitute the equivalent of *mean-field behavior* for percolation processes. It then becomes of interest to determine that behavior. Happily, it turns out that the absence of closed loops simplifies life so enormously (only one path connects any two points) that it is possible to derive explicit equations which determine p_c and $P(p)$ for a Bethe lattice.

For all its apparent peculiarity, a Bethe lattice is actually the natural geometric framework for the description of some quite familiar phenomena (e.g., chain reactions, avalanche showers) which come under the heading of *cascade processes*. Realizing this permits us to write down the percolation threshold almost "by inspection." Think of a species in which each individual gives birth during its lifetime to two offspring, each of which has a probability p of surviving long enough to undergo reproduction itself. If p is less than $1/2$ the species is doomed, it faces certain extinction as the average number of offspring per parent ($= 2p$) is less than one. If p exceeds $1/2$ the species should be viable; a single individual has a nonzero chance of producing an immortal dynasty. These statements solve for p_c on the $z = 3$ Bethe lattice. A living individual corresponds to an unblocked bond, and the number of surviving-to-reproduction offspring (0 or 1 or 2 in this case) corresponds to the number of unblocked bonds which appear at the next fork. An immortal dynasty corresponds to an infinite cluster. Essentially the same analogy applies to site percolation.

For a Bethe lattice with coordination z, each fork presents $z - 1$ independent opportunities for continuation. (The continuations are completely independent because, once embarked upon, their paths never cross. This powerful simplification is not available for ordinary lattices, in which the presence of closed loops permits innumerable crossings.) For the average number of realized continuations per fork to exceed unity (the prerequisite for percolation), $(z - 1)p > 1$. This condition applies equally to bond processes and site processes. In general, therefore, $p_c = 1/(z - 1)$ is the value of the percolation threshold on a z-coordinated Bethe lattice. A more formal (but less enlightening) demonstration of this result is contained in the following derivation for the full percolation probability $P(p)$.

Assume that, following a path of unblocked bonds on a Bethe lattice with coordination z, we arrive at a given vertex (site) and are faced with $z - 1$ new bonds, each of which initiates a branch. For a bond process on such a tree lattice, with p the fraction of bonds being unblocked, let $R(p)$ denote the probability that a branch is restricted to a finite size (i.e., is a dead end). For a branch to be finite, either the first bond of the branch is blocked (which happens with probability $1 - p$), or, if it is unblocked (probability p), then *all* of the $z - 1$ branches which lead out from the far end of the first bond must be dead-ended. But the probability of the latter event, $z - 1$ independent branches being finite, is R^{z-1}. Therefore

$$R = 1 - p + pR^{z-1}. \tag{4.18}$$

Equation (4.18) determines $R(p)$. It also determines p_c (for readers requiring a more mathematical argument than that already given). For p less than $1/(z - 1)$, $R = 1$ is the smallest positive root of Eq. (4.18). $R = 1$ means that there is *no* chance of finding a branch which is *not* finite, i.e., no percolation. For p larger than $1/(z - 1)$, Eq. (4.18) possesses a positive root smaller than unity, which permits percolation. Hence, $(z - 1)^{-1}$ is confirmed as the percolation threshold.

We wish to know $P(p)$, the probability that a randomly chosen bond belongs to an infinite cluster. For it to do so, the bond must be unblocked (probability p) and *at least one* of the branches that lead outward from the two ends of the bond must be infinite. Now there are $2(z - 1)$ outward-bound branches, and the chance of their *all* being blocked is simply R^{2z-2}. Thus the probability is $1 - R^{2z-2}$ that at least one branch is unblocked. Therefore the percolation probability is given by

$$P(p) = p(1 - R^{2z-2}). \tag{4.19}$$

Knowing $R(p)$ from Eq. (4.18), we have determined $P(p)$.

For concreteness, consider the simplest Bethe lattice, the $z = 3$ case illustrated in Fig. 4.9. [The $z = 2$ case corresponds to the linear chain, a nonpercolating situation with $p_c = (z - 1)^{-1} = 1$.] Equation (4.18) becomes a simple quadratic equation, which has the solution $R = (1 - p)/p$ for $p > p_c = \frac{1}{2}$. Substituting this for R in Eq. (4.19) yields for the percolation probability $P = p(1 - R^4) = p[1 - (1 - p)^4 p^{-4}]$, for $p > \frac{1}{2}$.

Note that, for the closed-form solution available for this example, $P(p)$ rises from zero with finite slope at $p = p_c$. This corresponds to $\beta = 1$ in $P \sim (p - p_c)^\beta$. This result, which obtains in general for percolation on trees, constitutes the mean-field value for this exponent. It was the value entered in the $d \geq 6$ column of Table 4.3. The other entries in that column also represent results derived for these "infinite-dimensional" entities, Bethe lattices.

The analysis of site percolation on a Bethe lattice is quite similar. In fact the equation for the site analog of R is identical to Eq. (4.18) for the bond case, showing that p_c is indeed the same for site processes on a tree as for bond processes. However, the functional form of $P(p)$ is different in the site case, because the site analog of Eq. (4.19) replaces the power $2z - 2$ (the number of bonds adjacent to one bond) by z (the number of sites adjacent to one site). Thus the percolation probability for site percolation on, for example, the $z = 3$ tree is $p[1 - (1 - p)^3 p^{-3}]$. Note that, throughout $p_c < p < 1$, $P^{\text{site}}(p) < P^{\text{bond}}(p)$; though their percolation thresholds are equal, it remains true that bond percolation is "easier" than site percolation.

It is worthwhile to pause momentarily to take note of the structure of Eq. (4.18). When we set about the task of constructing an expression for the quantity $R(p)$, which was done on route to a determination of the order parameter $P(p)$, we found that the resulting expression—the right-hand side of Eq. (4.18)—itself contained R. This *self-consistent* structure is characteristic of mean-field theories. Such an implicit equation [of which genre Eq. (4.18) is a very simple example] determines, for example, the Weiss molecular field in the mean-field theory of magnetic phase transitions.

Figure 4.11 portrays an idealized version of a class of physicochemical phenomena which provides a fine vehicle for the interplay of percolation ideas, most notably the competition between the percolation model and its mean-field (Bethe-lattice) counterpart with respect to suitability for the description of a subtle and complex experimental situation. It also represents an application of percolation ideas which anticipated, by about 15 years, Hammersley's explicit formulation of "the percolation problem." This historical aspect will be briefly described below.

When sodium metasilicate—Na_2SiO_3—is dissolved in water (forming "waterglass"), the dissolution may be considered to proceed via the reaction

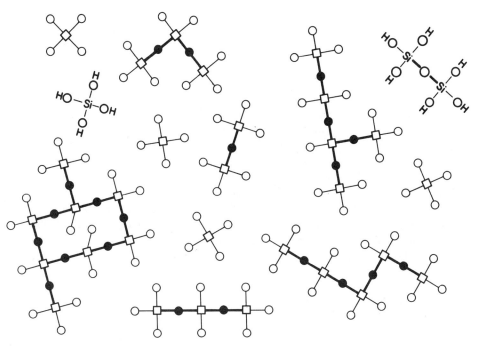

Figure 4.11 Schematic representation of the polymerization reaction by which silica gel is formed from a soup initially containing molecules of Si(OH)$_4$.

$Na_2SiO_3 + 3H_2O \leftrightarrows 2NaOH + H_4SiO_4$. The second product, more appropriately written in the form Si(OH)$_4$, is called monosilicic acid and is a tetrahedral molecule consisting of a silicon atom bonded to four OH groups. For pictorial convenience, Si(OH)$_4$ tetrahedra are schematically represented in Fig. 4.11 as planar (square) units. In the solution, two molecules of monosilicic acid may combine to form a larger molecule represented at the upper right of Fig. 4.11. Two OH groups on nearby Si(OH)$_4$ molecules have reacted to form a Si—O—Si bridge, with the release of a water molecule. (No H$_2$O molecules are shown in Fig. 4.11; the aqueous solvent is simply understood to be present and to occupy the "empty space" which separates the silicic molecules shown.)

This polymerization reaction can continue, building up larger molecules as indicated in the figure. Each initial Si(OH)$_4$ "monomer" can form as many as four "reacted bonds," oxygen bridges to silicon atoms originating from other molecules. Although in actual fact the polymerization process is known to occur through the mediation of ionic variants of these species, and not through the simple combining of uncharged silicic acid molecules, the picture painted in Fig. 4.11 captures the essential feature of this *condensation* phenomenon. Because of the stability of the silicon–oxygen bond, we may view this particular condensation process as essentially irreversible: Once the Si—O—Si cross-links form, they remain.

What comes next has certainly been guessed by the reader. As the reaction goes on and more cross-links form, larger and larger molecules appear. Eventually, and abruptly, at a critical stage in the condensation process, an "infinitely extended" molecule appears, a huge molecule whose extent is limited only by the size of the vessel in which the reaction is taking place. The macroscopically extended 3d-network molecule is called the *gel macromolecule*, the process of its formation is called *gelation*, and its abrupt appearance is called the *sol–gel transition* or the *gel point*. The term *sol* applies to the solution containing only finite molecules, while the term *gel* applies to the system containing the extended network.

In the sol phase, only atomic-scale molecules are present, and the material is a quite conventional liquid. Its viscosity increases as the sol–gel transition is approached, and close to the transition this physical parameter goes through the roof (becomes unmeasurably large) as many physical properties change spectacularly. Above the transition, with the gel present, the system is no longer liquid. The material can now resist shear stresses, and it deforms elastically like an isotropic solid. The system is now neither a conventional solid nor a conventional liquid, since an extended and continuous solid component (the gel macromolecule) is microscopically coexistent with a liquid component (some sol remains and is intimately intermixed with the gel, above the sol→gel transition). This complexity makes it difficult to give a clean definition of a gel, a point which has been aptly described by Henisch (1970): "A gel, for instance, has been defined as a 'two-component system of a semisolid nature, rich in liquid,' and no one is likely to entertain illusions about the rigor of such a definition."

For us the essential point to keep in mind about a gel is that, despite its complexity, it clearly contains a solid component provided by the three-dimensional covalently bonded framework that comprises the gel macromolecule. Solidity is manifested by the appearance of mechanical rigidity at the gel point. The rigidity increases as cross-linking continues beyond that point. The material which results from the specific condensation reaction described above and illustrated in Fig. 4.11 is called silica hydrogel. With the solvent still present (a "swollen" gel) so that the material is mostly water, silica hydrogel is a soft solid easily cut with a knife. With the solvent largely removed (a "dry" gel), the material is a brittle solid which is the familiar desiccant known as silica gel.

Silica gel is an amorphous solid, a not-too-distant relative of fused silica. It differs from the SiO_2 crn structure of Section 2.4.5 in that not all oxygens are bridging (Si—O—Si) but many are dead-ended in hydroxyl groups (Si—OH). It is also evident that silica gel is much less dense than silica. It is very porous (a property important for the application of dry gels as catalysts), with pore sizes on the order of 20–100 Å in diameter. One way to visualize the structure of silica gel is as a kind of highly diluted, loosely connected, continuous random network. (A related material, in a sense intermediate between fused silica and silica gel, is Vycor glass, a porous glass formed by dissolving away the soluble phase of an initially two-phase borosilicate glass. The pores in Vycor are coarser than in silica gel, and there is a sharper distinction between the "empty space" of the pores and the surrounding sponge of "bulk material"—which is close in structure to a standard, dense, SiO_2-type crn.)

One way to phrase the difference between the three-dimensional random network of silica gel and that of fused silica is in terms of their respective ring statistics. It is clear that, whatever the quantitative details, the low density and fine-grained microporosity of silica gel means that its network structure must be deficient in small rings relative to the network of SiO_2 glass.

A connection between percolation and the sol→gel transition was appreciated early on (Frisch and Hammersley, 1963), but was first analyzed in depth in a pair of independent 1976 papers by de Gennes and Stauffer. This connection is laid out in Table 4.5, which is reasonably self-explanatory. A few terms in the gelation column call for clarification. In the context of Fig. 4.11, the extent of reaction (cross-linking probability p) would be the fraction of oxygen atoms shown which belong to Si—O—Si bridges. The gel fraction G is the ratio of the molecular content of the gel macromolecule to the total molecular content (that of the gel macromolecule plus that of all of the sol molecules) of the system. G is the gelation analog of the percolation probability P. (Just as, above p_c, the infinite cluster coexists with finite ones, so, above the gel point, the gel macromolecule coexists with remaining sol molecules.) Functionality is the chemical term for the number of reacted bonds that each monomer can form, four in the case of our example (Fig. 4.11). Its percolation analog is the lattice coordination z. The last entry on the right of the table is the key one for our discussion.

Flory–Stockmayer theory (henceforth, FS theory) is the standard theory used in physical chemistry to describe the sol→gel phase transition and the molecular size distributions which accompany it. It dates from a series of classic papers written by Flory in 1941 and Stockmayer in 1943. The basic assumption of the theory is that gelation may be modeled as a branching process; FS theory is essentially a theory of *dendritic polymerization*.

Table 4.5 Percolation as a model for gelation

Percolation	*Gelation*
Percolation threshold	Gel point
Connected (unblocked) bond	Formed crosslinking chemical bond
Bond-connectedness probability p	Fraction of formed crosslinks (extent of reaction)
Finite cluster	Sol molecule
Mean cluster size s_{av}	Average molecular weight
Infinite cluster	Gel macromolecule
Percolation probability P	Gel fraction G
Coordination number z	Functionality z
Resistor-network conductivity	Elastic shear modulus
Bethe-lattice approximation	Flory–Stockmayer theory

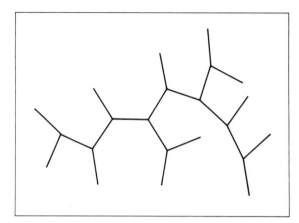

Figure 4.12   Schematic of a trifunctionally branched molecule (after Flory, 1941).

Figure 4.12 is from the first of Flory's three 1941 papers. It shows the skeleton of a possible molecule formed by the condensation of "trifunctional monomers" ($z = 3$). We immediately recognize this as one of the many possible bond clusters on a $z = 3$ Bethe lattice (compare Fig. 4.12 to Fig. 4.9e). FS theory assumes that *all* molecules created by the reacting monomers have such a treelike form. Flory's formulation of the condensation reaction as a branching process, and his quest for a quantitative theory which would account for "the existence of a sharply-defined gel point...which occurs when a critical number of intermolecular linkages has been exceeded," led him to derive a key connection between the critical "extent of reaction" (at which gelation occurs) and the functionality z of the reacting monomers. This key relation is, in effect, $p_c = 1/(z - 1)$; we have already met it as the value of the percolation threshold on a z-coordinated Bethe lattice. *The Flory–Stockmayer theory of gelation is equivalent to the exact solution of the percolation problem on a Bethe lattice.* Of course, percolation theory "was not around" in 1941. Yet gelation really provided the first application of percolation ideas to a phase transition, well in advance of the formal (and "official") introduction of the topic by Hammersley in 1956. Here is a good example of the physicist's/chemist's pragmatic invention of the mathematical machinery needed for the task at hand.

The correspondence between gelation and percolation seems irresistible. Consider the following quote: "We seek the answer to the question: Under what conditions is there a finite probability that an element of the structure, selected at random, occurs as part of an infinite network?" Evidently the question phrased here makes an inquiry about the percolation probability $P(p)$. But these lines are, in fact, from the beginning of the gelation chapter of Flory's 1953 classic on polymer chemistry!

FS theory is successful in several respects. It predicts a sharp gel point, as is observed experimentally, and the critical extent of reaction at which it occurs is given reasonably well by the FS value $p_c = 1/(z - 1)$. The very steep viscosity increase just prior to the gel point is also naturally explained, because the weight-averaged molecular weight diverges here [for a Bethe lattice $s_{av} \sim (p_c - p)^{-1}$, as shown on the right of Table 4.3] and viscosity is known to be closely tied to average molecular weight. Measurements of viscosity as a function of extent of reaction yield results which resemble the $s_{av}(p)$ curve of Fig. 4.5.

But there is an obvious problem. We have seen that percolation theory developed for a Bethe lattice constitutes a mean-field theory which should not work in dimensionalities lower than six. One way to see the problem is to note the fractal dimensionality of four (for large clusters near p_c) listed for Bethe-lattice percolation on the right of Table 4.3. This has the interpretation $l(s) \sim s^{1/4}$ for the radius $l(s)$ of clusters of size s as $s \to \infty$. In three dimensions, the average density of large clusters would then go as $(s/l^3) \sim s^{1/4}$, increasing without limit as s increases! This unphysical result (the density "blows up" for large clusters) reflects the overcrowding problem inherent in a ring-free branching structure. Inevitable overcrowding was evident in Figs. 4.9e and 4.10, somewhat disguised in Fig. 4.12.

The classical FS theory can therefore be considered to be "only" an important first approximation to gelation in three dimensions. The assumption of the FS theory which, on the one hand, allows it to be solved exactly (equivalent to Bethe-lattice percolation) and, on the other hand, sets limits to its realism and applicability to experiment, is *the absence of closed rings*. This prohibition against closed loops or cycles of bonds (one such ring was included in Fig. 4.11, at the lower left) is probably not serious at early stages in the condensation reaction, but must become important at later stages. The gel macromolecule itself possesses a *network* structure with a great many rings, as attested to by its observed rigidity. Thus, the treelike picture inherent in the classical theory has difficulty in squaring with the gelation process near the gel point and above it.

A model of gelation which automatically includes the presence of rings is bond percolation on a three-dimensional lattice. Each lattice site is considered to be occupied by a monomer, and each monomer may react with—and establish a covalent bond to—any of its z nearest neighbors.

Let us compare some of the predictions of the classical FS model with the corresponding predictions of a lattice bond-percolation model for $z = 4$, the functionality appropriate to the case of silicic acid gelation represented in Fig. 4.11. Qualitatively, the predictions are similar, since percolation underlies both theories. Each predicts a sharp gel point, a diverging average molecular weight (and, thereby, a diverging viscosity) below p_c, and an increasing elastic rigidity above p_c. Quantitatively, there are definite differences. The critical extent of reaction in the FS model is $p_c = 1/(z - 1) = \frac{1}{3}$. Taking a diamond structure as the three-dimensional, four-coordinated lattice for the lattice-percolation model, $p_c = 0.39$. More significant are the differences in the critical exponents. The gel fraction increases near threshold as $(p - p_c)^{0.4}$ in the lattice model, much faster than the linear $(p - p_c)^1$ increase predicted by the FS theory. Similarly, the divergence of the average molecular weight as the sol→gel transition is approached is more rapid for the 3d-lattice model than for the FS model (compare values for γ in Table 4.3). Experimental determinations of these gelation exponents are very difficult, and unfortunately it appears that no definitive test of the differing predictions is yet available.

Just as Bethe-lattice percolation, the classical FS theory of gelation can be criticized as a physical model on the grounds of its failure to take into account the undoubted occurrence of closed rings when large molecular structures are formed (a serious criticism at and above p_c), so also 3d-lattice percolation model has difficulties of its own. Obviously, real monomers in solution are not arranged on a regular lattice. More seriously, where are the solvent molecules in this model? As specified above, the lattice-percolation model is really a model for gelation in the *absence* of solvent. A more realistic model, which incorporates solvent molecules, is based on the generalization known as site-bond percolation (discussed in Section 4.8). Critical exponents for site–bond percolation have been shown to be the same as for ordinary percolation.

Earlier, when we introduced the analogy between percolation and magnetism that is expressed in Table 4.2, we noted there a correspondence between the resistor-network conductivity σ of the percolation model and a magnetic ("exchange-stiffness") version of an elastic constant which applies to spin waves in the ferromagnet. At that time a promise was made to deliver a connection between σ and a bona fide *mechanical* elastic constant. The time has come. This conductivity–rigidity connection was evident in a qualitative way when Fig. 4.1 was discussed (the screen *falls apart* at the same moment that *conductivity ceases*). The quantitative argument, given here for a gel, is due to de Gennes (1976, 1979). In a gelation process, the elastic shear modulus vanishes for the sol (a liquid), begins to grow at the sol→gel transition, and continues to increase as the cross-linking increases in the gel (solid) phase. What will now be shown is that the behavior of the elastic modulus, as a function of the cross-linking concentration, is the same as that shown for $\sigma(p)$ in Fig. 4.5.

We consider a lattice model with monomers at all sites and with some fraction p ($> p_c^{\text{bond}}$) of the mathematical bonds between nearest-neighbor sites corresponding to formed (reacted) chemical bonds between monomers. The elastic energy of this gel network may be written as

$$E_{\text{elastic}} = \frac{1}{2}\sum_{ij} k_{ij} (u_i - u_j)^2. \tag{4.20}$$

Here u_i is the displacement of monomer i from its equilibrium position at lattice site i, and k_{ij} is the force constant of bond ij. [To simplify the expression, the subscript specifying the Cartesian component of u_i has been suppressed. The sum in Eq. (4.20) should be interpreted as extending over the three orthogonal components.] The spring constant k_{ij} is taken equal to k if bond ij is connected (probability p); otherwise (probability $1 - p$), k_{ij} is zero.

For monomer i to be in equilibrium, the forces exerted upon it by the bonds joining it to its neighbors must cancel to zero,

$$\sum_{j} k_{ij}(u_i - u_j) = 0. \tag{4.21}$$

Now consider an electrical network that is isomorphic to the gel network, having a bond conductance g_{ij} equal to g when the force constant k_{ij} of the corresponding gel bond is k, and equal to zero when k_{ij} is zero. When subject to an applied voltage drop, the equilibrium voltage v_i at each node (site) i must satisfy the requirement (Kirchhoff's law) that there be no net current into the node,

$$\sum_{j} g_{ij}(v_i - v_j) = 0. \tag{4.22}$$

The equivalence between Eqs. (4.22) and (4.21) provides the sought-for correspondence, on a microscopic level. Macroscopically, the consequence of the bond conductances in the electrical case is the conductivity, while in the gel-network case the bond springs result in the elastic moduli.

An amorphous solid, being isotropic, possesses only two independent elastic moduli. One of these may be chosen to be the inverse compressibility or bulk modulus $B = -V(\partial P/\partial V)$, which is a measure of the elastic stiffness against volume decrease enforced by hydrostatic pressure. The other is the shear modulus which is a measure of the stiffness against volume-conserving change in shape. In interpreting the conclusion of the previous paragraph, we must bear in mind that the lattice bond-percolation model for gelation leaves out the solvent molecules. Below the gel point, in the sol phase, the material is a liquid. Its shear modulus, by definition, is zero. But the bulk modulus is finite; the solvent molecules are not infinitely compressible. Therefore, the correspondence to the resistor-network conductivity really applies closely only to the shear modulus, the measure of the elastic stiffness against shear, the rigidity which opposes distortion.

We conclude, therefore, that the elastic shear modulus of a gel should depend upon the fraction of formed cross-links in exactly the same way as the conductivity of a random-resistor network depends upon the fraction of conducting bonds. The functional form should closely resemble the $\sigma(p)$ curve of Fig. 4.5, starting from zero at the gel point and increasing gently beyond that. In particular, close to the gelation threshold, the shear modulus should increase as $(p - p_c)^t$, with t being the $d = 3$ conductivity exponent of Table 4.3. *This prediction of the 3d-lattice percolation theory of gelation is indeed borne out experimentally.* Note that the observed discrepancy between the slow increase in the elastic rigidity and the swift increase in the gel fraction has the same root cause as the corresponding discrepancy between $\sigma(p)$ and $P(p)$. Cross-links which belong to the gel macromolecule but which lead to dead-ended or dangling molecules do *not* contribute to the solid's shear stiffness. Only the backbone of the percolation path (that is, the infinite cluster with *all loose ends removed*) supports the electric current in the resistor-network case or the elastic rigidity in the mechanical-network case.

It should by now be quite clear that connections or correspondences or parallels or analogies, such as those given in Tables 4.2 and 4.5 of this chapter, are very instructive and helpful in providing insight. Such *mappings* of different fields onto each other are, of course, part of an old tradition in physics. Percolation provides particularly nice examples of such mappings. The connection to magnetic phase transitions (Table 4.2) has imbued percolation with much of the elegant and powerful machinery (scaling, universality, renormalization) which pervades phase-transition theory. Similarly, the correspondence between percolation and the theory of gelation has allowed the latter to take advantage of, for example, the results of the extensive computer simulations that have been used to determine the exponent t for random resistor networks. Putting two and two together (magnetism $\leftrightarrow$ gelation connection, via percolation), it is also true that phase-transition concepts such as scaling apply to the sol $\rightarrow$ gel transition.

The sol $\rightarrow$ gel transition can be viewed, in the context of the concept of network dimensionality which was introduced in Section 3.1.1, as a transition between a zero-dimensional-network situation and a three-dimensional-network situation. In the gelation process that produces silica gel, the initial liquid state contains atomic-scale (0d-network) molecules such as $Si(OH)_4$, while the final solid state is supported by a 3d-network macromolecule akin to the SiO_2 random network. This 0d $\rightarrow$ 3d feature accompanies percolation processes in three dimensions. (In general, in d dimensions, 0d-network elements merge to form a d-dimensional connected network at the percolation threshold.) There is another type of chemical cross-linking process, called vulcanization, which essentially a 1d $\rightarrow$ 3d connectivity transition.

In vulcanization, a dense melt made up of individual, flexible, organic polymer chains is transformed, by the action of a cross-linking agent, into a single giant molecule. Thus the process resembles gelation except that the initial building blocks are 1d-network molecules rather than 0d-network molecules. Chemically, the prototype vulcanization process involves a cross-linking mechanism which is not very different from that indicated in Fig. 4.11 for our gelation example. Representing the fourth column of the Periodic Table we now have, instead of silicon, the carbon atoms of the hydrocarbon polymer chains. From column six, in the role of the twofold-coordinated bridging oxygens of Fig. 4.11, we now have bridging sulfur atoms contributed by a sulfur-containing species added to the polymer melt. C—S—C cross-links between chains play the part that corresponds (in Fig. 4.11) to the Si—O—Si cross-links between monomers.

Each of the N carbons of the chain backbone provides an opportunity (with probability p) for cross-linking, so that for each chain the effective coordination number z equals N for this very nonstandard percolation problem. Because the chains are long and $N \gg 1$, the effective coordination number is enormous, rings are unimportant, and FS theory is valid: $p_c = 1/(z - 1) = 1/N$. Critical cross-linking occurs when there is, on average, one cross-link per chain (the equivalent of one offspring per parent in a cascade process).

The above result, namely a very small p_c for an assembly of long chains, is relevant to the structure of the chalcogenide glass $Se_{1-x-y}As_xGe_y$ which was schematically sketched in Fig. 3.7. Ge and As atoms in this system provide, in effect, cross-linking points for chains of Se atoms, and we see that only a very small concentration $(x + y \approx N^{-1})$ of such cross-linking atoms need be present to stitch together a 3d-network macromolecule. This point has practical importance in the use of these materials as photoconductors, as will be mentioned in Chapter Six.

4.7 CONTINUUM PERCOLATION AND THE CRITICAL VOLUME FRACTION

Along with its relevance to a rich variety of physical phenomena, the percolation model possesses a transparency that gives it great educational value. Figure 4.13 illustrates a homey experiment which is beginning to find its way into undergraduate laboratories as a pedagogic device for conveying ideas about phase transitions and critical phenomena. The procedure is typically as follows. Crumpled aluminum foil is pressed into the bottom of a glass beaker and is electrically connected to a battery. A mixture of small plastic and

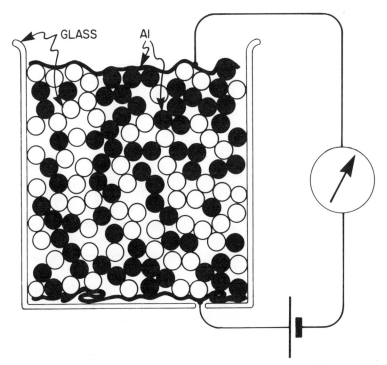

Figure 4.13   The experiment carried out by Fitzpatrick, Malt, and Spaepen (1974) as part of a freshman physics course at Harvard. Aspects simultaneously present in this simple arrangement include random close packing, percolation, and random-resistor network.

metallic balls of equal size is now poured into the beaker and "pressed down and shaken together" (following Section 2.4) in the ancient recipe for random close packing. A second crumpled-foil electrode is pressed onto the top of the mixture, and the circuit is completed through an ammeter to the other side of the battery. The current, if any, that is flowing in the circuit is measured by the ammeter and recorded.

The above experiment associates a value of electric current I with another number that specifies the overall composition of the system, the mix of conducting (metal) and nonconducting (plastic) balls. We may choose, as the composition parameter p, the fraction of the balls in the beaker which are metallic. The procedure described above determines, say, the current I_1 that corresponds to composition p_1. The beaker is now emptied and refilled with a different metal/(metal-plus-plastic) mix p_2, and the current I_2 is measured. A series of such measurements, repeated with different mixtures in the beaker, maps out a function $I(p)$. Question: What is the form of $I(p)$ and, in particular, what is the critical composition—the minimum fraction of metal balls needed to support across-the-beaker conductivity—that separates the setups for which current flows from those for which no current is seen?

The experimental situation of Fig. 4.13 simultaneously incorporates several distinct aspects: random resistor network, percolation, and *an underlying structure which is topologically disordered*. We recognize the particular form of topological disorder that is present here, i.e., that which characterizes the array of spheres within the beaker, as the random close-packed structure which was described in Chapter Two in connection with the atomic-scale structure of amorphous metals. The aspect of *underlying* topological disorder is very important. It distinguishes this percolation process from all of those that have been discussed thus far.

In conventional percolation on a lattice, the structural starting point is that most regular of geometric objects—a periodic lattice. Disorder is introduced by *superimposing* on the sites or bonds of such a lattice a randomly assigned two-state property (filled/empty or unblocked/blocked). The bimodal statistical variable, imposed on a regular geometry, results in a stochastic-geometry situation. (Incidentally, the use of the term *stochastic geometry* dates back at least as far as the 1963 review on percolation by Frisch and Hammersley.) But the situation depicted in Fig. 4.13 represents a *higher order of stochastic geometry* because the disorder-generating statistical variable (metal/insulator) is superimposed on a structure that is itself topologically disordered.

If the balls of Fig. 4.13 had been arranged, not in random close packing, but, for example, in a face-centered-cubic crystalline array, then the answer to the question asked about the critical composition (metal-ball fraction) for current flow could have been plucked from the fourth column of Table 4.4: p_c^{site} (fcc) = 0.198. But how do we handle the analogous site-percolation problem on a noncrystalline lattice such as rcp? Or deal with situations in which no underlying lattice whatever may be defined? A pragmatic approach to this problem invokes a stochastic-geometry construct (Scher and Zallen, 1970) which has come to be called the *critical volume fraction for site percolation*.

The notion of critical volume fraction is most simply defined in the context of a specific example. Figure 4.14 shows the situation for site percolation on the 2d honeycomb lattice. With each lattice site we associate a circle of radius equal to half the nearest-neighbor separation. In Fig. 4.14, open circles surround empty sites, shaded circles surround filled sites, and clusters of connected sites are indicated by the heavy lines joining adjacent filled sites. Now while p is the fraction of the circles that are shown shaded in Fig. 4.14, the volume fraction ϕ is defined as *the fraction of space* that is taken up by the filled (shaded) circles. The relationship between ϕ and p is dictated by simple geometry: $\phi = vp$, where v is the filling factor for the lattice, the fraction of space occupied by *all* of the circles of Fig. 4.14. The critical volume fraction ϕ_c is the value of ϕ which applies at the site-percolation threshold: $\phi_c = vp_c^{\text{site}}$.

For the two-dimensional example of Fig. 4.14, the volume fraction is, of course, an area fraction. The term "volume fraction" should be interpreted as appropriate to the dimensionality at hand. ("Space fraction" or "content fraction" are alternatives that could be used to avoid a 3d connotation, but these terms are less transparent than volume fraction.) In general, for the simple (connections only between nearest neighbors) lattice j in d dimensions, the critical volume fraction is $\phi_c(j,d) = v(j,d)p_c(j,d)$, where $p_c(j,d)$ is the site-percolation threshold and $v(j,d)$ is the filling factor of the lattice corresponding to the packing of equal, touching, nonoverlapping, d-dimensional spheres centered on the lattice sites.

Thus far, for site percolation on any given lattice, all we have done is to convert the percolation threshold from a critical site-occupation probability p_c

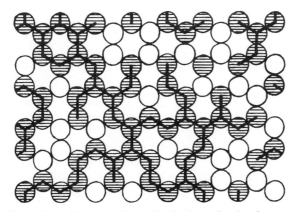

Figure 4.14 Illustration of the idea of the critical volume fraction for percolation, in the context of the two-dimensional honeycomb lattice. Shaded circles indicate filled sites, open circles indicate empty sites, and linked clusters are shown connected by heavy lines (Scher and Zallen, 1970).

P.176

to a critical volume fraction ϕ_c. What have we gained by this? The benefit that results may be seen by comparing the fourth and eighth columns of Table 4.4, which list p_c and ϕ_c values for various 2d and 3d lattices. While the variability ("nonuniversality") of p_c from lattice to lattice has already been remarked upon, what we find for ϕ_c is, for all lattices of the same dimensionality, *a remarkable insensitivity to lattice structure*. In the notation of the previous paragraph, $v(j,d)p_c(j,d) \approx \phi_c(d)$, the lattice-specifying index j may be dropped, to good approximation, from ϕ_c. To within a few percent, the critical volume fraction is 0.45 in two dimensions, 0.16 in three dimensions.

The critical volume fraction constitutes *an approximate dimensional invariant for the threshold in site-percolation processes*. As such it is an extremely valuable construct, useful in a wide variety of applications. One such application is the experiment which has concerned us here, that pictured in Fig. 4.13. We can now answer the question posed earlier. For the random packings typical of such experiments, the filling factor v is close to 0.60 (somewhat lower than the optimum rcp value of 0.637 discussed in Chapter Two). Using $\phi_c(d = 3) = 0.16$ in conjunction with $v = 0.6$, we obtain $p_c = 0.27$, predicting that current flow occurs when the fraction of metal balls in the mixture exceeds 27%. This is what is observed. Above this threshold, the current-versus-composition characteristic follows a curve similar to $\sigma(p)$ of Fig. 4.5.

A microscopic counterpart of the basic situation portrayed in Fig. 4.13 is provided by a class of solids, usually prepared in thin-film form at low temperature, that are atomically mixed dispersions of alkali atoms in rare-gas solids. Examples are $Rb_x Kr_{1-x}$ and $Cs_x Xe_{1-x}$. These alloys behave as "paradigms of percolative processes on an atomic scale" (Phelps and Flynn, 1976). The insulator→metal transition in these solids occurs, as a function of composition, when the proportion (x) of alkali atoms corresponds to an alkali-atom volume fraction of 16% ($= \phi_c$).

Many macroscopic two-phase systems are controlled by the critical volume fraction for percolation. One example is the required quantity of photoconducting material which, when dispersed in an insulating "binder," is adequate for the composite material to exhibit photoconductivity. Another example is the composition at which the normal→superconducting transition occurs for a disordered system consisting of Nb_3Sn filaments embedded in a copper matrix. In such systems the stochastic geometry resembles a situation (glance ahead briefly to Fig. 4.16) that may be characterized by the expression *percolation on a continuum*, which we are nearly ready to discuss.

The empirical basis for ϕ_c, the observed near-constancy of vp_c^{site} as exhibited by the numerical values in the last column of Table 4.4, is given a graphic setting in Fig. 4.15a for the three-dimensional case. The reciprocal p_c^{-1} of the site-percolation threshold is shown plotted against the lattice packing fraction or filling factor v. The slope of the straight line is ϕ_c^{-1}. Shown in Fig. 4.15b is an analogous plot for a *different* dimensional invariant (also, as ϕ_c, an approximate one) that was noticed early in the game for *bond* percolation and for which numerical evidence was included in the next-to-last column of

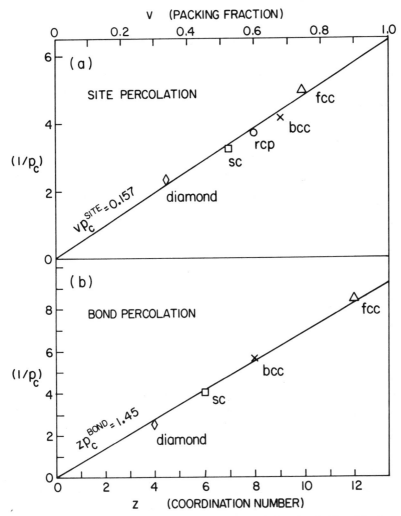

Figure 4.15 Two empirical correlations between percolation threshold and lattice connectivity for various three-dimensional structures. For site percolation, the correlation between the reciprocal of the percolation threshold and the lattice packing fraction (or filling factor) is shown in *a*. For bond percolation, the correlation between the reciprocal of the threshold and the lattice coordination is shown in *b*. The correlation demonstrated in *a* provides an empirical basis for the usefulness of the approximate dimensional invariant known as the critical volume fraction.

Table 4.4. Plotted here, for 3d lattices, is the reciprocal of the bond-percolation threshold as a function of the lattice coordination z. The significance of the straight line in *this* plot is that zp_c^{bond} is a dimensional near-invariant applicable to bond processes. In three dimensions, for regular lattices with nearest-neighbor bonds, zp_c^{bond} is close to 1.5. This may be paraphrased as follows: At the bond-percolation threshold for 3d lattices, an average site sees one-and-a-half unblocked bonds.

Although the sensitivity of the value of the percolation threshold to the specifics of the lattice structure has been repeatedly pointed out, we now have found that there exists, for both site percolation and bond percolation, a restatement of the threshold which, to a useful level of accuracy, is independent of the details of lattice geometry. For site percolation, that determinant of the composition at threshold is the critical volume fraction ϕ_c; for bond percolation, it is the critical bond number zp_c—the critical value of the average number of connected bonds per site. These constructs restore a respectable degree of universality to the position of the percolation threshold.

Note a highly suggestive parallelism of form between the two parts of Fig. 4.15. In each instance, a measure of the *ease of percolation* (the inverse threshold) is seen to be proportional to a measure of the *connectivity of the structure*. For bond percolation, that measure of connectivity is our familiar standard for this purpose, the coordination number z. But for site percolation, z is replaced by v, the filling factor based on the packing of equal spheres. For site-percolation processes, the proper measure of connectivity is the efficiency with which the structure (viewed as a packing) fills space. This distinction between bond and site percolation is analogous to the graph/froth or crn/rcp dichotomy of Chapter Two. On an atomic-scale bonding level, z is the proper measure of connectivity for covalent solids, while v is the measure of choice for metals.

These dimensional invariants, useful in $d = 2$ and $d = 3$, may also be estimated for dimensionalities up to $d = 6$ by virtue of computer experiments on the $d > 3$ counterparts of simple cubic lattices (''hypercubic'' lattices). The data are given in Table 4.4, and $\phi_c(d)$ was given earlier in Fig. 4.8 along with those more traditional dimensional invariants, the critical exponents.

Enter now the unruly geometry of Fig. 4.16, and the topic of continuum percolation. (Another example of this genre will appear, as Fig. 5.15, in the discussion of electron states in amorphous semiconductors.) There is no lattice, no discrete structure, which serves as a substrate for the disorder shown here. Yet we are interested in the connectivity properties (e.g., Does black percolate left-to-right in Fig. 4.16?); hence, we are dealing with percolation on a continuum. One reason for having hope in the face of such a wild scene as that of Fig. 4.16 is the percolation criterion provided by the critical volume fraction, because *this concept dispenses with the need for a lattice!* Where p_c is the occupation fraction for a *discrete* set of locations (the lattice sites), ϕ_c is the occupation fraction for a *continuous* set (spatial positions). In addition to the empirical justification of $\phi_c(d = 3) = 0.16$ provided by Fig. 4.15a, computer

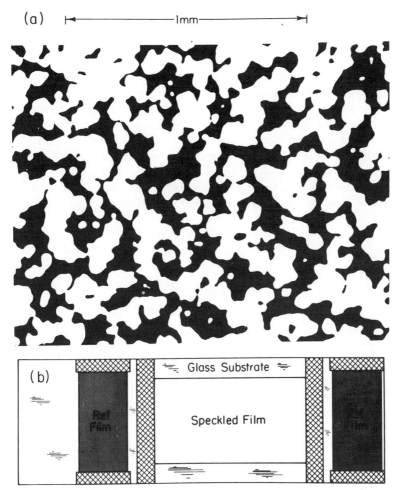

Figure 4.16 Sample geometry in the electrical experiments of Smith and Lobb (1979). The stochastic-geometry conductor–insulator films were photolithographically generated from laser speckle patterns. In *a*, which shows a small section of a speckled film, the black areas are metal. In *b*, the cross-hatched regions are thick metal contacts.

simulations of continuum percolation (Webman, Jortner, and Cohen, 1976) support the same result.

The two-dimensional pattern that appears in Fig. 4.16 was generated from a laser speckle pattern by a photographic procedure (Smith and Lobb, 1979). In this experiment, green light from an argon laser (λ = 5145 Å) was incident upon a diffusely scattering surface, and the speckle pattern formed by the scattered light was recorded on high-contrast film. The developed film shows no gray scale; it contains a pattern of clear regions and opaque regions whose relative area fractions depend on the laser intensity and the exposure time. Using the developed film as a mask for contact printing, the pattern was then photolithographically reproduced as a vapor-deposited metal thin film on a glass slide. In the small sample of a typical pattern that is shown in Fig. 4.16, the black areas are metal; they correspond to the dark parts (light intensity below a cutoff level determined by film exposure) of the original speckle pattern.

Figure 4.17 shows the results of electrical measurements on a series of films made in this way. Current was measured between opposite edges of each film, using thin-film electrodes indicated by cross-hatching in Fig. 4.16b. In Fig. 4.17, normalized conductivity σ/σ_0 is plotted against area fraction ϕ. (ϕ was determined by measuring the optical transmission of the film). The $\sigma(\phi)$ curve, between threshold and full coverage, is not badly approximated by a linear behavior. The reason for this is that the critical exponent t in $\sigma \sim (\phi$ –

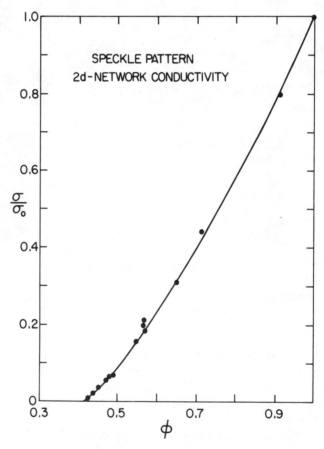

Figure 4.17   Conductivity measurements on films of the type shown in Fig. 4.16, as a function of metal area fraction (Smith and Lobb, 1979).

$\phi_c)^t$ is close to unity: t is about 1.2–1.3 for the data of Fig. 4.17. This is not significantly different from the value (based on computer experiments for random-resistor networks on two-dimensional lattices) listed earlier in Table 4.3: $t(d = 2) = 1.1$. Critical exponents for continuum percolation seem to be similar to those for conventional (lattice) percolation.

Although the critical volume fraction for site percolation on 2d lattices is $\phi_c(2d) = 0.45$, it will be shown in the next chapter (Section 5.6) that for 2d continuum percolation in systems characterized by black/white symmetry, ϕ_c is ½. This symmetry is not present in Fig. 4.16 (black regions tend to be concave with cuspy exteriors, white regions tend to be convex with rounded exteriors), which is why ϕ_c is different from ½ in Fig. 4.17.

4.8 GENERALIZATIONS AND RENORMALIZATIONS

Continuum percolation, as discussed in the previous section, is an example of a generalization of conventional or classical percolation theory—one that happens to provide (via the critical volume fraction) an extremely useful extension to situations in which the geometry is quite irregular. In this section, several other generalizations of percolation are discussed. While many such generalizations are imaginable, those mentioned here are ones with high applicability to physical systems. In addition, the powerful theoretical technique known as the renormalization group is illustrated in this section in a percolation context.

Probably the simplest generalization of classical percolation is *site-bond percolation*, in which both sites *and* bonds on a lattice are randomly occupied (unblocked, connected) with probabilities p^{site} and p^{bond}, respectively. First proposed by Frisch and Hammersley in 1963, this hybrid model was not extensively developed until 15 years later when Stanley and others adopted it as a model for gelation. In the plumbing analogy of Fig. 4.3, a site–bond percolation process has valves placed both at the joints *and* within the pipes of the flow network.

In the last section's treatment of bond percolation as a model for gelation, monomer molecules occupied *all* of the sites of a lattice, and reacted chemical bonds between adjacent monomers were present with probability p^{bond}. The shortcoming of this pure bond-percolation model is its failure to include the solvent molecules, whose presence has the effect of diluting the concentration of reactant monomer molecules. Such *effects of random dilution* are readily treated by percolation theory, and the remedy prescribed in this instance is to have each lattice site occupied by *either* a monomer molecule (with probability p^{site}) *or* an inert molecule (with probability $1 - p^{site}$). In this system, connected clusters (condensation-reaction products) consist of continuous sequences of occupied sites (monomers) linked by occupied (reacted) bonds. Thus, site-bond percolation is a very natural model for gelation.

With both sites and bonds occupied at random (but with, in general, different probabilities: $p^{site} \neq p^{bond}$), there will occur occupied bonds which have either or both endpoint sites empty. In a model for gelation, occupied bonds of this type are unphysical; they do not represent reacted chemical bonds, which must join pairs of monomers (both endpoint sites filled). Fortunately, such bonds do not contribute to cluster connectivity since they are either dead-ended (one endpoint site empty) or completely isolated (both endpoints empty). Therefore, they do not influence such matters as the presence or absence of an infinite cluster (gel macromolecule), and they do not spoil the model. But the presence of some occupied but unphysical bonds means that the proper measure of cluster size is the number of sites it contains rather than the number of bonds. Occupied sites, even if isolated (solitary singlets) or dead-ended (on the "outside surface" of a cluster), retain their ability to represent monomer molecules in a physical model.

In a site–bond percolation process, the composition of the system is specified by two independent variables, p^{site} and p^{bond}. For a percolation path to occur, both p^{site} and p^{bond} must be large. How large each must be depends on the other; they are interdependent. Figure 4.18 displays, for site–bond percolation on the square lattice, the large-p high-density portion of the phase

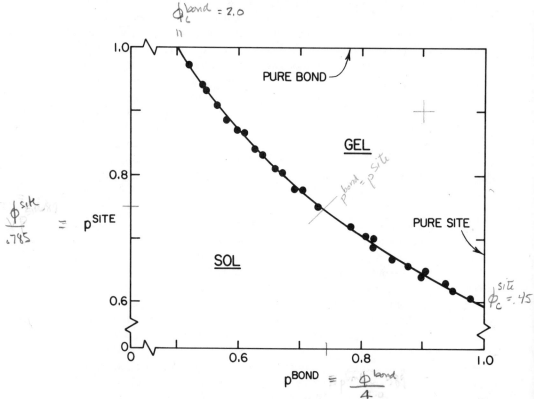

Figure 4.18   Phase diagram for site–bond percolation on the square lattice. The solid line separates the percolating regime (labeled "gel") from the nonpercolating regime (labeled "sol") (after Agrawal, Redner, Reynolds, and Stanley, 1979).

diagram in (p^{site}, p^{bond}) composition space (Agrawal et al., 1979). The solid line is the phase boundary that separates the percolating regime (labeled "gel" in the figure) from the nonpercolating regime (labeled "sol"). For compositions in the "gel" region, an infinite cluster occurs; for compositions in the "sol" region, all clusters are finite.

The upper border of Fig. 4.18 is the line along which p^{site} is unity while p^{bond} varies. This horizontal line represents those situations in which all sites are connected, but only some bonds are. It corresponds to pure bond percolation, and the sol ↔ gel transition line intercepts it at (p^{site}, p^{bond}) = (1, p_c^{bond}), the bond-percolation threshold. Similarly, the right-hand boundary of the figure represents pure site percolation, and this vertical line is intercepted by the sol ↔ gel phase boundary at the site-percolation threshold (p_c^{site}, 1). The fact that the two points which represent the thresholds for the two "pure" limiting cases are connected by a smooth and continuous phase-transition curve, the locus of percolation thresholds that comprises the sol ↔ gel boundary of Fig. 4.18, has a special significance discussed below.

Figure 4.19 contains two diagrams, both of which, like Fig. 4.18, show the two-dimensional (p^{site}, p^{bond}) composition field for site–bond percolation on the square lattice. (Unlike Fig. 4.18, p^{site} is plotted horizontally and p^{bond} vertically in Fig. 4.19. Also, the full field is shown in each diagram of Fig. 4.19, not just the part near the transition line.) Both diagrams, published independently and near simultaneously by Shapiro (1979) and by Nakanishi and Reynolds (1979), represent the *renormalization-group flow diagram* for site-bond percolation. Although any attempt at a proper development of renormalization-group theory is well beyond the scope of this book, percolation provides so fine an illustration of such a flow diagram that it deserves to be mentioned. Also, as with other concepts in the theory of phase transitions, the percolation model serves as a lucid setting for displaying the ideas of the renormalization group. Before discussing Fig. 4.19, a simpler example is now presented for the renormalization-group concepts of *flow diagram* and *fixed point*.

The basic idea of the renormalization group is that of a continuous family of transformations on the length scale of the system. Consider a political analogy. Let Fig. 4.20*a* represent a small portion of the two-dimensional political map of the United States. There are 81 voters in this region, and we assume total two-party dominance, with all individuals voting either for the Democratic (**d**) or Republican (R) candidate. In the election indicated in the figure, the Republicans have carried the region 48 to 33, that is, with about 59% of the vote. Using the Republican fraction to specify the composition p of the system, p = (48/81) = 0.59.

Now suppose that this same region contains nine election districts of equal size (nine voters), as shown in Fig. 4.20*b*. On the district level, the Republicans have taken the region seven districts to two. Viewed on *this* scale, in which the smallest discernible unit is an election district (with each district credited, in winner-take-all fashion, to the party having the majority of votes

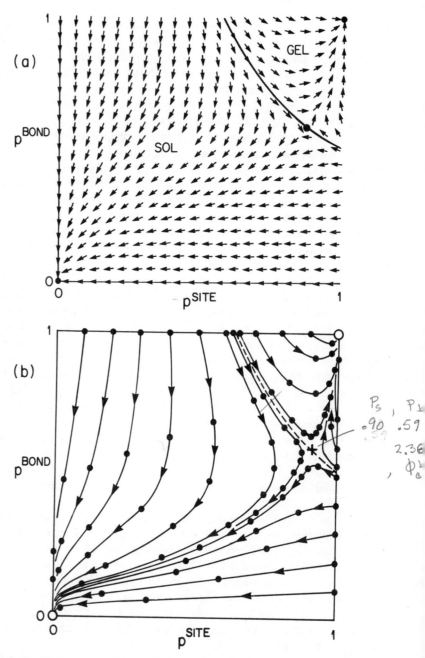

_b=2_

_b=2_

P_S , P_b
.90 .59
.59
2.36
, ϕ_c^b

Figure 4.19 Two representations of the renormalization-group flow diagram for site–bond percolation on the square lattice. Diagrams _a_ and _b_ are, respectively, from the work of Nakanishi and Reynolds (1979) and Shapiro (1979).

194

each trajectory represents increasing number of scalings by a scaling factor of b^2. After ∞ steps the system evolves to one of two stable (under scale transformation) "fixed" points. The other scale-transformation-invariant point is the unstable critical point...

within it), the political map of the region looks like Fig. 4.20c, and the composition of the system is now $p' = (7/9) = 0.78$. In going from a to c in Fig. 4.20, the unit of length has dilated by a factor of $b = 3$, and the original composition p has been "renormalized" via this scale transformation to the value p'. In general, p' depends on both p and b: $p' = p'(p,b)$. The scale factor b can be expressed as $N^{(1/d)}$, where N is the number of original units that transform into one elementary unit of the renormalized system and d is the dimensionality. For our example, $N = 9$ and $d = 2$.

This process can be repeated, as indicated in Fig. 4.20d. Assume that nine election districts make up one county, and award each county to the party that carries the majority of its nine districts. In our sample region, which contains only a single county, the renormalized composition after a second $b = 3$ scale transformation to the county level is $p'' = (1/1) = 1$. In a much larger area, p'' would not be unity but, say, $p'' = 0.97$. Thus, in the two stages of scale transformation which have been carried out, p has gone from 0.59 at the voter level to 0.78 at the district level to 0.97 at the county level. For the renormalization function $p'(p,b)$: $p'(p = 0.59, b = 3) = 0.78$ and $p'(p = 0.78, b = 3) = 0.97$.

The bottom panel of Fig. 4.20 graphically places these compositions on a linear composition field. The effect of successive scale transformations is to generate a flow diagram in composition (or density) space. For $p > \frac{1}{2}$, as in our example, the composition flows to the right (appropriately enough for a Republican victory) in Fig. 4.20e, approaching the $p = 1$ high-density limit. For $p < \frac{1}{2}$, the flow is to the left, toward the $p = 0$ low-density limit. There are three special points in this flow field in composition space, each of which corresponds to a composition or density that *remains invariant under scale transformation*. These are the renormalization-group *fixed points*, points in parameter space that correspond to states of the system which, in some sense, "look the same" at all scales of length. (The concept of self-similarity, discussed in Section 3.2.5 in connection with Brownian motion and Fig. 3.13, is also useful here.) Of the three fixed points in Fig. 4.20e, two of these are relatively uninteresting. These are the two "sinks" of the flow, the endpoints at $p = 0$ (the completely empty limit) and at $p = 1$ (the completely filled limit). These are "stable" or "homogeneous" fixed points; renormalization drives compositions toward these "pure" limits in which the minority component entirely disappears. In our political analogy, renormalization to large units tends to produce unanimity. American presidential elections under the electoral college system are one-stage renormalizations from the voter level to the state level ($N \approx 10^6$), and reasonably close popular-vote results (say, $p = 0.45$ or $p = 0.55$) are often transformed into electoral-college "landslides" ($p' < 0.1$ or $p' > 0.9$).

The key point in the flow field is the "source" of the flow, the *unstable fixed point* at $p = \frac{1}{2}$. This corresponds to a phase transition of the system. It separates the regions of the phase diagram that scale toward different single-component limits. Critical phenomena, in renormalization-group theory, are

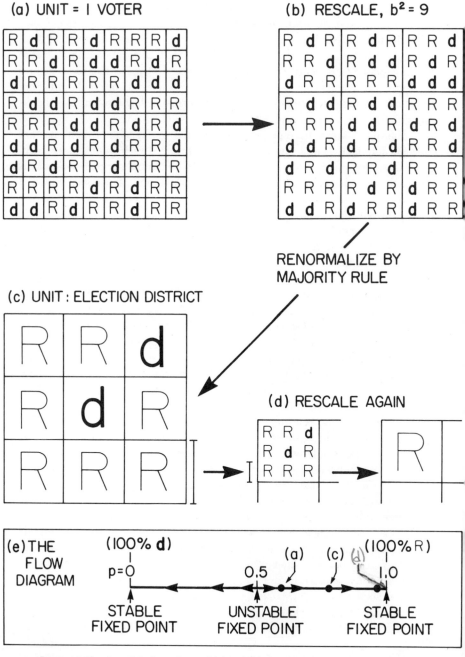

Figure 4.20 A political analogy which exhibits renormalization-group concepts.

controlled by the properties of the flow diagram in the immediate neighborhood of the unstable fixed point.

In Fig. 4.20, the criterion for deciding the state (R or **d**) of the renormalized element in *c* is based on majority rule applied to the corresponding cell of elements of the original system in *b*. For percolation, the criterion for deciding the state (occupied or empty) of the renormalized element is based on whether or not the corresponding cell of the original system percolates, that is, exhibits connectivity between two opposite faces (in $d = 3$) or edges (in $d = 2$). There are some subtleties and ambiguities in implementing this "spanning-cluster" criterion when the scaling factor *b* is small, but for large b ($b \gg 1$) all reasonable criteria converge to the same limit.

We can now interpret Fig. 4.19, which shows two different renormalization-group implementations (both for $b = 2$) for percolation on the square lattice. Unlike the one-dimensional flow diagram of Fig. 4.20e, the flow diagram for site–bond percolation occurs in the plane $(x, y) = (p^{site}, p^{bond})$. The two sets of results in Fig. 4.19 agree well with each other, although they are presented in two somewhat different formats. Figure 4.19a shows the direction of flow throughout the phase diagram, while Fig. 4.19b shows a set of distinct trajectories. Their results along the lines $p^{bond} = 1$ and $p^{site} = 1$ may be used to gauge the accuracy of these calculations. The known thresholds (Table 4.4) for the square lattice are $p_c^{site} = 0.593$ and $p_c^{bond} = 0.500$. By comparison, the estimates for p_c^{site} are 0.57 and 0.63, and for p_c^{bond} are 0.51 and 0.52, for Figs. 4.19a and 4.19b, respectively.

Like the political flow diagram of Fig. 4.20e, the percolation flow diagram of Fig. 4.19 contains two homogeneous fixed points and one critical fixed point. The homogeneous fixed points occur at compositions (0,0) and (1,1), corresponding to the dull cases in which all bonds and sites are empty or occupied, respectively. The critical fixed point occurs at $(p^{site}, p^{bond}) = (0.90, 0.59)$. Viewing the flow diagram as a drainage pattern on an actual terrain, the critical fixed point is a saddle point, the low point in a ridge line that connects the percolation thresholds for pure site percolation $(p_c^{site}, 1)$ and pure bond percolation $(1, p_c^{bond})$. This ridge line is the continental divide of the terrain. The watershed lying to the south and west of this line drains to the empty lattice fixed point at (0,0); the watershed lying to the north and east drains to the fully occupied fixed point at (1,1). The southwest watershed corresponds, of course, to the nonpercolating region of phase space; compositions in this region renormalize under rescaling to lower and lower density. The northeast watershed corresponds to the percolating regime, embracing all (p^{site}, p^{bond}) values that renormalize toward the high-density limit. Just as in the political case, *the scaling-induced flow is away from criticality* (away from the phase-boundary ridge line in Fig. 4.19, away from the point representing a 50–50 vote split in Fig. 4.20).

In our discussion of scaling behavior in Sec. 4.5, the statement that site-percolation processes and bond-percolation processes both belong to the same universality class (have the same critical exponents) was asserted as empirically established, as based on experiment. With regard to this important point there is special significance in the demonstration, via Fig. 4.19, that the critical fixed points for the two basic types of percolation processes are both controlled by (renormalize to, flow into) the same critical fixed point in the phase space for the hybrid process. This demonstration augments the empirical evidence with a degree of theoretical underpinning: In the context of renormalization-group theory, its interpretation *assigns site percolation and bond percolation to the same universality class.*

The approach exemplified by Figs. 4.19 and 4.20 is called direct or *real-space* renormalization. A picturesque way to express the underlying viewpoint of this approach is in terms of a declaration that captures the philosophy of the renormalization-group theory of phase transitions: At the critical point, *all scales of length coexist.* The characteristic length for the system goes to infinity, becoming arbitrarily large (macroscopic) with respect to atomic-scale lengths. For magnetic phase transitions, the measure of the diverging length scale is the correlation length ξ. For percolation, the diverging length scale is set by the connectedness length l_{av}. The presence of a diverging length scale is what makes it possible to apply to percolation the elegant program of real-space renormalization.

There is another form of renormalization-group theory, one which operates on the Hamiltonian constructed for the statistical mechanics of the physical system at hand. This is the form which yields the "ϵ expansion," in which the results of mean-field theory (valid for $d = d^*$, the marginal dimensionality) are expanded in a power series in $\epsilon = d - d^*$. The epsilon expansion explicitly exhibits the renormalization group as a next stage, beyond mean-field theory, in the development of the theory of phase transitions.

For completeness, it is worthwhile to point out that there exists, for the percolation model, such a Hamiltonian (or momentum-space) formulation of the renormalization group. The basis is essentially the percolation ↔ magnetism analogy given previously in Table 4.2. A generalization of the Lenz–Ising model for magnetism, known as the "*s*-state Potts model," generates bond percolation as a particular limiting case.

Though it might seem an unlikely match, it happens that the combination of the elegant ideas of renormalization-group analysis with the brute-force methods of computer simulations results in a rather productive marriage. Figure 4.21 shows a set of Monte Carlo calculations (Kirkpatrick, 1979) for bond percolation on the three-dimensional simple cubic lattice. The horizontal axis shows p, the fraction of bonds that are connected in a sample cube containing b^3 bonds. Plotted along the vertical axis is the fraction p' of such samples that percolate (have opposite faces connected). Five such "experimental" curves are shown in Fig. 4.21, each of which corresponds to a given cube size b^3. The five values for the cube edge b are $b = 10, 20, 30, 50,$ and 80, and each data point shown represents a large number of cases that ranges from 20,000 for $b = 10$ to 500 for $b = 80$.

Naturally, in the limit $b \rightarrow \infty$, the function $p'(p)$ becomes a step function

Potts model [margin note]

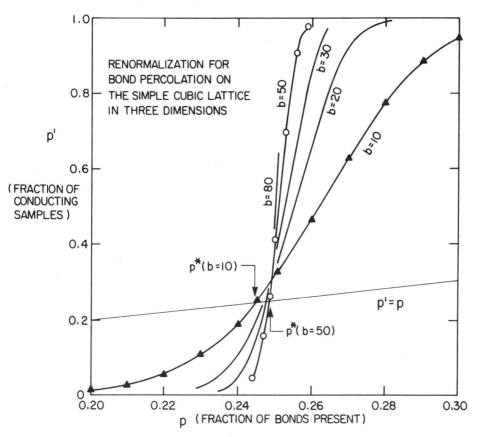

Figure 4.21 Computer calculations of scaling properties for bond percolation on the three-dimensional simple cubic lattice. When p is the fraction of connected bonds, $p' = p'(p, b)$ is the fraction of cubic samples of edge length b that contain a continuous path of connected bonds (a spanning cluster) which links opposite faces of the sample (after Kirkpatrick, 1979).

that jumps abruptly from $p' = 0$ to $p' = 1$ at the percolation threshold $p = p_c$. For finite b, the transition from $p' \approx 0$ to $p' \approx 1$ is smeared out over a finite range of p. Now in the spirit of the renormalization group, we interpret $p'(p,b)$ as the renormalized probability corresponding to an initial probability p and a lattice rescaling by a scale factor b. Then, for a given b, a critical fixed point $p^*(b)$ is determined by finding the value of p that transforms into itself under that rescaling

$$p'(p^*,b) = p^*. \tag{4.23}$$

The solution of Eq. (4.23), for each value of b, may be obtained graphically from Fig. 4.21 by locating the intersection of the curve $p'(p)$ with the straight line $p' = p$. We see that, though $p^*(b)$ slightly underestimates p_c (which is

0.247 for bond percolation on the sc lattice), p^* closely approximates p_c even for b as small as 10. By $b = 80$, p^* is within 0.001 of $p^* (\infty) = p_c$.

Besides providing an alternate route to p_c, a computer-experiment implementation of the renormalization-group philosophy also yields accurate information about the correlation-length exponent ν. In terms of the rescaled bond-connectedness probability $p'(p,b)$ illustrated in Fig. 4.21, the news about ν is contained in the steepness of each curve near p^*. If, for a scale factor b, the slope at $p = p^*$ is denoted by $(dp'/dp)_{p^*}$ it is a result of scaling theory that ν is given by

$$\nu = \frac{\ln b}{\ln (dp'/dp)_{p^*}} \tag{4.24}$$

We present a rough argument for the plausibility of Eq. (4.24). Let $l \sim (\Delta p)^{-\nu}$ and $l' \sim (\Delta p')^{-\nu}$ denote the connectedness lengths l and l' in the initial and rescaled systems, respectively. $\Delta p \equiv p - p_c$ and $\Delta p' \equiv p' - p_c$ denote the initial and rescaled distance from threshold, which is assumed to be small. Since $(l/l') = b$ because the rescaled unit of length is b times the original unit, then $(\Delta p'/\Delta p)^{\nu} = b$. Interpreting $\Delta p'/\Delta p$ as dp'/dp at $p = p^* \approx p_c$ and solving for ν yields Eq. (4.24).

Figure 4.22 shows the determination of ν according to the scaling-law expectation expressed in Eq. (4.24) and the numerical data shown in Fig. 4.21. The slope of $p'(p)$ near p_c is plotted against b on a log–log plot. The straight line that closely fits these data corresponds to a value of $\nu = 0.85$ for the connectedness-length critical exponent in three dimensions.

The bond-percolation computer experiments whose results yield the data displayed in Figs. 4.21 and 4.22 provide a straightforward and concrete realization for renormalization-group concepts. It is similarly straightforward to obtain analogous results for site percolation. Here, a block of b^3 sites on the original lattice would transform into a single site on the rescaled lattice. The renormalized probability $p' = p'(p,b)$ would now state the probability for that site to be occupied, given p as the occupation probability on the initial lattice. This situation closely resembles the magnetic "site-spin→block-spin" transformation that was introduced by Kadanoff in 1966 in the theoretical development that provided the intuitive conceptual precursor of the formal renormalization-group approach to phase transitions.

We close with a few final generalizations of percolation theory. Figure 4.23 illustrates two such generalizations: *polychromatic percolation* (Zallen, 1977) and *extended-range percolation* (Domb and Dalton, 1966). Conventional percolation is a problem in black and white. The statistically assigned state function takes on either of just two values (e.g., filled or empty), and our attention is in fact normally focused on only one (e.g., the filled sites) of the two species present. In polychromatic percolation, the state function is permitted to take on three or more discrete values. In polychromatic site percolation, each site is occupied by a "particle" of some color, and adjacent sites are connected if

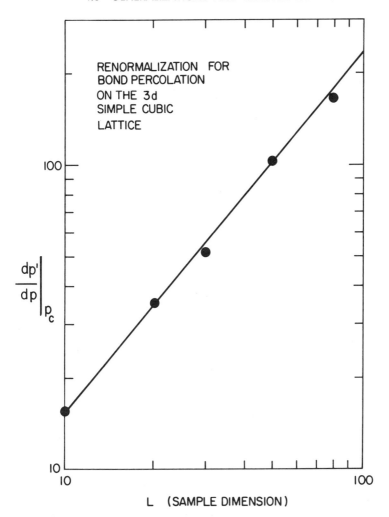

Figure 4.22 Plot for determining the critical exponent ν from Eq. (4.24) and the results of Fig. 4.21. (Kirkpatrick, 1979).

they are occupied by particles of the same color. Similarly, in polychromatic bond percolation, each bond is permeable to a "fluid" of some color, and adjacent bonds of (permeable to the fluid of) the same color are linked.

Figure 4.23 represents a site-percolation process with three species (colors) present, i.e., trichromatic site percolation. The three states or particle types are denoted by solid circles, open squares, and crosses, and are present here in concentrations of roughly 45%, 45%, and 10%, respectively. Since p_c^{site} is 0.59 for the square lattice, none of the species percolate. Polychromatic percolation becomes interesting when it is possible for two or more species to

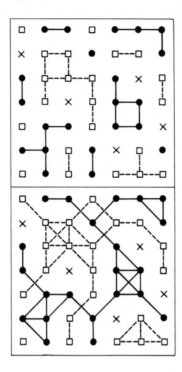

Figure 4.23 Polychromatic percolation with three species (colors) present. In this example, increasing the interaction range to second neighbors permits two colors to percolate, while none percolate when only nearest-neighbor interactions exist.

simultaneously percolate. This cannot happen for standard site percolation in two dimensions, since p_c is never less than ½. For site percolation in two dimensions, an unbounded black sponge cannot coexist with a white one. But in three (or higher) dimensions, p_c can be quite small and several sponges (infinite clusters) of different colors can coexist. Thus, for site percolation on the $d = 3$ sc lattice, with $p_c = 0.31$, both the circles and the squares (with $p = 0.45$) would percolate, forming a pair of extended and intertwined sponges.

Such interleaving three-dimensional percolation paths are hard to show in a two-dimensional illustration. Fortunately, there is another route to a multiple percolation situation that is much easier to illustrate. The basic prerequisite for a low threshold is a high connectivity. While connectivity increases with increasing dimensionality, high connectivity can also be achieved for lattices of low dimensionality by dropping the limitation (adhered to up to this point) to simple lattices with bonds only between nearest-neighbor sites. By adding direct connections *beyond* nearest neighbors, we can encounter quite low percolation thresholds in three dimensions—and even in two.

Multiple percolation via increased range of interaction in two dimensions is demonstrated in Fig. 4.23. While $z = 4$ and $p_c = 0.59$ for the simple square lattice shown in the upper half of the figure, extending the interaction range to next-nearest neighbors increases z to 8 and lowers the site-percolation threshold to $p_c = 0.41$. While, for the configuration shown, none of the species percolate on the simple lattice, both the circles and the squares percolate ($p >$ p_c) on the more highly connected lattice shown in the lower half of Fig. 4.23.

A simple setting for envisioning polychromatic percolation is a resistor network composed of three types of connectors: perfect insulators, unit conductors, and superconductors, present with probabilities p_0, p_1 and p_∞, respectively. Depending upon the values of these probabilities, the macroscopic network is either insulating, conducting, or superconducting. If p_∞ exceeds p_c, the system is superconducting, whatever the values of p_0 and p_1. If p_1 exceeds p_c while p_∞ does not, the system is a normal conductor. It is also normally conducting if the combined network of conductors and superconductors percolates, even though neither component percolates on its own: $p_1 < p_c$, $p_\infty <$ p_c, but $p_1 + p_\infty > p_c$. Only if the union of conductors and superconductors fails to percolate ($p_1 + p_\infty < p_c$ or, equivalently, $p_0 > 1 - p_c$) is the system insulating. These conditions carve up the composition field into three regimes which constitute the phase diagram for this polychromatic percolation system.

The last generalization of percolation that will be mentioned now to close this chapter is one which will be useful, in the following chapter, in the treatment of the free-volume theory of the glass transition. It is called *high-density percolation* (Reich and Leath, 1978, Cohen and Grest, 1979). In the site-percolation version, each site is occupied (p) or not ($1 - p$) as usual, but an occupied site is considered to belong to a cluster *only* if it is adjacent to m or more occupied sites. Thus, isolated ($m = 0$) filled sites are ignored. If m is chosen to be 1, the situation remains essentially similar to standard site percolation. For values of m lying between $m = 2$ and $m = z$, high-density percolation yields a percolation process that focuses on the appearance of dense, well-connected clusters. For $m = 2$, p_c is the same as for standard ($m = 1$) percolation, since only the inessential dangling bonds at the outskirts of the infinite cluster are eliminated by the $m = 2$ requirement. But for $m \geq 3$, the percolation threshold is shifted to higher densities. Critical exponents remain the same as for ordinary percolation.

REFERENCES

Agrawal, P., S. Redner, P. J. Reynolds, and H. E. Stanley, 1979, *J. Phys. A* **12**, 2073.

Cohen, M. H., and G. S. Grest, 1979, *Phys. Rev. B* **20**, 1077.

de Gennes, P. G., 1976, *J. Phys. (Paris)* **37**, L1.

de Gennes, P. G., 1979, *Scaling Concepts in Polymer Physics*, Cornell University Press, Ithaca.

Domb, C., and N. W. Dalton, 1966, *Proc. Phys. Soc. (London)* **89**, 859.

Fitzpatrick, J. P., R. B. Malt, and F. Spaepen, 1974, *Phys. Lett.* **A47**, 207.

Flory, P. J., 1941, *J. Am. Chem. Soc.* **63**, 3083, 3091, 3096.

Flory, P. J., 1953, *The Principles of Polymer Chemistry*, Cornell University Press, Ithaca.

Frisch, H. L., and J. M. Hammersley, 1963, *J. Soc. Indust. Appl. Math.* **11**, 894.

Hammersley, J. M., 1957, *Proc. Cambridge Phil. Soc.* **53**, 642.

Harrison, R. J., G. H. Bishop, and G. D. Quinn, 1978, *J. Stat. Phys.* **19**, 53.

Henisch, H. K., 1970, *Crystal Growth in Gels*, The Pennsylvania State University Press, University Park, p. 41.

Kadanoff, L. P., 1966, *Physics* **2**, 263.

Kirkpatrick, S., 1973, *Rev. Mod. Phys.* **45**, 574.

Kirkpatrick, S., 1976, *Phys. Rev. Letters* **36**, 69.

√ Kirkpatrick, S., 1979, in *Ill-Condensed Matter*, edited by R. Balian, R. Maynard, and G. Toulouse, North-Holland, Amsterdam, p. 321.

Leath, P. L., and G. R. Reich, 1978, *J. Phys. C* **11**, 4017.

Mandelbrot, B. B., 1977, *Fractals: Form, Chance, and Dimension*, Freeman, San Francisco.

Nakanishi, H., 1980, private communication.

Nakanishi, H., and P. J. Reynolds, 1979, *Phys. Lett.* **A71**, 252.

Phelps, D. J., and C. P. Flynn, 1976, *Phys. Rev. B* **14**, 5279.

Porter, C., 1935, lyrics* from the song "Just One of Those Things" from the musical comedy *Jubilee:*

> It was just one of those nights,
> just one of those fabulous flights,
> a trip to the moon on gossamer wings,
> just one of those things.

Reich, G. R., and P. L. Leath, 1978, *J. Stat. Phys.* **19**, 611.

Scher, H., and R. Zallen, 1970, *J. Chem. Phys.* **53**, 3759.

Shapiro, B., 1979, *J. Phys. C* **12**, 3185.

Smith, L. N., and C. J. Lobb, 1979, *Phys. Rev. B* **20**, 3653.

Stanley, H. E., 1983, *Introduction to Phase Transitions and Critical Phenomena*, second edition, Oxford University Press, London.

Stauffer, D. J., 1976, *J. Chem. Soc. Faraday Trans. II* **72**, 1354.

Stockmayer, W. H., 1943, *J. Chem. Phys.* **11**, 45.

Straley, J. P., 1977, *Phys. Rev. B* **15**, 5733.

Thouless, D. J., 1979, in *Ill-Condensed Matter*, edited by R. Balian, R. Maynard, and G. Toulouse, North-Holland, Amsterdam, p. 1.

Toulouse, G., 1974, *Nuovo Cimento B* **23**, 234.

Webman, I., J. Jortner, and M. H. Cohen, 1976, *Phys. Rev. B* **14**, 4737.

Wilson, K. G., and M. E. Fisher, 1972, *Phys. Rev. Letters* **28**, 240.

Zallen, R., 1977, *Phys. Rev. B* **16**, 1426.

CHAPTER FIVE

Localization ↔ Delocalization Transitions

5.1 LOCALIZED-TO-EXTENDED TRANSITIONS IN AMORPHOUS SOLIDS

There are two major topics in the physics of amorphous solids that involve changes from situations in which particles are localized in atomic-scale regions of space to situations in which macroscopically extended spatial regions are accessible to each particle. The first of these is the *glass transition*, phenomenological aspects of which have been described in the first chapter. Here the "particles" are the atoms themselves, and it is the atomic mobility which undergoes a dramatic change [$a \leftrightarrow b$ in Fig. 1.7] at the glass ↔ liquid transition. The second is the *Anderson transition*. Here the particles are electrons, and the effect is a subtle form of disorder-mediated metal ↔ insulator transition in which the electronic wave functions experience a crossover from extended to localized character. Theoretical approaches to these two phenomena, the glass transition and Anderson localization, form the main content of this chapter.

During the course of this chapter, concepts developed in the context of the percolation model will prove to be quite helpful. Percolation, as detailed in the preceding chapter, is a fundamental model for portraying a *localized ↔ extended transition* in a disordered system. Although it must be applied with caution, we will see that percolation illuminates ideas relevant to these phenomena.

To illustrate the above remarks, and also to provide a bridge to the next sections which deal with theories of the glass transition, two examples of connections with percolation are pointed out here. The first connection is the one between percolation and gelation, as spelled out previously in Table 4.5. The sol → gel transition (the sol phase which precedes the formation of silica gel was schematically shown in Fig. 4.11) is beautifully modeled as a percolation process. Site–bond percolation, as indicated in Fig. 4.18, serves as a revealing model for this phenomenon. Now, there are many resemblances between gelation and the glass transition. The sol→gel transition is, in fact, a form of liquid→glass transition. It differs in proceeding as an apparent chemical reaction rather than as a thermally driven phase change. The second example is a connection implicit in a "bond-lattice" model for the glass transition put forward for covalent liquids and glasses by Angell and Rao (1972). This model essentially resembles a bond-percolation picture of the transition which occurs when temperature traverses T_g.

In both of these examples, the appearance of long-range connectivity at the percolation threshold signals the arrival at the solid side of the glass transition. In Section 5.3 of this chapter, we will meet a quite different approach in which the above viewpoint is turned *inside out*. The free-volume model of the glass transition focuses, not on the structural (solidlike) elements of the system but on the mobile (liquidlike) elements. In that view, the appearance of the percolation path (connecting regions of mobile material) signals arrival at the *liquid* side of the transition.

5.2 DYNAMIC MODELING: MONTE CARLO SIMULATIONS OF THE GLASS TRANSITION

The liquid ↔ glass transition is of central importance in the physics of noncrystalline condensed matter. The experimental facts about this fundamental phenomenon were spelled out in Chapter One in Figs. 1.1, 1.2, 1.8, 1.9, and 1.10. It would be nice to present, at this point, "the theory of the glass transition." Unfortunately, this is not yet possible. No definitive theory exists that satisfactorily accounts for the bulk of observations. Still, illuminating models for simple glasses have been developed, and some of these are offered in this section and the next.

Perhaps the most direct approach exploits the ever-increasing speed and power of modern computers to attack the dynamics of simple models of many-particle systems. Such *computer simulations* are discussed in this section. The straightforward, if brute force, tack of these "computer experiments" can be very helpful and may point the way to deeper theories possessing greater insight and intuitive content. Computer simulations are most applicable to solids and dense liquids composed of spherical "molecules," such as the closed-shell inert-gas atoms Ne, Ar, Kr, and Xe. Amorphous solid forms of these elements have not yet been prepared. Thus far, the physical systems to which computer-simulation glass transitions most closely correspond are simple metallic glasses whose structure is describable in terms of hard-sphere models similar to the rcp model of Section 2.4. Indeed, as described below, some computer simulations are fruitfully carried out for motion in hard-sphere systems. This work can be viewed as *dynamic modeling*, a kinematic counterpart of the *static* modeling (e.g., random close packing) of Chapter Two.

The conceptual model that envisages a solid or dense fluid as made up of

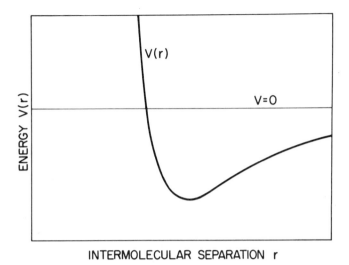

Figure 5.1 General form for the interaction energy of a pair of molecules.

hard spheres (i.e., perfectly rigid, noninterpenetrating, balls of uniform size) invokes a particle–particle interaction containing a repulsive potential that is infinitely steep. Usually no attractive potential is included, so that the form of the intermolecular potential-energy function $V(r)$ is

$$V(r) = 0 \quad \text{for } r > D,$$
$$= \infty \quad \text{for } r < D. \qquad (5.1)$$

D is the hard-sphere diameter and r is the separation between the mass centers of the two molecules. The hard-sphere potential of Eq. (5.1) is meant to simulate an actual intermolecular potential which has the form shown in Fig. 5.1. The main qualitative features of intermolecular interaction are very strong repulsion at small separations (owing to overlap of the electronic cores), weak attraction at large separations (owing to dipole fluctuations for closed-shell electrically neutral objects), and vanishing interaction at very large separations. The hard-sphere potential of Eq. (5.1), with D set at the zero crossing of Fig. 5.1, omits the weak mid-range attraction but retains (in fact, exaggerates) the strong short-range repulsion. It has been found that this approximation is not bad for dense fluids and solids, evidently because of the dominant role played at high density by the repulsive part of the intermolecular interaction.

Computer-simulation techniques treat the motions of the molecules by means of classical mechanics, which is acceptable if the molecule is not too light (as is helium). Quantum mechanics enters in the form of the interaction energy $V(r)$. The choice of this function, which is the primary input for the computation, is the first step. One extreme is to use the result (usually unavailable) of a full-fledged quantum-mechanical calculation for the interaction between a pair of the molecules in question. The other extreme is the choice of a severely simplified form, as the hard-sphere potential of Eq. (5.1). In between is the choice of a mathematically convenient yet reasonably realistic form such as the "Lennard-Jones 6-12 potential":

$$V(r) = 4\epsilon[(\sigma/r)^{12} - (\sigma/r)^6]. \tag{5.2}$$

The first term in Eq. (5.2) provides a steep repulsive interaction at short distances, while the second term provides a long-range attraction falling off as $1/r^6$ (as expected for a van der Waals fluctuating-dipole intermolecular interaction). The dimensional quantities σ and ϵ set, respectively, the length and energy scales characteristic of the $V(r)$ of Eq. (5.2). The distance σ, whose significance is similar to that of the hard-sphere diameter D in (5.1), is the value of the separation r corresponding to the zero crossing $V(r) = 0$. The energy ϵ is the depth of the minimum in $V(r)$. The minimum, $V(r) = -\epsilon$, occurs at a separation of $r = 2^{1/6}\,\sigma \approx 1.1\sigma$. Viewed more generally, ϵ is a measure of the strength of the interaction and σ is a measure of the size of the repulsive core. The form of $V(r)$ given in Eq. (5.2) is the one adopted for the example to be discussed below in connection with Figs. 5.2, 5.3, and 5.4.

Once the forces between molecules are given, either by a quantum-mechanical treatment or by some assumed form for their mutual interaction such as Eqs. (5.1) or (5.2), classical mechanics takes over. There are two distinct techniques of computer simulation; they go under the names of

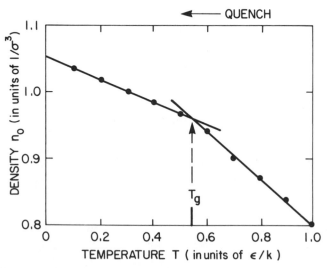

Figure 5.2 Monte Carlo calculation of density as a function of quench temperature for a system of simple spherical molecules interacting via a 6–12 potential. The quantities ϵ and σ are the characteristic energy and length, respectively, of the potential function (after Abraham, 1980).

molecular dynamics and *Monte Carlo* methods. The method of molecular dynamics executes a solution of the classical equations of motion for a system of interacting particles. The input is $V(r)$ and a set of initial positions and velocities. At each stage, $V(r)$ is used to calculate the force exerted on each molecule by its neighbors, and the evolution of the configuration of the system then proceeds by stepwise numerical integration of the equations of motion. Note that the use of the same $V(r)$ for every pair of particles (i.e., "pairwise additivity" of the intermolecular interactions) invokes the assumption that the force between two molecules is, to a good approximation, unaffected by the presence of other molecules. In principle, the molecular dynamics method is entirely deterministic and provides information about the time dependence of the configuration of the system. Particle trajectories similar to those shown earlier in Fig. 1.7, for solid (Fig. 1.7*a*) and liquid (Fig. 1.7*b*) phases of the system, are generated by molecular dynamics computer simulations.

In both Monte Carlo and molecular dynamics calculations, equilibrium properties are determined by averaging over the configurations obtained. In Monte Carlo simulations, a sequence of configurations is constructed by successive random displacements which derive, in roulette-wheel fashion, from a random-number generator. However, in order to simulate an actual thermodynamic system, it is algorithmically arranged that the probability of configurations $\{r_i\}$ appearing in the sequence is proportional to the Boltzmann factor $\exp[-V\{r_i\}/kT]$, where $V\{r_i\}$ is the potential energy of the configuration. Although Monte Carlo calculations cannot yield time-evolved trajectories, the Monte Carlo method has certain advantages vis-à-vis molecular dynamics. It is capable of dealing with more particles, and permits a larger number of "moves," in a simulation occupying a given amount of computer time. It is also well suited for computing thermodynamic behavior.

Figures 5.2, 5.3, and 5.4 display results obtained by Abraham (1980) in a Monte Carlo simulation of the liquid ↔ glass transition for a system of 108 spherical atoms or molecules interacting (within a cube of adjustable volume, replicated by periodic boundary conditions) via the "6-12" $V(r)$ of Eq. (5.2). These three figures show the result of (simulated) quenches in which the temperature was abruptly reset to a lower value and the system allowed to "equilibrate" with $\approx 10^6$ moves. The initial configuration preceding the quench was that of a well-equilibrated liquid phase at temperature T_0 and pressure P_0, where the reference temperature and pressure are established by the well-depth and size parameters of the potential function:

$$
\begin{aligned}
kT_0 &= \epsilon \quad, \\
P_0 &= \epsilon/\sigma^3.
\end{aligned}
\tag{5.3}
$$

Each of Figs. 5.2, 5.3, and 5.4 shows results of quenching (at constant pressure $P = P_0$) from temperature T_0 to temperature $T < T_0$, as a function of the final temperature T. In all of these figures, numerical values associated with the temperature scale denote the dimensionless ratio T/T_0.

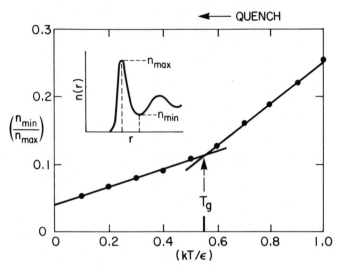

Figure 5.3 Microstructural signature uncovered by Abraham (1980) in his computer simulations of the glass transition in an interacting system of simple molecules. The function $n(r)$ is the number of atom centers per unit volume at distance r from an average atom and is closely related to the radial distribution function $4\pi r^2 n(r)$.

The calculated density-versus-temperature characteristic of Fig. 5.2 should be compared to Figs. 1.1 and 1.2 of Chapter One (recognizing, of course, that the specific volume plotted in the earlier figures is the reciprocal of the density plotted here). We identify the break with decreasing T, from a line of high slope (i.e., high thermal expansion) to one of low slope, as the glass transition T_g. Figure 5.2 indicates that, for $P = P_0$, T_g occurs at about $0.55 T_0$ in a temperature quench. In addition to studying liquid→glass transformations induced by abruptly decreasing temperature ("quench"), Abraham's Monte Carlo computations also explored the glass transition induced by abruptly increasing pressure ("crush") as well as by doing both simultaneously ("crunch" = "crush" + "quench"). For the pressure crush computer simulations at $T = T_0$, Abraham finds the transition at $P_g = 7P_0$. Taken together with Fig. 5.2, this implies that at $P = 0$, T_g is about $0.48 T_0$ (i.e., $0.48\epsilon/k$). $T_g = 0.48(\epsilon/k)$ thus specifies the connection between the glass transition temperature and the characteristic intermolecular-interaction energy ϵ for a model based upon the Lennard-Jones $V(r)$.

By averaging the kinetic and potential energies of the particles over many configurations of the system at each temperature, these computer experiments also yield the internal energy and the enthalpy as functions of T. The specific heat thus obtained exhibits a step at T_g, qualitatively similar to the $C_P(T)$ data of Figs. 1.8 and 1.9.

Thus far we have discussed Monte Carlo results for *thermodynamic*, i.e., macroscopic, manifestations of the glass transition. Computer simulations also

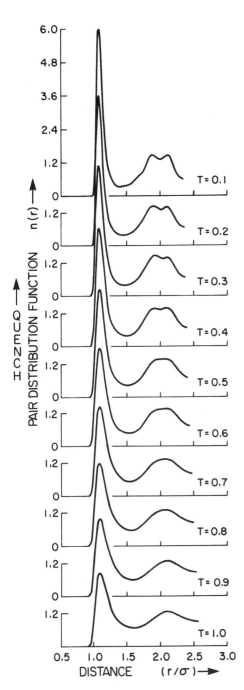

Figure 5.4 Evolution of the atomic-scale distribution for a series of Monte Carlo quenches from temperature T_0 to temperature T (expressed in units of T_0). The split second peak, which signals the random-close-packed structure of a simple glass, is evident in quenches to low temperatures (Abraham, 1980).

yield information about *microscopic* manifestations of the transition, and this aspect is the subject of Figs. 5.3 and 5.4. As discussed earlier in Sections 2.3.2 and 2.6, the radial distribution function is the best available experimental characterization of an amorphous solid. The radial distribution function $RDF(r)$ is $4\pi r^2 n(r)$, where $n(r)$ is the number density of atom centers per unit volume at distance r from an average atom. Abraham investigated the function $n(r)$ for the "Lennard-Jones glass" of Fig. 5.2 and, as shown in Fig. 5.3, found that a *microstructural signature* of the glass transition could be clearly discerned. Figure 5.3 plots the temperature dependence of the ratio (n_{min}/n_{max}) where, as indicated in the insert, n_{max} is the value of $n(r)$ at the nearest-neighbor peak and n_{min} is its value at the subsequent minimum. The kink in the behavior of this ratio, seen near T_g, is quite evident.

The evolution of the radial distribution function with decreasing temperature is presented in Fig. 5.4 for a sequence of quenches terminating at temperature T (T values on the figure are given in units of T_0). A remarkable feature of this family of computer-simulated $n(r)$ curves is the appearance, in the glass phase ($T < 0.5 T_0$), of a definite splitting in the second peak. This split second peak was encountered earlier. *It is the trademark of random close packing*, as discussed in Section 2.6. Figure 2.22 illustrated it {in a somewhat different format, for the function $4\pi r[n(r) - n_0]$}, and the second-neighbor configurations responsible for its two components were shown in Fig. 2.23. This confluence of the dynamic modeling of the present section and the static modeling described in chapters two and three is a gratifying outcome. It is especially pleasant here, for the case of an elemental glass based on equal spheres, because this model serves as the simplest conceivable construct for an elemental amorphous solid. The role played by the Lennard-Jones size parameter σ in the length scale of Fig. 5.4 is very close to that played by the hard-sphere diameter D in the length scale of Fig. 2.22.

5.3   THE FREE-VOLUME MODEL OF THE GLASS TRANSITION

The theory to be discussed now is the most widely used theoretical picture of the glass transition. It earns its appeal by virtue of its inherent conceptual transparency. The free-volume model is noted for its physical plausibility, mathematical simplicity, and its value for qualitatively predicting various experimental observations connected with the liquid ↔ glass transition. The basic idea is to approach the localization/delocalization issue for the atoms and molecules from the viewpoint of the volume available to each molecule, and to address the question of whether or not there is adequate room for molecular maneuver.

While the concept of free volume is intuitively appealing and extremely useful for an understanding of molecular-mobility phenomena, it is often handled in a vague way in the literature. The clearest visualization is in the context of the hard-sphere model introduced in the last section, that is, spherically symmetric classical molecules interacting via the step-function repulsion of Eq. (5.1). An illustration of free volume in a hard-sphere picture is given in Fig. 5.5.

Figure 5.5 shows a possible configuration of a *two-dimensional* hard-sphere system (i.e., an assembly of *hard disks*). The (2d) free volumes associated with each of three molecules, in this particular configuration, are represented by the shaded areas. These areas are well defined and are constructed in the following way. For a given molecule, say C in Fig. 5.5, consider all of the *other* molecules in the system to be frozen in position at their instantaneous locations in the configuration under consideration. Then determine the region of space accessible to the mass center of molecule C, subject, of course, to the constraint that the repulsive potential of Eq. (5.1) prohibits overlap with neighboring molecules. For C, the result is the small shaded region shown in the lower part of Fig. 5.5. Hemmed in by five neighbors, the free volume associated with molecule C is quite small. It is restricted to a snug cell formed by its nearest neighbors. (In molecular dynamics computer simulations, mentioned in the last section, the free volume of a particular molecule can be generated by assigning a very light mass to the particle in question and observing the portion of space swiftly swept out by its motion.)

Similarly constructed as for C, free volumes defined by the locus of points

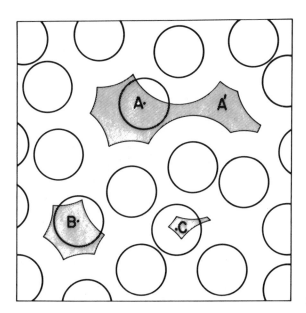

Figure 5.5 Visualization of the free-volume notion for a two-dimensional hard-sphere system. The shaded areas show the regions accessible, in this configuration, to the centers of three molecules. Molecule A is capable of taking a diffusive step to a ''new position,'' while B and C are confined within small cages.

accessible to the their centers are shown for molecules B and A in Fig. 5.5. For B, the free volume is larger than for C, but this molecule is still confined to a cage defined by its nearest neighbors (six in this case). For A, the free volume is greater still, and now something interesting has happened. Molecule A, at the position shown in the figure, is loosely surrounded by six nearest neighbors, But its free volume is sufficiently extended to permit A to have the opportunity to migrate to the position indicated as A′. Here the molecule is again loosely surrounded by six nearest neighbors, but four of them are different from the original ones it had at A. In effect, an A → A′ displacement represents a translational step or *diffusive excursion* to a "new position." This is a type of movement that is unavailable to molecules B and C. The latter two molecules, with their confined free volumes, are limited to oscillatory motions about centers defined by the cages formed by their nearest neighbors. A is free to "change partners" via a displacement comparable in magnitude to the molecular diameter.

The distinction being drawn here, between B and C, on the one hand, and A, on the other hand, is closely related to the vibrational/translational distinction illustrated earlier in Fig. 1.7. The point is that, for these three molecules within the configuration shown in Fig. 5.5., A has an opportunity to engage in a translational step representing a contribution to a diffusive (liquidlike) trajectory such as that sketched in Fig. 1.7b, whereas B and C have no such opportunity and are positioned only for oscillatory (solidlike) motion such as that sketched in Fig. 1.7a. This distinction is important, and will be elaborated further in the following development of the free-volume theory.

Although the hard-sphere model presents a drastically simplified view of condensed matter, Fig. 5.5 provides a clear setting for demonstrating the idea of free volume. We will use the symbol v_f to denote the free volume associated with a given molecule, and the symbol v to denote the total local volume associated with that molecule. The quantity v is simply the volume of the Voronoi cell surrounding the molecule, as illustrated earlier (for another two-dimensional configuration, in which the molecular centers are represented by heavy dots) by the shaded areas in Fig. 2.5. Note that v_f is *not* the same as the "empty" volume within v, that is: $v_f \neq v - v_{mol}$, where v_{mol} is the volume "occupied" by the molecule ($v_{mol} = \pi D^2/4$ for the hard-disk model of Fig. 5.5). One way to see this is to consider a random-close-packed array of hard spheres. As we saw in Section 2.4, the filling factor $\langle v_{mol}/v \rangle$ is 0.637 for the rcp structure, so that the average value of the fractional amount of empty space ($v - v_{mol})/v$ is about 0.36 in this configuration. But since this configuration is a close-packed one, each hard sphere is jammed in place and $v_f = 0$. Thus the free volume vanishes while the empty volume is still considerable, clearly revealing that, although a monotonic relationship certainly connects them (an increase in $v - v_{mol}$ must increase v_f), these are two different quantities.

The liquid ↔ glass transition, from the free-volume point of view, is a macroscopic manifestation of changes occurring in the microscopic *distribution of molecular free volumes*. Suppose the transition is approached from the liquid side, with decreasing temperature. At high temperature, sufficient free volume is present for diffusive-motion opportunities (such as that afforded to A in Fig. 5.5) to occur frequently, and fluidity results. With decreasing temperature, the total volume V of the material decreases, and with it, the total free volume V_f. At some point, V_f is reduced to a critical level, below which there is *inadequate room for molecular maneuverability and macroscopic fluidity*. The temperature at which this happens is T_g, and below this temperature, the material is an amorphous solid. *The glass transition occurs when the free volume is sufficiently squeezed out of the system.*

Implicit in the above phenomenological description is the involvement of thermal expansion, a point which will be returned to later. The intuitive appeal of the picture presented above can be augmented, at this juncture, by taking note of a very suggestive result obtained in the Monte Carlo simulations of the Lennard-Jones glass which were discussed in the last section. The volume at which the glass transition occurs in that study was determined, not only in temperature-quench simulations (Fig. 5.2), but also in pressure-crush simulations. The glass-transition volume was found to be the *same* for both computer experiments, which vary V in two quite different ways (lowering temperature versus raising pressure). Similar results have been found in molecular dynamics studies. Clearly, these findings provide support for the idea that volume is a central factor. The density at which the amorphous solid forms in these studies can be converted to an equivalent hard-sphere packing fraction, which is found to have a value of about 0.54, that is, about 85% of the packing fraction of the random-close-packed structure. Computer studies thus suggest that such a simple liquid locks itself into the glass state when its excess volume, relative to that of the rcp structure, has been reduced to about 18%.

The qualitative picture of the liquid ↔ glass transition as a maneuverability or free-volume crisis will now be addressed in more depth. The treatment which follows is based on the development of the free-volume model carried out in a classic series of papers by Cohen and Turnbull (1959, 1961, 1970) and later elegantly extended and combined with percolation ideas by Cohen and Grest (1979, 1981). The Cohen–Turnbull formulation of free volume was for simple molecular glasses of the type which we have focused on in this chapter, and it is of historical interest to note that their work led them to predict, in 1959, that "liquids of even the simplest structure would go through the glass transition if sufficiently undercooled." This was quickly proved correct, in the 1960s, with the discovery of metallic glasses formed by splat quenching (Fig. 1.4c) of the melt.

Four simple assumptions underlie the theory:

1. It is possible to associate a local volume v of molecular scale with each molecule.

2. When v reaches some critical value v_c, the excess can be regarded as free.

3. No local free energy is required for redistributing free volume among the molecules, and the free volume $v_f = v - v_c$ of each molecule fluctuates with the continual redistribution of the total free volume V_f.

4. Molecular transport occurs (as at A → A′ in Fig. 5.5) only when a molecule acquires sufficient free volume: $v_f > v_f^*$.

The first assumption invokes a cell picture of the atomic-scale structure, and we may identify v with the volume of the Voronoi polyhedron (Fig. 2.5) defined by the location of the molecule within the system's instantaneous configuration. Such a cell or cage picture is valid for *dense* phases such as glasses and dense liquids. The cell volume v is adopted as the key coordinate specifying the local configuration. Assumption 2 is a necessary simplification for the relation between the cell volume v and the free volume v_f, and supposes the simplest form: $v_f = v - v_c$. (From our previous discussion of hard spheres, we note that the critical volume v_c must be somewhat larger than the molecular volume, perhaps by a factor roughly given by the reciprocal of the rcp filling factor.)

The central physical conception of the mechanism of molecular transport in the free-volume model is embodied in assumptions 3 and 4. In the notion of free volume illustrated in Fig. 5.5 in a hard-sphere context, the adjective "free" naturally refers to the freedom of each molecule's movement within its cell. Assumption 3 stated above adds another crucial dimension to the use of the term *free* volume: V_f may be redistributed freely, i.e. *without cost in energy*, among the molecules in the system. The free-volume model is far more general than the hard-sphere picture of Fig. 5.5, and it assumes that the molecules interact with each other through a realistic intermolecular potential-energy function of the form shown in Fig. 5.1. Within this framework, the rationale for energy-conserving redistribution of free volume will now be given.

Figure 5.6 displays a function, the local free energy $f(v)$, which represents the contribution of a cell of volume v to the total free energy. This is roughly

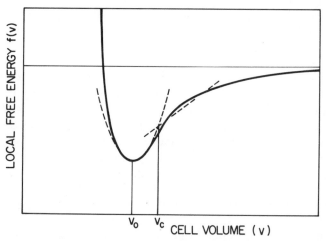

Figure 5.6 General form for the local free energy as a function of cell volume, and the approximate form (the dashed line) used by Cohen and Grest (1979, 1981) to distinguish between solidlike and liquidlike cells.

the negative of the work needed to remove the molecule from the cell to infinity, and the shape of $f(v)$ is necessarily similar, qualitatively, to the underlying intermolecular potential of Fig. 5.1. The free-volume view approximates this function by one indicated by the dashed segments shown in Fig. 5.6:

$$f(v) = f(v_0) + \tfrac{1}{2}K(v - v_0)^2 \quad \text{for } v < v_c,$$
$$= f(v_c) + L(v - v_c) \quad \text{for } v > v_c. \qquad (5.4)$$

The critical volume v_c, whose value is closely related to the location of the point of inflection in the "exact" $f(v)$, defines a demarcation between the quadratic oscillatorlike behavior at volumes not far from the minimum at v_0 and a linear regime which applies at larger cell volumes. K and L are constants which characterize the shapes of the quadratic and linear regions, respectively. It is the linear regime in this model $f(v)$ which permits the free exchange of free volume among neighboring cells. Two adjacent cells with volume in excess of v_c can transfer a volume increment between them, because the energy increase of the expanding cell is exactly balanced by the energy decrease of the contracting one. This argument justifies the third basic assumption of the free-volume model.

The fourth assumption contains the free-volume view of molecular transport: *Transport is attendant upon the occurrence of fluctuations in which the local volume exceeds a critical size.* Such requisite attenuations in the local density, in which voids of molecular size open up and opportunities for transport (A → A′ in Fig. 5.6) appear, occur when v exceeds $v_c + v_f^*$. (Here v_f^* is the minimum amount of free volume needed for a diffusive step.) At smaller cell volumes, the molecular motion is oscillatory; at larger cell volumes, the motion gains a diffusive contribution. Transport results from the fluctuation-induced occurrence of such open configurations.

The next step is to consider the *distribution* of cell volumes. At low temperatures, in the solid, $f(v)$ is confined to the quadratic regime close to the minimum at v_0. Because of the parabolic dependence on v, the total volume V is partitioned nearly equally among the cells because it is too energetically expensive to accumulate volume in cells of appreciably above-average size. At high temperatures, however, free volume appears when the portion of the curve above $f(v_c)$ comes into play. Free volume is partitioned randomly among the cells. Let N be the total number of cells (molecules), V_f be the total free volume, and let the range of v_f values be subdivided into discrete intervals with $n(i)$ being the number of cells having free volume $v_f(i)$ corresponding to interval i. N and V_f are conserved quantities; for each configuration, $\Sigma_i n(i) = N$ and $\Sigma_i n(i) v_f(i) = V_f$. The number of ways W in which distribution $\{n(i)\}$ may occur is $W = N!/\Pi_i n(i)!$. Maximizing W for given N and V_f and passing to the continuum limit [i.e., replacing $n(i)/N\delta(i)$, where $\delta(i) = v_f(i + 1) - v_f(i)$, by the continuous distribution $p(v_f)$ in the limit $\delta \rightarrow 0$] yields an exponential dependence for $p(v_f)$, where $p(v_f)dv_f$ is the probability of a cell having free volume between v_f and $v_f + dv_f$. (The derivation is a standard one fully analogous to that used to arrive at Maxwell-Boltzmann statistics. In the present case, V_f plays a role corresponding to that of the total energy in the MB case.) The result is $p(v_f) = (N/V_f) \exp[-(N/V_f)v_f]$.

The fluidity (η^{-1}, where η is the viscosity) now follows from the distribution $p(v_f)$ of free volumes among the cells. It is proportional to the probability of a molecule having access to a free volume in excess of v_f^*:

$$\eta^{-1} = \text{const.} \int_{v_f^*}^{\infty} p(v_f) dv_f$$

$$= \text{const.} \exp[-Nv_f^*/V_f]. \tag{5.5}$$

Equation (5.5) connects the fluidity with the total free volume V_f. It has the form $\eta^{-1} = A' \exp[-B'/V_f]$, and is usually called the Doolittle equation. If we now approximate V_f by $\alpha(T - T_0)$, where α is the thermal expansion coefficient in the vicinity of the glass transition and T_0 is the temperature at which the free volume vanishes, we then obtain an expression of the form $\eta = A \exp[B/(T - T_0)]$. This is the Vogel–Fulcher equation, which was mentioned earlier in Section 1.4. It describes the steep viscosity-versus-temperature characteristic of many glass-forming liquids at temperatures above T_g. The gradual decrease in free volume with decreasing T is converted into a precipitous decrease in molecular mobility, because V_f appears within the exponent of Eq. (5.5).

The thermal expansivity α which enters above is actually $\alpha_{\text{liq}} - \alpha_{\text{glass}}$, the difference between the expansion coefficients of the liquid and the glass. α_{glass} is considerably smaller than α_{liq}, as seen earlier in Figs. 1.1 and 1.2. In the solid, thermal expansion occurs as a consequence of the anharmonicity of the intermolecular interactions. For a perfectly harmonic solid (of which there are none), thermal expansion vanishes. For the free-volume form of the local free energy which was given in Eq. (5.4), it is evident that thermal expansion remains small until the temperature is high enough to appreciably populate the linear regime of $f(v)$. For values of local free energy below $f(v_c)$, the function $f(v)$ is symmetric about its minimum and harmonic (parabolic) in form. Asymmetry and pronounced anharmonicity set in when $v > v_c$ (when the molecules have enough energy to aggressively probe the soft large-separation part of the intermolecular potential of Fig. 5.1), so that the onset of the high thermal expansivity characteristic of the liquid corresponds to the formation of considerable free volume.

5.4 FREE VOLUME, COMMUNAL ENTROPY, AND PERCOLATION

As discussed thus far, the free-volume model presents a physical picture of the molecular mobility and yields a simple argument for the typical form of its temperature variation above T_g. In Chapter One, especially in connection with the vanishing excess entropy illustrated in Fig. 1.10, the viewpoint was expressed that the glass transition is reasonably interpreted as an equilibrium thermodynamic transition (though modified by kinetic effects). We now discuss the ambitious effort of Cohen and Grest (1979) to extend the free-volume theory to cover thermodynamic behavior and the glass transition itself. Their theory addresses the entropy of the system and makes important use of percolation theory.

Inherent in the segmented form used for the local free energy function $f(v)$ of Eq. (5.4) and Fig. 5.6 is a definite dichotomy between cells of volume $v < v_c$ and those of volume $v > v_c$. Cells with $v < v_c$ are in the quadratic oscillatorlike regime of $f(v)$ and are termed *solidlike*; cells with $v > v_c$ are in the linear regime and are termed *liquidlike*. Only liquidlike cells have free volume. The solidlike/liquidlike distinction drawn here ascribes to individual cells the previously discussed vibrational/diffusive distinction between panels *a* and *b* of Fig. 1.7 and hard-sphere molecules C and A of Fig. 5.5.

This division of the cells into two classes leads in a natural way to the introduction of percolation concepts. I illustrate this in a highly schematic manner in Fig. 5.7, using the geometric context of the 2d square lattice for simplicity. (A more realistic sketch would need to include the essential aspect of the distribution of cell sizes and shapes, and would resemble the Voronoi froth of Fig. 2.5.). Although there are three separate groups of liquidlike cells in Fig. 5.7, only *one* liquidlike *cluster* is identified in the figure. The reason for this is that a percolation process slightly different from classical percolation is involved here.

In order for the isolated liquidlike cell at the upper right of Fig. 5.7 to change its volume, it is necessary that there be a compensating net change in the volume of the surrounding solidlike cells. Since the liquidlike cell is in the linear regime of $f(v)$ while the solidlike cells are in the quadratic regime, this entails a change in the sum of the local free energies. Because the total energy changes when the volume of this cell changes, the isolated liquidlike cell is not free to exchange volume with its neighbors. Similarly, repartition of free volume between the two neighboring liquidlike cells at the lower left of Fig. 5.7 necessarily involves changes in the surrounding solidlike cells, and this generally entails an energy cost.

In order for a pair of liquidlike cells to engage in a free exchange of free volume with each other, they must be nearest neighbors *and* they must have enough neighboring liquidlike cells to ensure that no solidlike cells are forced to change also. This is the case for the two cells, shown connected by an arrow denoting such an exchange, in the center of Fig. 5.7. Such cells will be said to belong to the same *liquidlike cluster*. This situation, in which clusters must satisfy a connectivity requirement which exceeds that for ordinary percolation, is called *high-density percolation* and was discussed at the end of the last chapter. In a lattice of coordination z, an occupied site is considered to belong to a cluster only if it is adjacent to m or more occupied sites (where m > 1; $m = 1$ corresponds to ordinary percolation). The schematic example illustrated in

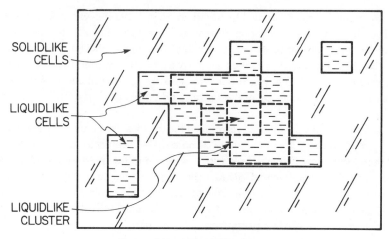

Figure 5.7 A liquidlike cluster.

Fig. 5.7 is for $z = 4$, $m = 3$. It exhibits one liquidlike cluster of eight cells, shown outlined by a dashed line. Within this cluster, repartition of free volume takes place freely because, owing to the linearity of $f(v)$, a gain in energy of one cell is exactly cancelled by a loss in energy of a neighboring cell.

As anticipated at the end of Section 5.1, the way in which percolation enters in the free-volume view of the glass transition is the *reverse* of the way that it has entered in, say, the gelation process discussed earlier. In gelation, attention is focused upon linked clusters of the structural units (the "particles" in the system), while in Fig. 5.7 we are interested in clusters of the low-density relatively empty liquidlike cells (roughly speaking, the "holes" in the system). Role reversals such as this appear often in physics. One nice example is T. D. Lee's model of quark confinement in a nucleon as the analog of an inside ↔ outside mapping of flux exclusion from a superconducting sphere.

Let p denote the fraction of liquidlike cells and p_c denote the percolation threshold for the appropriate percolation problem (high-density percolation, with $m \approx z/2$ a reasonable estimate). If $p(T)$ denotes the temperature dependence of the liquidlike fraction, the condition $p(T) \approx p_c$ may be roughly regarded to locate the glass transition. For $p < p_c$, only finite liquidlike clusters are present and the amorphous system is a solid. For $p > p_c$, an infinite liquidlike cluster pervades the system and provides opportunities for long-range molecular transport. However, as we know from the last chapter, the fractal wispiness of the percolation path just above threshold softens the character of the transition.

The intricate mathematical development of the Cohen–Grest free-volume thermodynamic theory of the glass transition cannot be treated here in depth. However, some features of their theory will now be sketched. One main point is an explicit calculation of the *communal entropy* associated with the liquidlike clusters.

Communal entropy is an entropy measure associated with *diffusional* (as opposed to vibrational) contributions to density fluctuations. It is defined as the difference in entropy between a system of molecules and a structurally related reference system in which the molecules are confined to individual cells. For a system of hard-sphere molecules such as that shown earlier in Fig. 5.5, the reference system is the random-close-packed amorphous solid and the cells are the Voronoi polyhedra of the rcp froth of Section 2.4. When the density of the molecular system in question is nearly as great as that of the reference solid, diffusional contributions are negligible and the communal entropy vanishes. When the density of the system is gradually reduced, communal entropy develops until, when the system is liquid, it becomes substantial. The development of the communal entropy is a basic aspect of the glass ↔ liquid transition.

In the Cohen–Grest theory, communal entropy enters via the liquidlike clusters. (Actually, in what follows, clusters termed "liquidlike" are those which satisfy an additional mild requirement: The total free volume within the cluster exceeds v_f^*. The quantity v_f^*, introduced earlier among the basic assumptions of the free-volume picture, is the minimum amount of free volume needed for a molecule to undergo a diffusive step.) Within a liquidlike cluster, such as the eight-molecule example indicated in Fig. 5.7, each molecule *moves in time throughout the entire cluster*. Each molecule has accessibility to the space of the full cluster, i.e., there is communal sharing of the cluster volume. The number of ways of distributing the free volume among the cells of the cluster represents its contribution to the communal entropy of the system. Clearly the communal entropy arises from the *delocalization* of the molecules out of their individual cells and their consequent movement throughout the communal volume of the cluster. The communal entropy thus depends on the distribution of liquidlike clusters, and an important contribution comes from the infinite (percolating) cluster when it is present.

Ingredients of the statistical mechanical treatment of Cohen and Grest (1979, 1981) include, *inter alia*, the local free-energy function $f(v)$ of Fig. 5.6, an analytic approximation for the percolation cluster-size distribution $n(s)$ [$n(s)$ is the number of liquidlike clusters of size s, $s = 8$ for the example of Fig. 5.7], and the percolation critical exponent β. Central roles are played by p, the fraction of liquidlike cells, and $N(v)$, the distribution of cell volumes. $N(v)dv$ is the number of cells having cell volumes in the range from v to $v + dv$. The relationship between p and $N(v)$ is

$$p = N^{-1} \int_{v_c}^{\infty} N(v)dv, \qquad (5.6)$$

where N is the total number of cells (molecules). An expression for the total free energy F is constructed which includes the term $\int N(v)f(v)dv$ as well as the term $-TS_{\text{com}}$, where S_{com} is the communal entropy. It is S_{com}, with its dependence upon the statistics [$n(s)$] of the liquidlike clusters, which incorporates the infor-

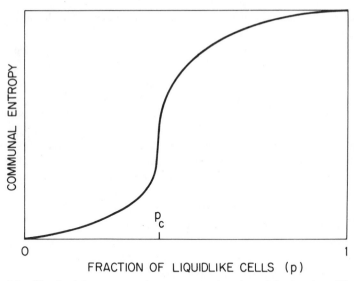

Figure 5.8   Sketch of the communal entropy as a function of the fraction of liquidlike cells in the Cohen–Grest percolation formulation of the glass transition.

mation from percolation theory. A sketch of $S_{com}(p)$, showing the qualitative behavior of the communal entropy as a function of the fraction of liquidlike cells, is shown in Fig. 5.8. There is a definite ''event'' at $p = p_c$. With decreasing p, S_{com} drops steeply when the percolating liquidlike cluster breaks up. The slope at this point depends on the percolation critical exponent β, and diverges for $\beta < 1$. For ordinary percolation, we know from the preceding chapter that β is 0.4 in three dimensions.

In order to proceed from $S_{com}(p)$ to an equilibrium theory of the glass transition, Cohen and Grest determine the temperature dependence $p(T)$ by performing a sophisticated optimization of the free energy F which yields a self-consistent condition for p. For the form of $S_{com}(p)$ indicated in Fig. 5.8, they find a first-order transition: with decreasing temperature, p decreases and then jumps discontinuously from a value above p_c to one below p_c at a critical temperature. This result is not entirely satisfying since, as described in Chapter One, the bulk of the thermodynamic behavior observed near T_g is suggestive of a second-order transition. One aspect of the resolution of this difference resides in the softening of the sharp distinction between liquidlike and solidlike cells, since the segmented form adopted for $f(v)$ is an approximation for the smooth curve of Fig. 5.6. This may have the effect of removing the infinite slope of S_{com} at p_c, resulting in a second-order transition. We also know that kinetic effects are important near T_g. Above T_g, p decreases continuously with decreasing T. Now in order for p to change, an exchange of volume must occur between liquidlike and solidlike cells, which is not a free exchange but instead involves an energy cost. Cohen and Grest therefore argue that $p(T)$ falls out of thermodynamic equilibrium below T_g, and they show that the inclusion of relaxation effects in their theory can successfully account for the observed thermodynamic properties and such effects as $T_g(\dot{T})$, the shift of T_g with quench rate. Their analysis finds that the observed dispersion of relaxation times is associated with the dispersion of surface-to-volume ratios of the liquidlike clusters.

5.5 ELECTRON STATES AND METAL ↔ INSULATOR TRANSITIONS

We now turn to the other fundamental localization ↔ delocalization phenomenon in the physics of amorphous solids. This concerns the motion, not of atoms, but of electrons. The topic of disorder-induced electron localization is even deeper than that of the atomic-motion case (glass transition) discussed in the preceding sections. Where the liquid ↔ glass transition could be visualized largely in a classical picture of the molecular movements (quantum mechanics entered mainly in determining the intermolecular interaction which underpinned the assumed form for the cell free energy), the electronic phenomenon is intrinsically quantum mechanical in character. Moreover, important features of this phenomenon represent conceptual departures from the traditional quantum-mechanical treatment of electrons in crystalline solids.

In order to introduce some of the main ideas, and to place the issue of disorder-induced localization in the context of other basic electronic phenomena, we compare in Table 5.1 three types of atomic-scale metal ↔ insulator transitions.

The first class of transition considered in Table 5.1 is one which is encompassed within the Bloch or band-theory framework for electronic states in crystalline solids. A brief reminder of basic elements of band theory is given in the next paragraph, phrased for simplicity in one-dimensional form. (The three-dimensional generalization is straightforwardly obtained; e.g., replace kx by $\mathbf{k} \cdot \mathbf{r}$, etc.)

Band theory is a one-electron independent-particle theory which assumes that there exists a set of stationary states available to any one electron, and that all of the electrons are distributed among these states according to Fermi–Dirac statistics. These states ψ_{nk} are given by the solutions of a Schrödinger equation $H\psi_{nk} = E_{nk}\psi_{nk}$ in which the Hamiltonian operator H includes, in addition to the electronic kinetic energy term $p^2/2m = -(\hbar^2/2m)(d^2/dx^2)$, a crystal potential term $V(x)$ which is intended to account for the interaction of one electron with all of the other particles in the crystal. Since V (and therefore H) is periodic in space with the translational periodicity of the crystal structure, i.e., $V(x + a) = V(x)$, where a is the lattice constant, it then follows that the wave functions $\psi_{nk}(x)$ are Bloch functions of the form $e^{ikx}u_{nk}(x)$. The function u_{nk} is periodic, $u_{nk}(x + a) = u_{nk}(x)$, and modulates the plane-wave part of the wave function e^{ikx}. The quantum numbers characterizing each wave function are the wave-vector k ($-\pi/a < k < \pi/a$), an integer band index n, and the energy eigenvalue E_{nk} or $E_n(k)$. The functional form of the dependence of $E_n(k)$ upon the crystal momentum ($\hbar k$) for each

Table 5.1   The Anderson transition in the context of other metal ↔ insulator transitions

| | Electron Wave Functions | | | | |
	Metal side of Transition	Insulator side of Transition	Characteristic Energies	Change at the M → I Transition	Criterion for Localization
Transition					
Bloch	Extended	Extended	Bandwidth B	Partly filled bands → all bands filled or empty	—
Mott	Extended	Localized	Electron–electron (e^2/r_{ij}) correlation energy U	Correlation-induced localization	$U > B$
Anderson	Extended	Localized	Width W of the distribution of random site energies	Disorder-induced localization	$W > B$

band (n) specifies the electronic energy band structure of a crystalline solid. Each band can accommodate $2N$ electrons, where N is the number of unit cells in the crystal and the factor 2 arises from the spin degeneracy.

In the Bloch (or, more properly, the Bloch–Peierls–Wilson) theory of electrons in crystals, a solid is an insulator if each band is either completely filled or completely empty, and it is a metal if at least one band is partly filled. At zero temperature (which is often assumed here for the purpose of making a perfectly sharp metal/insulator distinction), all states lying lower in energy than the Fermi energy E_F are occupied by electrons, while all states lying higher than E_F are empty. For a metal, the band structure and the number of electrons is such that E_F lies within a band, which is thus only partly filled. For an insulator, E_F lies between bands and there is an energy gap separating the highest-lying filled (valence) band and the lowest-lying empty (conduction) band.

The type of metal ↔ insulator (M ↔ I) transition envisioned in the first row of Table 5.1 is illustrated by the schematic band structure shown in Fig. 5.9. A crystalline material, composed of atoms with an even number of electrons sufficient to populate an integral number of bands, is close to the borderline case in which the bands bracketing the Fermi level either just overlap in energy (as in Fig. 5.9a) or just miss overlapping (as in Fig. 5.9b). A small change in pressure or temperature may then cause the crystal's band structure to cross over to the other situation. Such a band-overlap M ↔ I transition occurs in ytterbium, in which a variation in pressure produces a crossover between band structures of types a and b of Fig. 5.9. It is interesting to note that the familiar "semimetal"

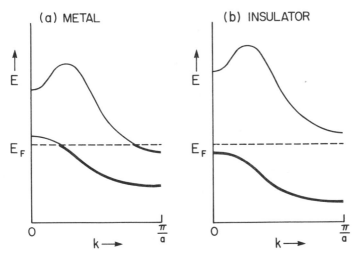

Figure 5.9   Within the Bloch theory for electrons in crystals, a metal → insulator transition can occur via a change in band structure which removes the overlap between filled and empty bands.

graphite lies, at STP, just on the metallic side of this border, but there is no way of removing the overlap to open up a gap as in *b*.

As indicated in the first row of Table 5.1, on *both* the metallic and the insulating sides of the above type of transition, the electronic wave functions in the vicinity of E_F correspond to *extended* states. Such wave functions have appreciable amplitude throughout the solid. For crystals, these are Bloch functions and have the form sketched in Fig. 5.10a. The solid line in the figure represents the real part (or imaginary part) of ψ, while the dashed line indicates the plane-wave envelope corresponding to the wave-vector eigenvalue k. For the particular Bloch function shown in Fig. 5.10a, k is approximately $0.3(\pi/a)$, corresponding to a wavelength $(2\pi/k)$ a bit under seven lattice constants.

For amorphous solids, k is not a good quantum number, since the validity of Bloch-function solutions depends upon the presence of crystalline long-range order and disappears in the absence of such order. Long-range order is, of course, totally absent in amorphous solids. $E_n(k)$ energy band diagrams, such as those indicated in Fig. 5.9, have no meaning for electronic states in glasses. This is an obvious way in which band theory breaks down for amorphous solids. Although extended electronic states are present and important in these solids, they cannot be characterized by quantized wave-vector values as in crystals.

The type of metal ↔ insulator transition most relevant to amorphous solids is the *Anderson transition*, which is listed in the last row of Table 5.1 and which will be described primarily in the following section. Here the insulating side of the transition corresponds, not to extended states in filled bands as in a Bloch-type transition in crystals, but instead to electron states which are *localized*. The meaning of a localized state is indicated by Fig. 5.10*b*. The wave function is concentrated near a center composed of just a few atoms, and has negligible amplitude elsewhere in the solid. Away from the small region that contains essentially all of its integrated probability $\int |\psi|^2 dr$, the amplitude spatially decays away *exponentially* with distance. This behavior is schematically shown by the dashed-line wave-function envelope in Fig. 5.10*b*, which falls off as $e^{-\alpha R}$ at large distances R from the localization center. The quantity α, an important parameter for a given localized state, is known as the *inverse localization length*, for clear physical reasons.

In crystalline solids, localized states are, of course, often introduced by chemical impurities. A familiar example is the case of donor impurities in semiconductor crystals such as phosphorus atoms in crystalline silicon, in which hydrogenlike bound states are associated with the impurity atoms at discrete energies within the host crystal's energy gap. These states are *extrinsic* to the host solid (the perfect crystal, without chemical impurities or structural defects), all of whose intrinsic states are extended Bloch functions. By contrast, we shall see that disorder-induced localized states are *intrinsic* to amorphous solids, and that their energy levels form a *continuous spectrum* rather than a discrete line spectrum as in extrinsic semiconductor crystals.

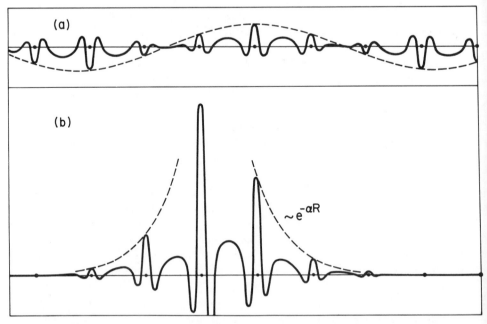

Figure 5.10   The distinction between extended and localized electron states. A Bloch-type extended-state wave function is illustrated in *a*; a localized-state wave function is illustrated in *b*.

Before proceeding to a discussion of disorder-induced localization and the Anderson transition, it is useful to introduce the concepts involved in the type of M ↔ I transition which is entered in the middle row of Table 5.1 and which is identified with a few of the many deep insights that have been contributed to condensed-matter physics by Sir Nevill Mott. The *Mott transition* (Mott, 1949, 1974) is both philosophically and physically related to the Anderson transition. Philosophically, both transitions are dramatic demonstrations of situations in which the conventional theory (Bloch functions, band structures, etc.) used for electrons in crystals breaks down. Band theory, often successful in explaining the electrical and optical properties of crystalline solids, fails completely (i.e., *qualitatively*) for these situations. Physically, the two are related in that there exist solids in which both effects are present. An interesting case for which this is so, phosphorus-doped silicon (Si : P), is described in the next section.

The Mott transition illuminates a regime in which the one-electron or independent-particle theory of solids fails, for clear physical reasons, within a crystalline context. Consider a monatomic solid composed of atoms which, when isolated, contain a single electron outside of a closed-shell core. Examples are the alkali metals. [A nice description of Mott-transition issues, using Na as an example, has been given by Adler (1982).] In a quantum-mechanical calculation for electronic states in a crystal, standard procedure is to assume that the atomic nuclei are fixed in position on the sites of a crystalline lattice. (Nuclear motions about the equilibrium lattice positions, i.e., phonon excitations, are put in later.) Carried out properly for an alkali, using the known bcc lattice constant a_0 of the actual crystal, a band-structure calculation reproduces reasonably well the properties of the metallic solid. The crystal is a metal because the highest band is only half filled, since there is only one electron per unit cell and thus only N electrons present to occupy the $2N$ states available in that band.

Mott (1949), however, asked a basic question about the implications of such a calculation carried out at *different* values of the lattice constant a. A gedanken experiment familiar to all condensed-matter physicists is a gradual condensation process in which the atoms are brought together ''from infinity'' to the observed solid structure, through a hypothetical set of structures having all possible lattice constants between $a = \infty$ and $a = a_0$. Suppose we set a at a very large value, say one meter, and repeat the calculation. As translational periodicity is retained, the one-electron solutions are still Bloch functions, the highest band is still half filled, and the material is still predicted to be a metal. But this is obviously silly. What we are clearly dealing with here is a set of *isolated* atoms, and the true solutions must be just the atomic solutions. Thus, using the example of sodium, the correct result should correspond to an assembly of neutral and noninteracting Na atoms, with the electrons of interest to us occupying the outermost ($3s$) orbitals—*one* on each atom. Configurations having *neither* or *both* $3s$ orbitals occupied on any atom, which necessarily appear if the solutions are Bloch functions, *cannot* occur. They would cost a large amount of energy, since such a transfer of an electron between a pair of isolated neutral atoms ($Na + Na \rightarrow Na^+ + Na^-$) requires an energy increase equal to $I - A$.

I and *A* are, respectively, the ionization energy required to remove the outer electron from a neutral atom (Na → Na$^+$ + e^-) and the electron affinity which is the much smaller energy regained by attaching the electron to another atom (Na + e^- → Na$^-$).

The above argument shows that band theory fails, in the atomic limit ($a \to \infty$), for a system with a half-filled highest band. The physical reason for this failure is disclosed with the aid of Fig. 5.11. Shown here are two electronic configurations which, in the independent-electron picture, have the same potential energy. In configuration *b* of Fig. 5.11, two of the outermost or valence electrons (the 3s electron contributed by each Na atom, in the sodium case) have been shifted from their positions in *a* to translationally equivalent positions in another unit cell. Because of the translational periodicity of the crystal potential $V(r)$, which approximates the *average* interaction of each electron with all of the 10^{24} other charged particles (*including* all of the other electrons) in the solid, configurations *a* and *b* have the same potential energy in the independent-particle picture. This is manifestly unreasonable. Certainly *b* has a substantially higher energy than *a* because of the repulsive Coulomb interaction (e^2/r_{12} for each pair, where r_{12} is the separation of electrons 1 and 2) among the valence electrons. This energy cost, associated with the electronic crowding shown in *b*, is a "many-body effect" which is left out of the independent-particle picture of one-electron theory.

One-electron theory is therefore unable to discriminate against configurations such as Fig. 5.11*b*, since it does not perceive their extra energy vis-à-vis

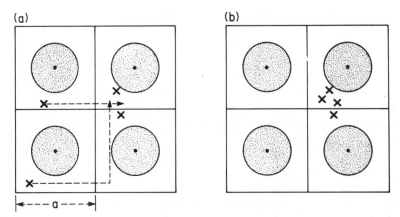

Figure 5.11 Two electronic configurations within a four-unit-cell region of a crystal. The shaded circles denote the ion cores; the crosses denote the positions of the valence electrons. Although configuration *b* certainly has higher energy than *a* because of its greater electron–electron Coulomb energy, both configurations are considered to have the same energy within the independent-electron approximation, in which all that matters is the position of each electron within its unit cell. Electrons in a real solid *correlate* their motions to reduce the frequency of occurrence of energetic configurations such as *b*.

configurations such as Fig. 5.11a. In the real solid, however, the electrons definitely do tend to *correlate* their movements in order to avoid such energetically unfavorable mutual proximity as in Fig. 5.11b. The amount by which band theory *overestimates* the ground-state energy of the system because of this *neglect of electron correlations* is called the *correlation energy*. The point embodied by the Mott-transition concept is that the correlation energy can, under certain conditions, cause a solid to have an insulating ground state when the neglect-of-correlation band picture would erroneously predict it to be a metal. Such a solid (an example is NiS_2) is called a *Mott insulator*, and the condition required for its occurrence is displayed in Fig. 5.12.

In this figure, the atoms are represented by potential wells and the single valence electron of each atom occupies, in the isolated-atom limit, a bound-state energy level which is indicated on each atomic well by a horizontal line. In the crystal, this level gives rise to an energy band of bandwidth B, as sketched on the left. Relative to the energy level of the free atom, the crystal band extends in energy from roughly $-B/2$ to $+B/2$. Since the band is only half filled, the average energy of a valence electron in the crystal is roughly $-B/4$. This lowering in energy, relative to the free atom, is responsible for metallic cohesion. It arises from a lowering of kinetic energy which is achieved by the delocalization of the electrons into extended states in the crystal. Delocalization smooths the wave-function oscillations and thus reduces the ∇^2 kinetic-energy

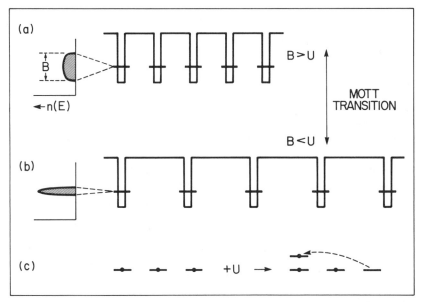

Figure 5.12 Schematic picture for the Mott transition. When the electron bandwidth B is decreased (by increased atom–atom separation) sufficiently to be smaller than the intrasite electron–electron energy U, correlation-induced localization takes place.

contribution to the total energy. This energy lowering is relatively ineffective when there is little overlap between orbitals on neighboring atoms, since then there is little opportunity for smoothing to be accomplished by interference between wave-function regions of opposite curvature which arise from different atoms. B thus decreases with increasing lattice constant a, as suggested in Fig. 5.12 by the difference between a and b. Of course, the bandwidth vanishes in the $a \to \infty$ atomic limit.

Now let us take into account the correlation energy which we know to be omitted from the band-theoretical viewpoint. The energy cost of doubly occupying an atom with a pair of valence electrons, which equals the average Coulomb energy $\langle e^2/r_{12} \rangle$ when both electrons 1 and 2 are in orbitals on the same atomic site, is denoted by the symbol U. Its significance is given in Fig. 5.12c. In Figs. 5.12a and 5.12b a single horizontal line at each well denoted the energy level of the valence electron of the isolated atom. This energy level is doubly degenerate, and can accommodate two electrons of opposite spin. The same level can be occupied by zero or one or two electrons. But this view *neglects* the e^2/r_{12} interaction. In the case when *two* electrons occupy the same site, we know that this repulsive interaction is present and raises the energy. This occupation-number-dependent feature introduced by e^2/r_{12} is represented by the equivalent two-level diagram of Fig. 5.12c. Two energy levels are associated with each site. The lower level, which corresponds to the level used in Figs. 5.12a and 5.12b, is available to the first electron to occupy the atom. The upper level, which really exists only when the lower level is filled, is lifted by the amount U and is available to the *second* electron to occupy the site. Of course, the assignment of "first" and "second" electrons when two are present is a bookkeeping artifact, since the positive energy U is contributed by their mutual e^2/r_{12} interaction. As discussed earlier, $U = I - A$ in the atomic limit.

In the independent-electron treatment which neglects the existence of U, the wave functions are Bloch functions so that each electron appears with equal probability at any site, independent of whether or not the other electron state at that site is filled. The probabilities of occurrence of empty, singly occupied, and doubly occupied sites are therefore $1/4$, $1/2$, and $1/4$, respectively. Recognizing that each doubly occupied site does indeed increase the energy by U, we see that the use of delocalized wave functions, and the consequent failure of the electrons to maximally avoid each other, introduces an average *potential-energy cost* per electron of $U/4$. Since we have seen that the average *kinetic-energy incentive* for band formation is about $B/4$, it follows that *the correlation cost exceeds the delocalization gain if $U > B$.*

This, then, is the condition for the occurrence of a Mott insulator:

$$U > B \quad \text{(correlation-induced localization).} \qquad (5.7)$$

If this inequality criterion is satisfied, the electrons are localized. Unlike the Bloch type of metal ↔ insulator transition (in which extended states apply to both sides), the Mott transition is a delocalization ↔ localization transition for the electron states. It occurs, as implied by $a \leftrightarrow b$, Fig. 5.12, when a change in interatomic separation causes a crossover in the relative importance of the two characteristic energies of the valence-electron system: the bandwidth B and the electron–electron correlation energy U.

The Mott transition is an exemplary embodiment of a recurrent theme in condensed matter physics: *Electron localization in the low-density limit signals the triumph of potential energy over kinetic energy in that regime.* An earlier version of this theme is the idea of the Wigner lattice. Wigner proposed that a low-density low-temperature electron gas would crystallize, with the electrons becoming localized near the sites of a lattice. Consider electrons moving within a uniform "jellium" background of neutralizing positive charge. If they are localized on a lattice of lattice spacing a, the average potential energy of each electron (negative because of the net binding which arises from the attractive interaction with the surrounding positively-charged jellium) is of order $-e^2/2a$. The kinetic-energy cost of localizing the electrons may be estimated from the Heisenberg uncertainty principle. A confinement length of order a implies a momentum spread of order $\hbar/a$, corresponding to a kinetic energy of order $(\hbar^2/2m)/a^2$. Since the potential energy scales with $1/a$ while the kinetic energy scales with $1/a^2$, the potential energy becomes the dominant factor in the $a \to \infty$ low-density limit. The kinetic-energy cost of localization becomes negligible in comparison to the Coulomb interactions, and the electrons form a lattice to maximally avoid each other in order to minimize the totally dominant potential energy.

The converse of the proposition stated above is that in the opposite *high-density* limit ($a \to 0$), it is the kinetic energy which dominates. Wave-function smoothing and extended states are the order of the day, and the neglect of correlation becomes exact with e^2/r_{ij} being of no importance. This is the rationale for the quest for a metallic solid form of hydrogen at very high pressures. Suspected to be an important constituent of the cores of the large planets, but thus-far unrealized (as of this writing) on earth, metallic hydrogen is the motivation for high-pressure research aimed at achieving the high-density regime.

5.6 DISORDER-INDUCED LOCALIZATION: THE ANDERSON TRANSITION

With the aid of the background provided by the preceding section, it should be possible to grasp the main aspects of disorder-induced localization, even though the mathematics involved in the derivation of Anderson's theorem is rather formidable and can only be presented here in outline form. In a format similar to that used in Fig. 5.12 for the Mott transition, an illustration of the Anderson transition is shown in Fig. 5.13. As in the case of the Mott and Wigner transitions, electron localization in the Anderson transition reflects the passage to a regime in which the potential energy wins out over the kinetic energy as the factor dominating the form of the wave functions. While the $a \to b$ delocalization $\to$ localization transition in Fig. 5.12 occurs via the weakening of kinetic-energy considerations with decreasing bandwidth at low densities, in the Anderson case of Fig. 5.13 it occurs via *the superposition of a disorder-induced potential energy of sufficient strength*. (In fact, as will be discussed shortly, there *is* an effective low-density aspect of Anderson localization which recovers a resemblance to the Mott transition.)

In Fig. 5.13*b*, the potential wells representing the atomic sites are no longer all the same. Instead, the well depths vary from site in a random way. Such a disordered potential is present in an amorphous solid. Because of the topological disorder characteristic of such a solid, all sites are different in a glass. Instead of a single well depth (and bound-state level) as in the crystalline case of Fig. 5.13*a*, there is a *distribution* of well depths (and corresponding levels) in the amorphous case schematicized in Fig. 5.13*b*. The width of this distribution, which specifies the energy range of the disorder-induced spatial fluctuations of the potential energy seen by an electron at the atomic sites, is denoted by *W*.

The competition between kinetic-energy and potential-energy influences on the electron states now resides in the ratio *W/B*. *W*, the magnitude of the random potential, and *B*, the (crystal) bandwidth in the absence of disorder, are the relevant characteristic energies. The essential point was mathematically demonstrated in his famous paper entitled ''Absence of Diffusion in Certain Random Lattices'' (Anderson, 1958). Anderson showed that when the dimensionless *disorder parameter W/B* is sufficiently large, *all* of the states in the valence

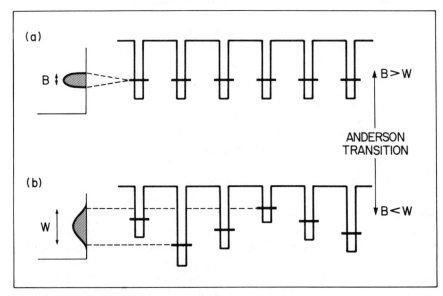

Figure 5.13 One-electron tight-binding picture for the Anderson transition. When the width *W* of the disorder exceeds the overlap bandwidth *B*, disorder-induced localization takes place.

electron band are localized. The criterion for localization listed in Table 5.1 omits a not-well-pinned-down numerical factor of order unity:

$$W > B \quad \text{(disorder-induced localization)}. \tag{5.8}$$

Anderson's model, as sketched further on, is a quantum-mechanical calculation that invokes a tight-binding independent-electron picture which corresponds to Fig. 5.13*b*. The *independent electron* aspect is important; the calculation employs a random potential, like that sketched in the figure, in a *one-electron* Hamiltonian. Unlike the Mott insulator, in which extended states predicted by an independent-electron picture are overturned when electron–electron correlation is introduced, *Anderson localization appears as a consequence of disorder in a purely independent-electron picture.*

The notion of disorder-induced localization was subsequently extended by Mott and others (Mott, 1968; Cohen, Fritzsche, and Ovshinsky, 1969) in a way that has strongly influenced current thinking about electronic states in amorphous semiconductors. When the disorder is great enough for W/B to satisfy Anderson's criterion, *all* of the states in the band become localized. Mott pointed out that, even for smaller degrees of disorder, *states in the tails of the band are localized.* This feature of electrons in glasses is indicated on the density-of-states diagram shown in the lower half of Fig. 5.14. The energies for which states are localized (the shaded regions) correspond to the tails at the top of the valence band and bottom of the conduction band. Within the main body of each band, the states are extended. The extended/localized distinction refers to the essential difference illustrated earlier in Fig. 5.10. In Fig. 5.14*b*, the demarcation energies separating regions of localized and extended states are referred to as *mobility edges.* The Anderson *transition* then refers to a localization ↔ delocalization transition in which a change (in composition, pressure, applied electric field, etc.) pushes the Fermi level *through* such a mobility edge.

The reason that states in the wings of the distribution of Fig. 5.13*b* are especially susceptible to localization may be seen in several ways. One way to see this (Thouless, 1974) is by, essentially, a percolation argument. In fact, the situation envisaged amounts to polychromatic percolation, the percolation generalization described near the close of Chapter Four. To make use of this argument, we need to anticipate a specific feature of quantum transport among the disordered sites. An electron may move with relative ease from one site to a nearby one only if the energy levels of the two sites differ by an amount that is less than a small fraction of the (overlap-determined) bandwidth, an amount roughly given by B/z, where z is the coordination number of the lattice. Sites differing in energy by a greater amount are effectively decoupled. Suppose we slice the distribution W into discrete energy ranges of width B/z, and label (or "color") each site by the energy range to which it belongs. Each site can communicate with a nearby site of the *same* energetically specified type (i.e., of the same color), but *cannot* communicate with any site of another type. The situation now resembles a polychromatic percolation process, such as that shown earlier in Fig. 4.23 for three colors, i.e., for three species of sites.

Let us use the case shown in the lower panel of Fig. 4.23 to illustrate the point, which is quite simple. For a species of site corresponding to an energy slice taken from the fat central part of the density of states, there is an abundance of available sites, their spatial concentration is high, and percolation is easy (squares and circles in Fig. 4.23). But for a color corresponding to a slice taken from a thin tail of the distribution, the available sites are sparse and are spread far apart in space. A percolation path is absent (crosses in Fig. 4.23), and localization is likely. This conveys the reason that disorder most readily localizes electrons in the extremities of the density-of-states distribution, as shown by the shaded region in Fig. 5.14. It is no accident that, interpreted in this way, Anderson localization appears as a *low-density regime* (distribution tail = low concentration in space) in which the potential energy (here, the site-decoupling disorder-induced potential) enforces electron localization. Viewed from this direction, the family resemblance to the Mott and Wigner cousins of the preceding section is plain to see.

Before leaving Fig. 5.14, the lower part of which has been used to show the meaning of mobility edges in amorphous semiconductors, we take the opportunity to compare a few overall aspects of the electronic structure of covalently bonded crystals and glasses. The upper part of this figure illustrates the form which the density-of-states function $n(E)$, the number of electron states per unit energy per unit volume, typically takes for a crystalline solid. Note, first of all, the sharp corners (discontinuities in slope) which are characteristic of a crystal's electronic spectrum. For a crystal, $n(E)$ is directly derived from the band structure $E(\mathbf{k})$ by simply counting the states in $\mathbf{k}$ space. It is an essentially geometric consequence of this counting procedure that sharp structure in $n(E)$ arises from the inevitable presence of special places in $\mathbf{k}$ space at which the gradient $\nabla_{\mathbf{k}} E(\mathbf{k})$ vanishes. This occurs whenever $E(\mathbf{k})$ has a local maximum or a minimum or a saddle point. Since $\mathbf{k}$ itself, as a quantum number labeling each electron eigenstate, depends for its validity upon translational periodicity, it follows that *the sharp structure in n(E) is a crystal property which requires long-range order for its existence.* In the absence of long-range order, these sharp features in $n(E)$ disappear. Hence the difference in this respect between the curves shown in the upper and lower parts of Fig. 5.14.

The emphasis bestowed above with the use of italics is intended, so to speak, to give the devil his due. Here is a case of a specific property of condensed matter which *cannot* be understood on the basis of short-range order alone, but which instead really requires long-range order for its explanation. (Other such properties of crystalline solids, mainly spectroscopic in character, are mentioned in the final chapter in connection with selection-rule breakdown in glasses.) It is, of course, a main theme of this book that short-range order suffices to account for *many* essential aspects of solids, both crystalline and amorphous. In the context of Fig. 5.14, it is therefore appropriate to point out that the *overall* form of $n(E)$ is indeed mainly determined by the short-range order. In covalent semiconductors such as crystalline and amorphous silicon, the average gap between the valence and conduction bands (i.e., the energy

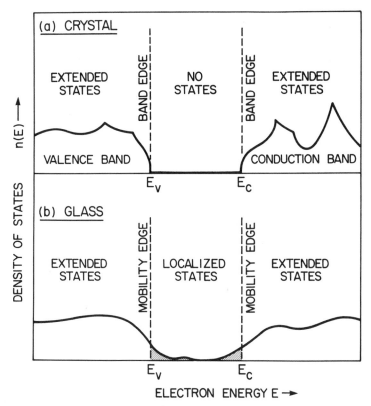

Figure 5.14 Schematic density-of-states diagram for a crystalline and an amorphous semiconductor, in the vicinity of the highest occupied and lowest empty states. $n(E)dE$ is the number of electron states, per unit volume, with energies in the interval from E to $E + dE$. With respect to many electrical properties, the mobility edges in the amorphous solid play a role analogous to that played by the band edges in the crystal.

difference between the centers of gravity of the two bands) is determined by the bonding–antibonding splitting of the states associated with the covalent bonds connecting nearest neighbors of the fourfold-coordinated network.

One final aspect of Fig. 5.14 requires comment. In the crystalline density of states of the upper diagram, there is an energy gap ($E_G = E_c - E_v$) between the top of the valence band (E_v) and the bottom of the conduction band (E_c). This energy gap (for a perfect crystal) is completely clear of states; it corresponds to a forbidden energy zone such as the one which separates the bands in Fig. 5.9*b*. In principle, these energies can be filled in for an amorphous solid, since the sharp band edges of the crystal (which are simply special cases of the sharp structure discussed above) need not persist in the presence of disorder. In the parlance, the gap of the crystal, in which $n(E) = 0$, is replaced in the glass by a "pseudogap" in which $n(E)$ is merely very small. It turns out that, for most continuous-random-network amorphous solids, the band edges are rather well defined and the disorder-induced tails (which are exaggerated in Fig. 5.14*b*) do not extend very far. One way to (literally) see this is simply to observe the outstanding transparency of optical glasses. (It should also be noted, however, that optical transitions between localized states are very weak, which lessens the optical impact of the tails.) Nevertheless, these disorder-induced density-of-states tails are present in glasses and contain, up to their mobility edges, localized states. The energy separation between the mobility edges of Fig. 5.14*b* is called the *mobility gap* (Cohen, Fritzsche, and Ovshinsky, 1969). It plays a role, with respect to the electrical properties of an amorphous semiconductor, which is similar to the role played by the energy gap in a crystalline semiconductor.

Anderson localization emerges from an intimate mix of quantum mechanics and strong disorder. Before returning to a Hamiltonian-based discussion of the full problem, let us attempt to isolate the aspect of disorder. Our favorite method of demonstrating consequences of disorder is, of course, the percolation model. We have already made use of one set of percolation ideas, in connection with the mobility edge. Polychromatic percolation, with the sites "colored" according to their well-depth energies in discrete intervals determined by overlap, was useful in aiding understanding of disorder-induced localization in the band tails. Now we consider a percolation picture which can be viewed as a *classical* limit of the Anderson transition (Zallen and Scher, 1971; Ziman, 1979).

To elucidate the topological element of the problem, we consider the motion of a classical particle in a random (but not wildly fluctuating) potential $V(\mathbf{r})$. Figure 5.15 represents a portion of a two-dimensional potential in the form of a topographic or contour map. The contour lines denote equipotentials of $V(\mathbf{r})$. For a (classical) particle of energy E, regions of space for which $V(\mathbf{r}) > E$ are completely forbidden. Allowed regions, for which $V(\mathbf{r}) < E$, are shown shaded in Fig. 5.15. Panels *a, b,* and *c* of the figure correspond to progressively larger values of the particle energy.

What happens with increasing E is immediately seen with the aid of a "great flood" analogy (Zallen and Scher, 1971; Zallen, 1979). Interpret each panel of Fig. 5.15 as an aerial photo of a part of a planet whose crust is so porous that all bodies of water have the same surface level (unseen channels interconnect them underground). $V(x,y)$ should now be thought of as the altitude of the planet's solid surface at position (x,y), while E becomes the universal water level. The allowed ($V < E$) regions shown shaded in the figure are now bodies of water, while the forbidden ($V > E$) regions left unshaded are the land areas. Each panel of Fig. 5.15 shows the same piece of planetary terrain. The sequence $a \rightarrow b \rightarrow c$ now shows how the map changes as the water rises.

Early, in Fig. 5.15*a*, we see a region which evidently belongs to a continental land mass. Isolated lakes are embedded in an infinite continent. Later, in *b*, the rising water level allows lakes to link up and grow larger. Finally, in *c*,

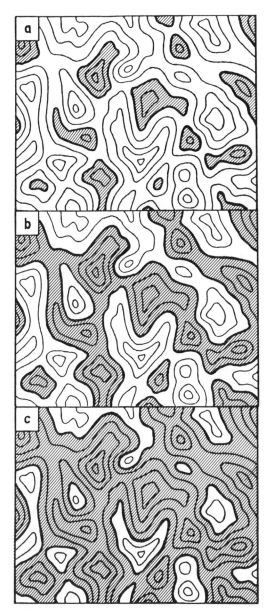

Figure 5.15 Percolation for a classical particle in a two-dimensional random potential (Zallen and Scher, 1971). The contour lines represent equipotentials of $V(\mathbf{r})$, and the shaded regions in a, b, and c indicate allowed ($V<E$) regions for three successively increasing values of E. In the "great flood" analogy described in the text, the lakes → ocean transition (a classical analog of the Anderson transition) is shown as the water level rises during the biblical forty days.

near the end of the biblical 40 days, we see a situation which is *topologically reversed* from the initial scene of *a*: In *c*, isolated islands are now embedded in an infinitely extended ocean. At some well-defined water level between *a* and *c*, somewhere close to *b*, the lakes → ocean transition takes place. From the viewpoint of ships able to move freely on the water surfaces, the lakes → ocean transition is a localization → delocalization transition. Although a ship placed in the ocean cannot go *everywhere* on the planet's surface (the land areas and the lakes are inaccessible to it), its motion can extend *indefinitely far* in any direction.

Reverting back to the original model of a classical particle of energy E in a random potential $V(\mathbf{r})$, it is then obvious that all states are localized for $E < E_c$ and that extended states appear for $E > E_c$. The critical energy E_c corresponds to the critical water level at which the lakes ↔ ocean transition occurs in the "great flood" analogy. This model of classical motion in a randomly varying potential thus generates a type of percolation process; it amounts to the problem of *percolation on a continuum* that we met previously in Section 4.7 of the last chapter. Continuum percolation is a generalization of site percolation, and the composition variable that plays the role of the site-occupation probability p is a quantity, which we denote $\phi(E)$, that specifies the fraction of space allowed to particles of energy E: $\phi(E) \equiv \int_{V(r)<E} d\mathbf{r} / \int d\mathbf{r}$. When the energy-dependent allowed-space fraction ϕ exceeds a critical value ϕ_c, the continuum-percolation threshold, an infinitely extended allowed region ("ocean") exists. The transition energy E_c, which amounts to the mobility edge of this classical model, satisfies

$$\phi(E_c) = \phi_c. \tag{5.9}$$

In two dimensions, for a broad class of random potentials, $\phi_c = \frac{1}{2}$. To see this, note that only *one* component (shaded *or* unshaded) can percolate in Fig. 5.15, because two extended networks cannot avoid intersection in 2d. It is also true that one component *must* percolate. Since the lakes → ocean transition is the topological mirror image of the continent → islands transition, it follows that $\phi_c = \frac{1}{2}$ for random potentials whose statistical distributions [$d\phi/dE$, i.e., $n(E)$] are symmetric about the average potential (Zallen and Scher, 1971). In one dimension, it is evident that $\phi_c = 1$, since there is no way to go *around* the mountain peaks of $V(\mathbf{r})$ as can be done in higher dimensions. [This is basically the same reason that $p_c(1d) = 1$.] In three dimensions, which naturally interests us most, no "rigorous" result is available. Nevertheless, it seems quite plausible to adopt the critical volume fraction discussed in Section 4.7 as a reasonable estimate for many situations, suggesting that $\phi_c(3d) \approx 0.16$.

The central panel of Fig. 5.15 vividly portrays a few properties of the percolation threshold that are worth mentioning here. One has to do with the number of lakes (clusters, in conventional lattice percolation). When the water level is very low, few lakes are present, and their number rises along with the rising water level. Later on, however, the number of lakes decreases as they merge to form large lakes or (when $\phi > \phi_c$) join the extended ocean. It turns out that the percolation threshold corresponds to the level at which the *number* of lakes is *dropping most steeply,* as they are rapidly consumed by the emerging ocean. Another point evoked by Fig. 5.15 is that the *total coastline is longest* at the percolation threshold. This feature follows from the fractal laciness of the large lakes near ϕ_c, as discussed for lattice percolation in the last chapter.

All of the discussion that has revolved around the potential of Fig. 5.15 has thus far been in terms of classical physics. But quantum-mechanical considerations are, of course, crucial for electrons. Real electrons can penetrate and tunnel through regions in which $V(\mathbf{r}) > E$, so that the sharp allowed/forbidden dichotomy of the classical problem disappears. Nevertheless, the *semiclassical approximation* involved in, say, the use of Eq. (5.9), is appropriate in certain situations in which the characteristic length scale of the topographic features (lakes, etc.) of $V(\mathbf{r})$ is much greater than a typical electron de Broglie wavelength.

Let us contrast the classical (continuum percolation) and quantum treatments in two interesting respects, before proceeding to discuss Anderson's analysis. The first comparison concerns the nature of states above the mobility edge. For $E > E_c$ in the classical case, lakes coexist with the infinite ocean. (In more traditional percolation terms, finite clusters coexist with the percolation path.) This suggests the possibility that localized states persist in the presence of extended states at the same energy. Quantum mechanics contradicts this: A localized wave function cannot avoid mixing with an enveloping extended state present at the same energy. Thus localized and extended states are cleanly separated (by mobility edges) in the energy spectrum, as shown earlier in Fig. 5.14.

The second comparison has to do with the dimensionality dependence of the two pictures. Both percolation and quantum treatments have all states localized by disorder in one dimension. A percolation transition is present for $d = 2$ and higher dimensions, while an Anderson transition occurs in $d = 2.001$ and higher (i.e., for $d > 2$). In the Anderson case, as discussed in the following section, $d = 2$ appears to be a borderline dimensionality.

The time has come to flesh out, at least to some extent, Anderson's quantum theory of electrons in disordered solids (Anderson, 1958). "The first problem was to create a model which contained only essentials" (Anderson, 1978). It is a tight-binding energy-disordered one-electron model embodied by the following Hamiltonian:

$$H = \sum_i E_i \, \mathbf{c}_i^\dagger \mathbf{c}_i + \sum_{i \neq j} T_{ij} \, \mathbf{c}_i^\dagger \mathbf{c}_j. \tag{5.10}$$

This Hamiltonian is expressed in "second quantization" notation in the site representation. E_i is the energy level of an electron at site i, T_{ij} is the matrix element of the Hamiltonian between sites i and j, and $\mathbf{c}_i^\dagger$ and $\mathbf{c}_i$ are operators that, respectively, create and annihilate an electron at site i. T_{ij} is called the transfer energy or transfer integral connecting sites i and j. (Other names for T_{ij} include overlap energy integral and "hopping" integral.) This energy is a measure of the coupling between a given pair of sites, and is a rapidly decreasing function of the distance separating sites i and j.

The essential feature of Eq. (5.10) is that E_i is randomly chosen for each i from a distribution of width W, as in Fig. 5.13b. [Note that the problem posed, while still immensely difficult, is still not as difficult (because it has been discretized) as solving the Schrödinger equation for some disordered "glass potential" $V(\mathbf{r})$ such as that sketched in Fig. 5.15.] This is the fundamental way in which disorder is introduced in the Anderson model: A distribution of site energies represents the distribution of atomic environments in the system. It is sometimes referred to as "diagonal" disorder, since it appears in the diagonal matrix elements of Eq. (5.10). "Off-diagonal" disorder (i.e., in the T_{ij} values) is not essential to the model.

The simplest way to discuss the model is in terms of tight-binding expansions of the form $\psi = \Sigma_i a_i \Phi_i$, where Φ_i is the atomic orbital centered at site i. If ψ is an eigenfunction of energy E, it satisfies $H\psi = E\psi$ with H given by Eq. (5.10), yielding a matrix equation for the amplitudes a_i:

$$Ea_i = E_i a_i + \sum_j T_{ij} a_j. \tag{5.11}$$

Equation (5.11) applies to a stationary state solution. For a general set of a_i's, the time-dependent equation of motion is

$$\frac{\hbar}{i} \frac{da_i}{dt} = E_i a_i + \sum_j T_{ij} a_j. \tag{5.12}$$

Suppose, at $t = 0$, an electron is placed at site m: $a_m(t = 0) = 1$, $a_i(t = 0) = 0$ for $i \neq m$. Assuming that it is possible to solve Eq. (5.12) for the subsequent evolution of the system, examine $a_m(t)$ in the limit $t \to \infty$. If $a_m(t \to \infty) = 0$, the electron has "diffused away," as expected for extended states. But if $a_m(t \to \infty)$ is finite, it has not diffused away but has only spread out over a finite neighborhood of site m. This corresponds to localization. Another way to look at this is in terms of the dependence upon the size of the system (which was assumed infinite in the preceding interpretation). Let the system contain N sites, and determine the (finite) value of $a_m(t)$ in the long-time limit for this finite system. Now increase N, and again determine $a_m(t \to \infty)$. If extended states apply, $a_m(t \to \infty)$ will continue to decrease without limit with increasing N, roughly as N^{-1}. But if localized states describe the system, the long-time amplitude at site m will asymptote with increasing N to a finite value, and will remain there as $N \to \infty$, decreasing no further no matter how large the system is made. Beyond a definite finite neighborhood, a localized electron does not "see" the other sites in the solid.

It is useful to look at Eqs. (5.10)–(5.12) in some limiting cases. First of all, since the essential stochastic element of the Anderson model resides in the E_i distribution, the model is often particularized to a simpler situation in which the sites are spatially arranged on a regular lattice and all transfer integrals beyond nearest neighbors are set equal to zero. Then Eq. (5.11) becomes

$$Ea_i = E_i a_i + T_{01} \sum_{\delta=1}^{\delta=z} a_{i+\delta} \tag{5.13}$$

where T_{01} is the transfer integral between nearest neighbors and the sum extends over the z nearest neighbors of site i.

One limiting case is that of vanishing disorder, i.e., $W = 0$, the crystalline case. Set $E_i = 0$ for all i in Eq. (5.13). To quickly remind the reader of crystalline "tight-binding bands," note the simplest case of Eq. (5.13) for $d = 1$ (linear chain):

$$Ea_n = T_{01}(a_{n-1} + a_{n+1}). \tag{5.14}$$

In Eq. (5.14), we use n instead of i as the site index. Now we "guess" a plane-wave solution for the amplitudes, $a_n = a_0 e^{in\theta}$, and find that this satisfies Eq. (5.14) if $E = 2T_{01} \cos \theta$. This corresponds to a crystal band of Bloch-function extended states at energies in the range $-2T_{01} < E < +2T_{01}$. The crystal bandwidth for the 1d case (for which the coordination is $z = 2$) is thus $B = 4T_{01}$, and in higher dimensions

$$B = 2zT_{01}. \tag{5.15}$$

Thus the bandwidth B, the characteristic kinetic-energy parameter of Fig. 5.13, is proportional to the coupling T_{01} and the connectivity z.

The opposite limit is to leave the E_i distribution intact in Eq. (5.13) and instead to set $T_{01} = 0$. With the coupling removed, the solutions are simply atomic orbitals at each site, that is, for $E = E_i$, $a_i = 1$ and $a_j = 0$ for $j \neq i$.

Having taken the problem apart to see the two opposite limits, $W = 0$ and $T_{ij} = 0$, we should now reassemble it. Anderson attacked the full problem [Eqs. (5.10)-(5.12)] with perturbation theory from both directions, using W as the perturbation in one case and T_{ij} as the perturbation in the other. Because the two limits are completely different, the execution of this program is a mathematical *tour de force* involving sophisticated techniques ("Greenian" operators, "propagator" series of perturbation-term diagrams in the plane-wave-like limit, "locator" series in the site-localized limit, etc.) which are well beyond the scope of our discussion. A central aspect of the analysis is to focus on statistical distributions rather than on average values.

In order to provide at least a rough argument for a critical value of the disorder parameter W/B, we give one here from the viewpoint of the strong-disorder limit $W \gg B$. Begin with the $B = 0$ localized limit and then turn on the transfer integral T_{01} and do perturbation theory with this coupling parameter as the perturbation. Consider an unperturbed state localized at site i ($a_i = 1, a_j = 0$ for $j \neq i$). First-order perturbation theory mixes this state with one on a neighboring site by an amplitude of order $T_{01}/(E_i - E_j)$, and higher orders of perturbation theory add terms containing higher powers of (basically) this quantity. The question is: How big can T_{01} be before localization is destroyed and extended states arise?

The site energies E_i and E_j are taken from a distribution of width W. Suppose we place E_i at the center of the distribution and assume that the E_j's of the z nearest-neighbor sites are uniformly spaced over the distribution at energy intervals of W/z. For this situation, the smallest energy denominator appearing in the perturbation parameter $T_{01}/(E_i - E_j)$ is $|E_i - E_j| = W/2z$, so that the largest (and therefore dominant) value occurring for this parameter is $2zT_{01}/W$. In order for the perturbation expansion, which is seen to be a power series in $(2zT_{01}/W)$, to converge, it is necessary that $(2zT_{01}/W) < 1$. If we identify the situation at which convergence breaks down with delocalization, then the localization ↔ delocalization transition occurs at $2zT_{01} = W$. This, via Eq. (5.15), is equivalent to the Anderson localization criterion $W > B$.

Long after Anderson's original paper, it was shown that his localization criterion applied exactly to the case of a Bethe lattice. Since a Bethe lattice (see Fig. 4.9) is equivalent in many respects to a lattice of very high dimensionality and since high dimensionality benefits delocalization, this result showed that the approximations used in his theory are conservative and actually underestimate the disorder-induced tendency to localization. The question of the dimensionality dependence of the Anderson transition brings us to the topic addressed in the closing section of the chapter.

5.7 SCALING ASPECTS OF LOCALIZATION

The significance of a disorder-induced localization edge, a mobility edge such as E_c in Fig. 5.14, is the following: Even if there does not exist a completely sharp cutoff (as occurs in a crystal) in the *distribution* of electron states in the energy spectrum, there does exist a sharp cutoff in the *character* of those states. Eigenfunctions change abruptly at E_c, even though the eigenvalue spectrum is continuous. A great deal of evidence on their optical and electrical properties supports this general picture for amorphous semiconductors, although much of the experimental evidence tends to be rather indirect.

A classic experimental demonstration of a localization ↔ delocalization transition for electrons in a disordered system occurs in crystalline silicon containing a carefully controlled quantity of phosphorus impurity. This transition is presented here both as a well-documented case in which disorder contributes to localization on the insulating side of the transition and, especially, as a splendid example exhibiting *scaling effects*—the main topic of this section.

Figure 5.16 displays the beautifully clean data of Hess, DeConde, Rosenbaum, and Thomas (1982) on phosphorus-doped silicon (Si:P, or $Si_{1-x}P_x$ with $x \sim 10^{-4}$) at very low temperature. In Si:P, phosphorus atoms replace a few of the silicon atoms on the fourfold-coordinated sites of the diamond-structure lattice. Four of the five outer electrons of each P atom enter into covalent bonds with the four neighboring Si atoms. The fifth electron cannot enter the crystal valence band and must go into a higher energy state, a "shallow donor" level just slightly below the conduction band edge. In this hydrogenic state, the electron is loosely bound to its P^+ ion in a large orbit of radius about 30 Å (its Bohr radius bloated by the large dielectric constant of the host crystal and the small effective mass of the crystal's conduction band). At very low concentrations, the bound donors (low temperatures are assumed here) are isolated and

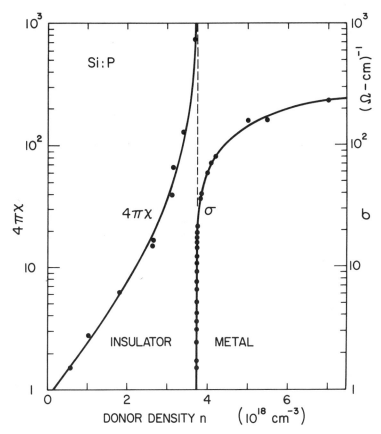

Figure 5.16   The clean experimental results of Hess, DeConde, Rosenbaum, and Thomas (1982) obtained at very low temperatures ($T \sim 10^{-2\circ}\text{K}$) on phosphorus-doped silicon in the composition range of the insulator ↔ metal transition. The divergence of the dielectric susceptibility $4\pi\chi$ as the transition is approached from the insulator side, along with the sharp but continuous conductivity threshold in $\sigma(n)$ on the metallic side, provide a striking manifestation of the metal ↔ insulator transition in a disordered system.

the system is an insulator. But at higher concentrations, the electron states become extended over the entire impurity system and the solid is a conductor. The critical concentration (sharply defined in Fig. 5.16) is $n_c = 3.7 \times 10^{18}$ cm^{-3}, corresponding to the composition $\text{Si}_{1-x}\,\text{P}_x$ of $x_c = 7.5 \times 10^{-5}$. Fewer than 0.01% of the Si sites are occupied by P atoms when the solid goes metallic.

The state of disorder present in Si:P is schematically indicated in Fig. 5.17. Since the Bohr radii of the donor wave functions so greatly exceed the lattice constant of the host crystal, the discrete nature of the sites available to the substitutional impurities is of no importance, and the situation is essentially that of spheres randomly placed in a continuum. The spheres (of radius given by the Bohr radius) may overlap and form clusters, and this percolationlike aspect of the problem is apparent in the figure.

The physics of the metal–insulator transition in doped silicon involves features of *both* the Anderson and the Mott transitions. While the Anderson transition (Fig. 5.13) and the Mott transition (Fig. 5.12) are *conceptually* quite distinct, they are often experimentally intertwined in specific instances. This is not too surprising. In any situation involving electrons in a noncrystalline setting, disorder introduces aspects of Anderson localization while electron–electron interactions are always present to exert some Mottian influence. It is important to emphasize that the two effects are allies in terms of enforcing a tendency to localization in the low-density (or low density-of-states) regime. In the case of Si:P, the Mott-transition aspect is especially plain because the M↔I transition is tracked as a function of concentration. Note also that the phosphorus donors essentially amount to the single-electron-outside-closed-shell atoms of Fig. 5.12. Anderson localization enters the physics of Si:P via the disordered potential experienced by the donor electrons in the field of the randomly distributed P^+ ions.

The experimental results of Fig. 5.16 are extrapolations to zero temperature from measurements taken down to extremely low temperatures ($T \sim 10^{-2}\,^\circ K$). Below the critical concentration n_c, the conductivity σ is undetectable; above n_c, $\sigma(n)$ rises sharply but continuously. The quantity monitored on the insulating

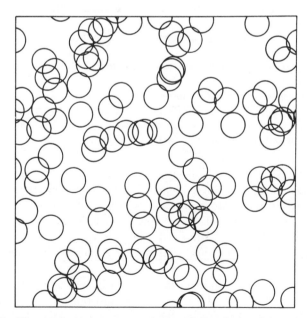

Figure 5.17   The stochastic geometry characteristic of phosphorus-doped silicon. The overlapping spheres represent donor wave functions.

side of the transition is χ, the contribution of the donor impurities to the dielectric susceptibility of the solid ($4\pi\chi$ is the contribution to the dielectric constant). This quantity diverges as the insulator→metal transition is approached. The solid lines in the figure are empirical fits using an assumed critical-exponent form for both $\sigma(n)$ and $\chi(n)$: For σ above n_c, $\sigma(n) \sim (n - n_c)^{0.50}$; for χ below n_c, $\chi(n) \sim (n_c - n)^{-1.2}$.

Called to mind by the overall form of Fig. 5.16, as well as by the form of the functions given above for the phenomenological description of the observed behavior of $\sigma(n)$ and $\chi(n)$, is the idea of a second-order phase transition. Compare Fig. 5.16 to Fig. 4.5, which displays the concentration dependence of key quantities in the percolation model (for the particular case of bond percolation on the square lattice). Bearing in mind the somewhat different format for the two figures (the vertical scale in Fig. 5.16 is logarithmic while that in Fig. 4.5 is linear), we immediately see the similarity between the two curves shown in Fig. 5.16 and, say, the two solid curves shown in Fig. 4.5. This suggests that scaling ideas, such as the renormalization-group approach developed in Section 4.8 in the context of the percolation model, may be appropriate to apply to the localization ↔ delocalization transition. This is indeed the case, and will now be discussed. First, however, it is necessary to make a brief comment on the above-mentioned qualitative connection between Figs. 4.5 and 5.16.

As described in the previous chapter, the critical exponents of the classical percolation model (Table 4.3) have been found to describe the M↔I transition which occurs in many types of disordered macroscopic systems. However, the percolation exponents do *not* fit the results of Fig. 5.16 for the transition in the microscopic, quantum-mechanical, Si:P system. In percolation, the analog of the polarizability or susceptibility χ is the mean cluster size s_{av} (as stated for the magnetic case in Table 4.2). The exponent describing the divergence of $s_{av}(p)$ as $p \to p_c$ is 1.7, as given in Table 4.3 for $d = 3$. This is significantly different from the value of 1.2 which applies to χ as $n \to n_c$ in Fig. 5.16. Nevertheless, the percolation model does provide at least a qualitative prediction of the "polarization catastrophe" (singularity in χ at the conductivity threshold) exhibited by Si:P.

More striking is the difference with respect to the threshold behavior of the conductivity itself. Comparing the σ curves of Figs. 5.16 and 4.5 reveals that the steep threshold seen for Si:P is quite different from the soft threshold for percolation conductivity. In critical-phenomena terms, this corresponds to the fact that the percolation exponent t in $\sigma \sim (p - p_c)^t$ is greater than unity, while the experimental exponent for $\sigma(n)$ near n_c is 0.5. This is why the conductivity curve of Fig. 5.16 resembles $P(p)$ rather than $\sigma(p)$ in Fig. 4.5. Percolation is, at best, only part of the story in any M↔I transition governed by quantum effects.

Having taken care to point out that the specific *detailed* predictions of percolation are inapplicable here, we may now proceed to apply the more *general* methods of scaling theory and renormalization group which were illustrated in the last chapter using percolation as the vehicle. By adopting the style characteristic of successful theories of critical phenomena, this approach exploits the

Table 5.2   The correspondence between the traditional and the scaling formulations of electron localization in disordered solids

The Anderson Model	*The Reformulation in Scaling Terms, by Thouless*
An individual atomic site i	A *block* of the solid containing many sites, in d dimensions a cube of side L, volume L^d
The spread W of the distribution $\{E_i\}$ of site energies	The average spacing $\Delta E \approx W/N$ of the N energy levels within the block L^d
The transfer energy or hopping integral T_{ij} which couples nearby sites	The energy shift δE caused by a change in the boundary conditions at the cube interface ($\sim \hbar/t_D$, where t_D is the time it takes for an electron to diffuse to the boundary)
$(W/B) \approx (W/2zT_{01})$, the dimensionless measure of the strength of the disorder	$(\Delta E/\delta E) \approx (1/g)$, where $g(L) = (\hbar/e^2)\sigma L^{d-2}$ is the *scale-dependent dimensionless conductance* which now serves as the disorder parameter

appealing analogy (Fig. 5.16 evokes Fig. 4.5) between localization and a second-order phase transition.

The simple scaling theory described here (Abrahams, Anderson, Licciardello, and Ramakrishnan, 1979) employs a reformulation (Thouless, 1974) of the Anderson model which we have summarized in Table 5.2. Instead of a single atomic site, the basic unit is now a *block* of volume L^d which contains many sites. (Characteristic of phase-transition theories is the phrasing in terms of arbitrary dimensionality d.) The solid is regarded as being built of such blocks, coupled to each other. W and B (or T_{01}), the two characteristic energies of the Anderson model, are replaced by characteristic energies which measure, respectively, the energy disorder within a block and the electronic coupling between adjacent blocks.

In place of W of Fig. 5.13b is ΔE, the average energy spacing between levels within one block. The qualitative connection between ΔE and W is straightforwardly given in terms of the density of states $n(E)$:

$$n(E) = 1/(L^d \Delta E). \tag{5.16}$$

In place of the site-to-site coupling T_{ij}, we now have an energy δE which is a measure of the sensitivity of a state within one block to a change in the boundary conditions at the interface with an adjacent block. An elegant heuristic argument, based on the uncertainty principle and outlined in Eqs. (5.17)–(5.20), connects δE to the conductivity σ in a macroscopic limit:

$$\delta E \approx \hbar/t_D, \tag{5.17}$$

$$t_D \approx L^2/D, \tag{5.18}$$

$$\sigma \approx e^2 D n(E), \tag{5.19}$$

$$\delta E \approx (\sigma \hbar/e^2)[L^2 n(E)]^{-1}. \tag{5.20}$$

Equation (5.17) contains the uncertainty-principle argument. In a macroscopic block of side L, an electron wave packet will diffuse to the boundary in a time t_D given by Eq. (5.18), where D is the diffusion constant. The uncertainty principle then implies that the influence of the boundary on the energy of the packet is about $\hbar/t_D$, as in Eq. (5.17). Equation (5.19) is a form of the Einstein relation between conductivity and diffusion. Combining the three equations then leads to the expression for δE given in Eq. (5.20).

As stated in Table 5.2, the ratio $\Delta E/\delta E$ is now adopted as the measure of the strength of the disorder, the analog of the ratio W/B in the "traditional" Anderson model. Extended states are sensitive to changes in boundary conditions ($\delta E > \Delta E$), while localized states do not "feel" the boundary and are quite insensitive ($\delta E < \Delta E$). The localization condition $\delta E < \Delta E$ may be interpreted as meaning that the energy mismatch between levels of adjoining blocks exceeds the extent to which the interface can shift them, making the alignment of levels (and electron transport) impossible.

The disorder parameter, denoted as g^{-1}, is now a *scale-dependent* quantity defined by

$$\frac{1}{g(L)} \equiv \frac{\Delta E(L)}{\delta E(L)}. \tag{5.21}$$

A convenient interpretation of the quantity defined in Eq. (5.21), the dimensionless, scale-dependent, (coupling-strength/disorder-strength) ratio $g(L)$, is provided by substituting expressions (5.16) and (5.20) for the two characteristic energies in Eq. (5.21):

$$g(L) = (\hbar/e^2)\sigma L^{d-2}. \tag{5.22}$$

Equation (5.22) applies to extended states in the macroscopic limit, since Eq. (5.20) relies on that limit. Since we recognize σL^{d-2} as the conductance of a (d-dimensional) cube of edge L and conductivity σ, we see that $g(L)$, the fundamental ratio of Eq. (5.21), may be viewed as a *generalized conductance* expressed in units of $e^2/\hbar$. It is the relevant parameter which determines the efficacy of the wave-function coupling from block to block when blocks of size L^d are fitted together, and it is the variable expressing the essential physics of the disordered electronic system.

The scaling theory examines the scale-length dependence of $g(L)$. Suppose that we know $g_0 = g(L_0) = \delta E(L_0)/\Delta E(L_0)$ for the system as composed of coupled cubes of size L_0^d. Scaling theory assumes that, given g_0 at length scale L_0, this determines g at a larger length scale $L = bL_0$ in which b^d of the original cubes have been combined to form new cubes of size $L^d = (bL_0)^d$. In the scaled-up situation in which the "granularity" of the system has been coarsened by a factor of b, the large cubes now have intracube levels spaced by $\Delta E(L)$ and the intercube coupling is $\delta E(L) = g\Delta E(L)$. Of course, both ΔE and δE are smaller at length scale bL_0 than at L_0, but the rescaled conductance $g(bL_0)$ can be either larger or smaller than $g(L_0)$. *The rescaling (or renormalization) of the relative coupling g is precisely analogous to the political rescaling illustrated earlier in Fig. 4.20 and the connectivity rescaling illustrated for percolation in Fig. 4.19.*

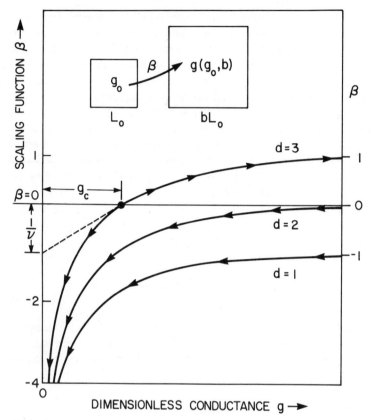

Figure 5.18 The qualitative behavior of $\beta(g)$ for one, two, and three dimensions, in the scaling theory of localization of Abrahams, Anderson, Licciardello and Ramakrishnan (1979).

At the new length scale bL_0, g is determined completely by the old value g_0 and the scale factor b: $g = f(g_0, b)$. If b is treated as a continuous transformation (i.e., $b = 1 + \epsilon$, where $\epsilon \ll 1$), then the scaling behavior can be specified by a differential scaling function $\beta(g)$:

$$\beta(g) = \frac{d \ln g(L)}{d \ln L}. \tag{5.23}$$

For positive β, g increases with increasing L; for negative β, g decreases with increasing L. The qualitative behavior of $\beta(g)$ is sketched in Fig. 5.18 for one, two, and three dimensions.

The curves shown in Fig. 5.18 were proposed by Abrahams et al. (1979), who realized that the behavior of $\beta(g)$ could be determined from the physics "by inspection" in the asymptotic limits $g \to \infty$ and $g \to 0$. For large g, macroscopic transport theory applies at large L so that Eq. (5.22) may be used and thus

$$\lim_{g \to \infty} \beta(g) = d - 2. \tag{5.24}$$

Therefore $\beta(\infty)$ is $+1$ in three dimensions, 0 in two dimensions, and -1 in one dimension, as indicated at the right-hand side of Fig. 5.18. For small g, the weak-coupling strong-disorder limit in which $g \ll 1$, Anderson's theorem tells us that the electron states are localized and decay exponentially at large distances as in Fig. 5.10*b*. At the boundary of a block of linear dimension L, the amplitude of the wave function of an electron localized within the block is of order $e^{-\alpha L}$, where $1/\alpha$ is the localization length. The block-to-block coupling therefore also decays exponentially with L, from which it follows that

$$\lim_{g \to 0} \beta(g) = \ln g. \tag{5.25}$$

Therefore $\beta(g)$ approaches $-\infty$ as g approaches zero, irrespective of dimensionality. Assuming that $\beta(g)$ varies smoothly and monotonically between Eqs. (5.24) and (5.25) yields the curves sketched in Fig. 5.18.

Figure 5.18 displays an aspect like that of the renormalization-group flow diagrams shown earlier for percolation in Fig. 4.19 and for politics in Fig. 4.20. This is shown by the arrowheads on the three curves of Fig. 5.18, which represent the direction in which g changes with increasing L: toward large g for $\beta > 0$, toward small g for $\beta < 0$. For the curve representing three dimensions in the figure, we recognize an *unstable fixed point* similar to those shown in Figs. 4.19 and 4.20. It is located by the zero crossing of the $d = 3$ $\beta(g)$ curve. At this critical value g_c of the dimensionless conductance, the criticality condition

$$\beta(g_c) = 0 \tag{5.26}$$

is satisfied. The significance of g_c is that it represents the special value of g for which this key parameter *remains invariant under change of scale*. It separates those values ($g > g_c$) that scale to the strong-coupling limit ($g \to \infty$) from those values ($g < g_c$) that scale to the decoupled limit ($g \to 0$). In Figs. 4.19 and 4.20, the unstable fixed points signalled the arrival of phase transitions, the percolation threshold in Fig. 4.19 and the party changeover in Fig. 4.20. (In connection with Fig. 4.20, it should be noted that a curve, which is quite similar to the curves describing the approach to $g = 0$ at the left of Fig. 5.18, describes the vanishing chances of the minority party as the electoral units scale to larger size.) In the case of the $d = 3$ $\beta(g)$ curve of Fig. 5.18, the *unstable fixed point signals the occurrence of the Anderson transition.*

For the $d = 1$ curve of Fig. 5.18, $\beta(g)$ is always negative so that *all g* values ultimately rescale to ("eventually flow to") the $g = 0$ localized limit. This agrees with theories which show (Mott, 1974) that in a static, disordered, one-dimensional system, all electron states are localized by even weak disorder. But it is a surprising (and somewhat controversial) result of the scaling theory that all g values also flow toward $g = 0$ for two dimensions as well, as implied by the $d = 2$ curve of Fig. 5.18. However, this apparently startling result appears to be similar to other theorems for two dimensions which, while theoretically rigorous, turn out to be inapplicable to most realizable systems because of considerations of relevant magnitudes. There is a famous theorem due to Peierls which states that, strictly speaking, there is no long-range order (and, therefore, no phase transition) in two dimensions. Experiments, such as those on the behavior of submonolayer helium adsorbed on solid surfaces, definitely contradict this. The reason is that the calculated correlation length (distance over which order is maintained), while finite, turns out to be enormous with respect to experimentally achievable sample dimensions. A similar situation may very well apply to the $d = 2$ scaling result of Fig. 5.18, especially since the $\beta(g)$ function lies so close to the $\beta = 0$ axis [so that $g(L)$ decreases *very* slowly with increasing L] at large g. The localization length may greatly exceed the sample dimensions in many experimental situations involving "2d" disordered systems. This may also explain why some computer simulations on $d = 2$ systems evidently exhibit a localized ↔ extended Anderson transition, even though the scaling theory [as a consequence of Eq. (5.24)] requires $d > 2$ for the transition to occur.

For three-dimensional disordered systems, most notably, amorphous solids, the scaling theory clearly predicts the existence of Anderson transitions. To connect the $d = 3$ unstable fixed point at $(g,\beta) = (g_c, 0)$ with an Anderson transition occurring as the Fermi energy passes through a mobility edge such as that at E_c in Fig. 5.14b, we need to note that the discussion of Eqs. (5.16)–(5.20) leading up to the definition of g in Eq. (5.21) refers to electrons at a given energy E. We have considered the scale-dependent behavior of g at fixed E. Now let us restrict our attention to g_0, the generalized conductance at a fixed microscopic length scale L_0 of the order of interatomic (or intersite) spacings. This quantity depends parametrically on the electron energy, $g_0 = g_0(E)$. A smooth variation of g_0 with E locates the mobility edge E_c at the energy at which g_0 crosses the critical conductance g_c: $g_0(E_c) = g_c$.

In addition to identifying the *zero* of $\beta(g)$ in Fig. 5.18 with the localization ↔ delocalization transition, the scaling theory analyzes the *slope* of the β curve as it crosses $\beta = 0$ at g_c to extract information about the critical behavior in the vicinity of a mobility edge. How this is done will not be detailed here, but it is closely related to the way in which the percolation exponent ν was extracted from the renormalization-group computer-simulation data of Figs. 4.21 and 4.22 using Eq. (4.24). For percolation, the exponent ν described the divergence of the average linear dimension of the clusters as the percolation threshold was approached from below: $l_{av} \sim (p_c - p)^{-\nu}$. For localization, the corresponding exponent (the same symbol is used) describes the divergence of the wave-function localization length (α^{-1} of $\psi \sim e^{-\alpha R}$) as the mobility edge is approached from the region of localized states:

$$\alpha^{-1} = l_{loc} = \frac{1}{(E_c - E)^{\nu}} \qquad (5.27)$$

The result, as indicated in Fig. 5.18, is that $(\nu g_c)^{-1}$ is given by the slope of $\beta(g)$ at $g = g_c$. Similarly, the conductivity exponent t of $\sigma \sim (E - E_c)^t$, describing the conductivity growth above the mobility edge, can also be determined. It turns out, for $d = 3$, that $t = \nu$. Estimates obtained from computer simulations and from epsilon expansions indicate that $t = \nu \approx 1$.

In comparing the scaling result for the conductivity exponent with the experimental results of Fig. 5.16 for the Si:P system, we find that while $t = 1$ fits better than the percolation value of $t = 1.65$ to the observed value of $t = 0.5$, it still corresponds to an insulator $\leftrightarrow$ metal transition that is much less sharp than the observed one. The scaling theory, as is clear from Table 5.2, is an independent-electron theory ultimately based on the original Anderson model, and neglects electron–electron correlation effects. It is such correlation effects, which were discussed in Section 5.5, that are believed to account for this difference.

REFERENCES

Abraham, F. F., 1980, *J. Chem. Phys.* **72**, 359.

Abrahams, E., P. W. Anderson, D. C. Licciardello, and T. V. Ramakrishnan, 1979, *Phys. Rev. Letters* **42**, 673.

Adler, D., 1982, in *Handbook on Semiconductors, Vol. 1,* edited by W. Paul, North-Holland, Amsterdam, p.805.

Anderson, P. W., 1958, *Phys. Rev.* **109**, 1492.

Anderson, P. W., 1978, *Rev. Mod. Phys.* **50**, 191.

Angell, C. A., and K. J. Rao, 1972, *J. Chem. Phys.* **57**, 470.

Cohen, M. H., H. Fritzsche, and S. R. Ovshinsky, 1969, *Phys. Rev. Letters* **22**, 1065.

Cohen, M. H., and G. S. Grest, 1979, *Phys. Rev. B* **20**, 1077.

Cohen, M. H., and D. Turnbull, 1959, *J. Chem. Phys.* **31**, 1164.

Grest, G. S., and M. H. Cohen, 1981, *Adv. Chem. Phys.* **48**, 455.

Hess, H. F., K. DeConde, T. F. Rosenbaum, and G. A. Thomas, 1982, *Phys. Rev. B* **25**, 5578.

Mott, N. F., 1949, *Proc. Phys. Soc. (London)* **A62**, 416.

Mott, N. F., 1968, *Phil. Mag.* **17**, 1259.

Mott, N. F., 1974, *Metal-Insulator Transitions,* Taylor and Francis, London.

Thouless, D. J., 1974, *Physics Reports* **13**, 93.

Turnbull, D., and M. H. Cohen, 1961, *J. Chem. Phys.* **34**, 120.

Turnbull, D., and M. H. Cohen, 1970, *J. Chem. Phys.* **52**, 3038.

Zallen, R., 1979, in *Fluctuation Phenomena,* edited by E. W. Montroll and J. L. Lebowitz, North-Holland, Amsterdam, p. 177.

Zallen, R., and H. Scher, 1971, *Phys. Rev. B* **4**, 4471.

Ziman, J. M., 1979, *Models of Disorder,* Cambridge University Press, Cambridge.

CHAPTER SIX

Optical and Electrical Properties

6.1 LOCAL ORDER AND CHEMICAL BONDING

There is a vast and rapidly expanding literature on the physical properties of amorphous solids, particularly amorphous semiconductors. Several excellent books and reviews, which provide a window onto this active field of experimental condensed-matter physics, are included in the reference list at the end of this chapter. Foremost among these is the classic book by Mott and Davis (1979), which extensively reviews many of the electronic phenomena observed in these solids.

The purpose of this chapter is to present the reader with a small set of hors d'oeuvres from the field of optical and electrical properties of amorphous solids, in the hope that his or her appetite may be stimulated enough to encourage a foray into the current literature on the field. While the selection included in this brief sampler is naturally biased by the author's background and taste (if any), the choice has also been influenced by opportunities to illustrate certain general principles and/or to connect with topics touched upon elsewhere in this book. In particular, properties relevant to some of the applications mentioned in Chapter One will be discussed in that regard. More generally, an attempt will be made to illustrate the ways in which the optical and electrical properties of amorphous solids differ (or, in some respects, do *not* differ) from the properties of crystalline solids.

A good place to begin is with the comparison, presented in Fig 6.1, of the optical properties of the three main condensed phases of germanium (Tauc, 1974). Shown here is the reflectivity R as a function of the photon energy $h\nu$ throughout the ultraviolet portion of the spectrum, the region of the spectrum associated with electronic excitations. The first thing to notice about these spectra is that liquid Ge displays a low-frequency behavior quite different from

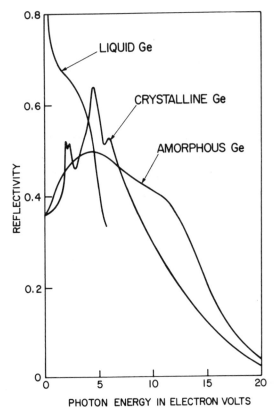

Figure 6.1 The fundamental reflectivity spectra, in the electronic-excitation regime, of crystalline, amorphous, and liquid germanium (Tauc, 1974).

that of the two solid forms. For liquid Ge, $R(h\nu)$ approaches 100% as $h\nu \to 0$; while for both solid forms of Ge, $R(0)$ is about 36%. Both crystalline and amorphous germanium are transparent in the infrared, and the value $R(0) = 0.36$ indicates that the long-wavelength refractive index of both solids is about $n = 4.0$. (The relation between R and the complex refractive index $n^c = n - i\kappa$ is $R = |(n^c - 1)/(n^c + 1)|^2$. For a transparent material, the extinction coefficient κ is small: $\kappa \ll 1$.] Their optical behavior at long wavelengths reflects the fact that both Ge crystal and Ge glass are semiconductors. Liquid Ge, on the other hand, exhibits characteristic metallic reflectivity at long wavelengths. In fact, liquid germanium *is* a metal.

This striking difference, between amorphous and crystalline Ge, on the one hand, and liquid Ge, on the other hand, provides a strong demonstration of the *primacy of the short-range order* in determining the basic nature of a material. Amorphous and crystalline Ge share the same four-coordinated local order, as described in Chapter Two. But in liquid Ge the short-range order changes markedly to approximately $z = 8$, and the electronic structure changes completely.

In contrast to the case of germanium, an analogous comparison carried out for crystalline, amorphous, and liquid selenium reveals that all of the forms of Se are semiconducting. The short-range order is the same ($z = 2$) in the various condensed phases of selenium.

Return to Fig. 6.1 and now focus on the optical comparison between amorphous and crystalline Ge. The sharp features in the crystal spectrum are consequences of the long-range order present in the crystalline state. As discussed in the previous chapter in connection with Fig. 5.14, sharp structure in the electronic density of states $n(E)$ arises in a crystal as a band-structure consequence of special places ("critical points") in $\mathbf{k}$ space at which the gradient $\nabla_{\mathbf{k}} E(\mathbf{k})$ vanishes. Because the first-order-allowed electronic transitions which dominate the ultraviolet region of the optical absorption spectrum of a crystalline semiconductor are $\mathbf{k}$ conserving "direct" transitions (this selection rule essentially amounts to momentum conservation), that spectrum mirrors the joint density of states $n_{cv}(E)$. Since $n_{cv}(E)$ is determined by $E_{cv}(\mathbf{k}) \equiv E_c(\mathbf{k}) - E_v(\mathbf{k})$ and thus has structure at energies corresponding to $\mathbf{k}$ values for which $\nabla_{\mathbf{k}} E_{cv} = 0$, the crystal band structure similarly gives rise to structure in the optical absorption spectrum which, in turn, is mirrored in $R(E)$ as shown for c-Ge in Fig. 6.1. These critical-point (or "van Hove") spectral singularities, which are especially sharp in the crystal spectrum at low temperatures, are specific consequences of translational periodicity ($\mathbf{k}$ as a good quantum number). This being so, similar fine structure is absent in the glass spectrum, as demonstrated by amorphous germanium in Fig. 6.1.

Ignoring the fine structure present for c-Ge and absent for a-Ge, there yet remains some overall similarity between the electronic spectra. For both semiconductors the reflectivity is largest in the vicinity of 5 eV, an energy which is a rough measure of the position of the first fundamental absorption band and is, also roughly, the average energy separation ("average gap", denoted E_{av}) between the highest valence band and the lowest conduction band in the solid. The similarity between their average gaps also reveals itself in the similarity of their infrared refractive indices n, since dielectric theory shows that $n^2 - 1$ scales as $(1/E_{av})^2$. Absorption-edge measurements made to locate the energy threshold E_g for valence-band to conduction-band electronic transitions (E_g denotes the bandgap for c-Ge, the mobility gap for a-Ge, as in Fig. 5.14) also find that even E_g is nearly the same (just under 1 eV) for both materials.

The observations of the preceding paragraph, namely that amorphous and crystalline germanium are semiconductors with strong similarities in their electronic structure (both their average gaps and minimum gaps are nearly the same), is a consequence of these two covalent solids having the same short-range order. Now, it should be remembered that to traditional solid-state physicists raised on the Bloch–Wilson–Peierls theory of electron states in crystals, a bandgap (such as that shown in the band structure of Fig. 5.9b) is intimately associated with the *translational symmetry properties* of the crystal. From this conventional condensed-matter-physics viewpoint, the fact that E_{av} and E_g of amorphous Ge are experimentally well defined and scarcely different than the corresponding energies in crystalline Ge is, to say the least, mysterious. (To say the most, it is embarrassing.) Most *chemists*, however, have no problem with that set of observations. Chemists are accustomed to focusing upon the *bonding structure* associated with the local order in a material. This chemical viewpoint, which automatically emphasizes the short-range order and is un-concerned about questions relating to the presence or absence of translational periodicity, is equally valid for *both* crystals and glasses.

Physicists have increasingly adopted the chemical viewpoint as a firm base from which to begin the analysis of the electronic structure of amorphous solids. Good discussions of this approach, and its relevance to covalent semi-conductors, have been given by Mooser and Pearson (1960), Kastner (1972), and Adler (1980). Energy-level diagrams illustrating this approach for two ele-mental solids which are prototypical of the two main classes of semiconductors (namely, tetrahedrally bonded semiconductors and chalcogenide semiconduc-tors) are shown in Figs. 6.2 and 6.3. Figure 6.2 illustrates the chemical-bonding

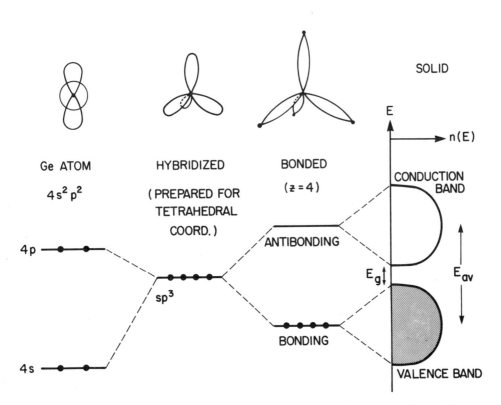

Figure 6.2 Bonding schematic for the electronic structure of a tetrahedrally coordi-nated covalent solid (crystalline or amorphous), illustrated for germanium.

picture of germanium, Fig. 6.3 provides a bonding picture of selenium, and each diagram applies to *both* the crystalline and amorphous form of these semiconducting solids.

Of the 32 electrons in an atom of germanium, 28 occupy low-energy orbitals making up the first three shells of states which are tightly bound to the Ge nucleus. These "core electrons," essentially the same in the solid as in the free atom, do not take part in the chemical bonding. It is the outer four electrons, which occupy the two $4s$ states and two of the $4p$ states in the atom, that bond to become significantly altered in the solid state, as visualized in steps in Fig. 6.2. In the first step, four equivalent sp^3 orbitals are constructed by hybridization. Since the hybridization mixes three p states with one s state while the ground-state atom has two s and two p states occupied, this step requires one $s \rightarrow p$ "promotion energy" (which is about 6 eV for Ge and Si). That energy cost is more than regained in the next step, in which four neighboring Ge atoms (similarly "prepared for bonding" by sp^3 hybridization) are brought in to enclose the initial atom in a configuration like that of the local order in the solid. The two atomic sp^3 orbitals on neighboring atoms which face each other to overlap along the Ge—Ge bond axis now interact to form two new orbitals, a bonding orbital and an antibonding orbital.

The antibonding orbital is the combination of sp^3 orbitals which has a node midway along the bond axis, while the bonding orbital is the spatially smoother combination in which the initial orbitals add constructively along this axis. The bonding orbital is lowered in energy with respect to the atomic sp^3 orbitals because of the reduction in kinetic energy that results from the delocalization of the wave function along the bond; the electron now is spread out over the vicinity of two atoms rather than just one. The antibonding orbital, which is much more oscillatory in space because of the zero crossing introduced at the bond midpoint, is raised in energy (vis-à-vis the atomic sp^3 orbitals) by a similar amount. The bonding–antibonding splitting is about 5 eV for Ge.

Along each bond, the bonding orbital can accommodate two electrons of opposite spin, as can the antibonding orbital. Since two electrons are available for each bond, one contributed by each of the involved Ge atoms, only the bonding orbitals are filled. Hence there is a net lowering of energy which is, of course, the reason that the covalently bonded solid forms in the first place. The bonding-energy arithmetic is roughly as follows. One $s \rightarrow p$ promotion (to form sp^3 hybrids from an s^2p^2 atom) costs 6 eV, but since all four hybrids drop into bonding states when the bonds form, that introduces a change of 4 × (-2.5 eV) so that the net change is -4 eV per atom (or -2 eV per bond).

In the solid, interactions between bonds broaden the bonding and antibonding levels into bands, as shown on the right-hand side of Fig. 6.2. E_{av} and E_g are the average gap and the bandgap. The point to be emphasized is that the overall aspect of this density-of-states sketch is similar for *both* the crystalline and the amorphous solids, since the overall electronic structure arises from the short-range order in the way indicated in the rest of the figure. E_{av} re-

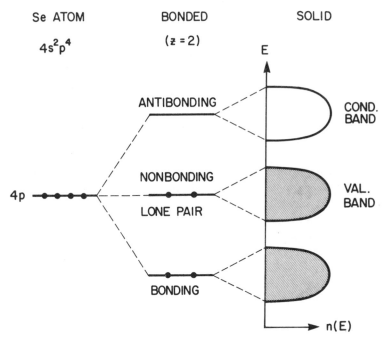

Figure 6.3 Bonding schematic for the electronic structure of solid selenium.

flects the bonding–antibonding splitting. The effects of long-range disorder in the amorphous form (such as the tailing of a finite density of states into the pseudogap region spanned by E_g) do *not* wash out the main features. This conclusion is even more evident in the case of the chalcogenides, now discussed via Fig. 6.3.

A selenium atom, having 34 electrons (as opposed to germanium's 32), has six electrons outside of closed shells: two $4s$ and four $4p$ electrons. The prevailing view holds that the $4s$ electrons lie sufficiently deep in energy for them to be considered as part of the atomic core, and thus chemically inert and out of the picture (as, indeed, they are in Fig. 6.3) as far as bonding is concerned. The chemical action takes place among the four $4p$ electrons. There are three orthogonal p-function orbitals, each available to up to two (opposite-spin) electrons. One of these orbitals is indeed occupied by two paired electrons of opposite spin. These electrons are referred to as *nonbonding* or *lone-pair* electrons. Their energy is nearly unaffected by, and they do not participate in, the chemical bonding. Bonding is effected by the other two p-orbitals, each occupied by one electron, when two neighboring Se atoms are brought nearby with singly occupied p-orbitals aligned to overlap and interact with those of the original atom (in analogy with the case of Ge sketched in Fig. 6.2, but now $z = 2$ instead of $z = 4$). The electronic structure in the solid is then as indicated at the right of Fig. 6.3.

As in the case of Fig. 6.2 for Ge, the conduction band in Se also originates from antibonding levels. However, as seen in Fig. 6.3, the highest valence band in solid Se is *not* formed from bonding states but instead from nonbonding or lone-pair states. For this reason, chalcogenide semiconductors (defined as having S, Se, or Te as the major constituent, as in Se, GeSe$_2$, and As$_2$S$_3$) are sometimes referred to as "lone-pair semiconductors" (Kastner, 1972).

The nonbonding/bonding division within the valence-band density of states, sketched for Se in Fig. 6.3, shows up in chalcogenides as a characteristic double-peak feature of their ultraviolet spectra. This feature, absent in the Ge-family tetrahedral semiconductors, is discussed in the following section on optical properties. At this junction, it is instructive to take note of another interesting difference in the physical behavior of tetrahedral and chalcogenide semiconductors, one which can be conveniently discussed in terms of the qualitative $n(E)$ features compared in Figs. 6.2 and 6.3.

Figure 6.4 displays the results of a set of optical experiments, carried out at high hydrostatic pressure, on a thin sample of As$_2$S$_3$ glass. Also included in the figure are results observed, during the same experimental run, on a thin ZnTe crystal. (The ZnTe sample was included in the pressure cell to serve as a pressure gauge during this experiment, but its presence also serves to illustrate the comparison now described.) Each curve in Fig. 6.4 shows the behavior (at a specific pressure) of the optical transmission as a function of photon energy in the vicinity of the optical absorption edge, the energy threshold for light-induced electronic transitions from the top of the valence band to the bottom of the conduction band. The cutoff in optical transmission, which occurs with increasing photon energy (note that $h\nu$ increases to the *left* in Fig. 6.4), provides a spectroscopic measure of the energy gap E_g.

The results shown in Fig. 6.4 disclose that when the solid is compressed, the bandgap E_g *increases* for the tetrahedrally bonded semiconductor ZnTe and *decreases* for the chalcogenide glass As$_2$S$_3$. In fact, this *qualitative distinction in the response to pressure* is generally observed: For tetrahedrally bonded germanium-family semiconductors, $dE_g/dP > 0$ is usually found, while for chalcogenide lone-pair semiconductors, $dE_g/dP < 0$ is the rule. *This distinction cuts across crystalline and amorphous lines.* The behavior of crystalline As$_2$S$_3$ under pressure is similar to that shown in Fig. 6.4 for amorphous As$_2$S$_3$, and amorphous tetrahedral semiconductors also usually behave similarly as their crystalline counterparts. Thus this dichotomy must have to do with the underlying bonding distinction between tetrahedral and chalcogenide covalent solids.

We can understand the pressure-response-of-the-bandgap dichotomy with the help of Figs. 6.2 and 6.3 and the structural concept of network dimensionality discussed previously in Chapter Three. Recall that the network dimensionality is defined as the number of dimensions in which the covalently bonded molecular unit is macroscopically extended. As shown for the crystalline examples listed in Table 3.2, Ge, As$_2$S$_3$, and Se represent 3d-network, 2d-network, and 1d-network solids, respectively.

In 3d-network solids, such as the four-coordinated semiconductors (both

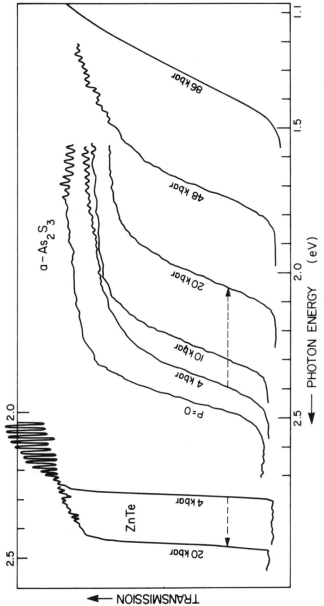

Figure 6.4 The opposite effect of pressure on the optical bandgaps of a tetrahedrally bonded and a chalcogenide semiconductor (Weinstein, Zallen, and Slade, 1980).

crystalline and amorphous) of the germanium family, the application of pressure necessarily compresses the covalent bonds in the material. As the bonds shorten, the bonding–antibonding splitting increases because of the increased interactions between bonded atoms. Therefore, from Fig. 6.2, E_{av} increases. The bandgap usually changes in the same way as the average gap, so that E_g increases as well, and dE_g/dP is positive.

In 1d- and 2d-network solids, as exemplified by the chalcogenide semiconductors, pressure does *not* (to first approximation) shorten the covalent bonds. These stiff bonds are insulated from the effects of compression by the presence of the soft intermolecular bonds in these essentially molecular solids. It is the mushy intermolecular volume that is primarily compressed under pressure, while the internal structure of the covalently bound molecular unit is relatively little affected. Thus, in selenium, the Se_N chains are pushed significantly closer together, but, within a chain, the Se—Se bond length scarcely changes. Because of this, the bonding–antibonding splitting is little affected by pressure and, from Fig. 6.3, this also holds for the nonbonding–antibonding splitting (which constitutes the average gap E_{av} for a chalcogenide).

Although there is relatively little effect on the energy separation E_{av} between the centers of gravity of the (nonbonding) valence band and the conduction band, pressure does have a significant influence on the *width* of these bands in the chalcogenides. Because of the enhanced intermolecular interactions (e.g., chain-chain interactions in Se) enforced by the crowding together of the molecular units in the compressed solid, both bands broaden under pressure. The broadening causes the top of the valence and the bottom of the conduction band to approach each other. Hence E_g decreases, and dE_g/dP is negative.

The above argument accounts for the prevalence of the behavior displayed in Fig. 6.4: a pressure-induced blue shift ($dE_g/dP > 0$) for the optical absorption edge in a tetrahedral semiconductor; a pressure-induced red shift ($dE_g/dP < 0$) for the absorption edge in a chalcogenide semiconductor. The argument involves network topology and chemical bonding; the presence or absence of translational periodicity is irrelevant. Note that lone-pair chalcogenide semiconductors, being molecular solids (network dimensionality less than three), are quite compressible. The effect of high pressure on the electronic structure can be considerable. The As_2S_3 glass sample observed in the experiment recorded in Fig. 6.4 is yellow-orange at zero pressure, deep red at 50 kilobars, and opaque (to the eye, though the sample still transmits in the infrared) at 90 kilobars. By varying the pressure over this range, the optical bandgap is "tuned" over a range of energy exceeding an electron volt.

6.2 OPTICAL PROPERTIES

The interaction with light provides a powerful means for probing the electronic and vibrational structure of a solid. In a first-order optical absorption process, in which a single elementary excitation (excited electron state, phonon mode, etc.) of the solid is involved, the energy of the absorbed photon equals the energy of the created excitation. Such experiments directly address the solid's characteristic energy spectrum. Optical properties of amorphous solids have already appeared in Figs. 6.1 and 6.4. In this section we present several further spectroscopic examples, and attempt to treat this important topic somewhat more systematically. In addition to what they teach us about the nature of these materials, the spectra also bear upon interesting applications that rely on their optical characteristics (e.g., transparency, for optical fibers, and its opposite, for solar cells).

First a few general notes about optical properties. The optical response function of a solid can be specified in different ways, each of which generally involves a pair of spectral functions (two are needed, to describe the amplitude and phase of the response) which are typically taken to be the real and imaginary parts of a frequency-dependent complex quantity. For example, the complex dielectric constant $\epsilon^c(\nu) = \epsilon_1(\nu) - i\epsilon_2(\nu)$ connects $E(\nu)$, the electric-field amplitude of an incident light wave of frequency ν, to $P(\nu)$, the complex amplitude of the polarization wave induced in the solid, by $P(\nu) = (1/4\pi)[\epsilon^c(\nu) - 1]E(\nu)$. The complex refractive index $n^c(\nu) = n(\nu) - i\kappa(\nu)$ is related to $\epsilon^c(\nu)$ by $[n^c(\nu)]^2 = \epsilon^c(\nu)$, and the complex reflectance amplitude coefficient $R^{1/2}e^{i\theta}$ is given by $(n^c - 1)/(n^c + 1)$. Here $R(\nu)$, a real quantity, is the reflectivity, the experimentally observed ratio of the *intensity* of the reflected beam to that of the incident beam (observed at near-normal incidence, for a sample arranged so that no light returns from rear-surface reflections). Because of the general requirement of any response function that cause must precede effect, there are integral equations (the Kramers–Kronig relations) which relate the real and imaginary components of each of these complex quantities. Knowledge of one component over a sufficiently wide frequency range permits, via the use of the appropriate integral transform, the other component to be determined with reasonable accuracy. Thus, if the reflectivity $R(\nu)$ is known for ν over a broad spectral regime, the conjugate phase-factor component $\theta(\nu)$ may be determined and (from the relations mentioned above for transcribing between the sets of optical functions) $n^c(\nu)$ and $\epsilon^c(\nu)$ as well. Just this procedure is often followed. Finally, for situations in which the light transmitted through a sample can be observed (as in Fig. 6.4), the optical absorption coefficient $\alpha(\nu)$ may be measured. This quantity describes the attenuation of the intensity of a beam $[I(x) = I(0)e^{-\alpha x}$, where $I(x)$ is the intensity after a path length $x]$ propagating within the solid. Unlike the other optical quantities mentioned, $\alpha(\nu)$ is not dimensionless but instead has dimensions of reciprocal length. In the narrow spectral region of an absorption edge, $\alpha(\nu)$ and $\epsilon_2(\nu)$ are approximately proportional to each other via the relation $\alpha = (2\pi\nu/nc)\epsilon_2$, where c is the light velocity. $\epsilon_2(\nu)$, the imaginary part of the dielectric constant, is a dimensionless measure of absorption and is the optical function most closely related to microscopic properties of the solid such as electronic and vibrational densities of states.

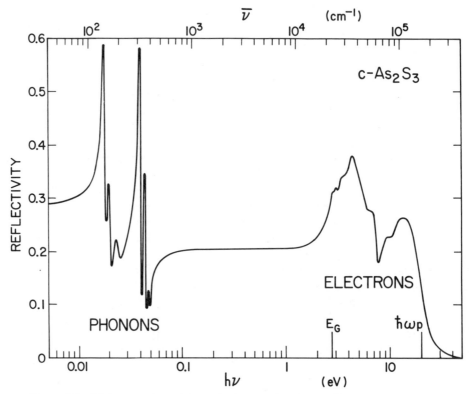

Figure 6.5 Global overview of the optical properties of crystalline As_2S_3. The reflectivity is shown for one light polarization over a range of photon energies from the far infrared to the far ultraviolet. (Zallen and Blossey, 1976).

Figure 6.5 presents a broad overview of the optical response of the crystalline form of the chalcogenide As_2S_3 (Zallen and Blossey, 1976). Relatively few solids have been experimentally characterized over such a wide spectral range; four decades of photon energy are spanned in Fig. 6.5. Since the amorphous form of As_2S_3 is an exemplary glass, which is used at several places in this book to illustrate typical properties of the amorphous solid state, its crystalline counterpart provides a convenient point of departure. The reflectivity function $R(h\nu)$ of c-As_2S_3 is shown in Fig. 6.5 for photon energies from the far ultraviolet down to the far infrared (from right to left in the figure).

A reflectivity spectrum such as that of Fig. 6.5 should be read from right to left since it tends to be cumulative with *decreasing* frequency. At enormously high frequencies the reflectivity is vanishingly small and the solid optically resembles free space, since little in it is capable of responding to such rapid electromagnetic excitations. Well off to the right of Fig. 6.5 there occur nuclear excitations in the gamma-ray region, and closer in there occur core-electron excitations in the X-ray region; both make negligible contributions to R on the scale shown. (In any case, these high-frequency excitations are properties of the free atom; no difference can exist between crystal and glass.) Starting at roughly 20 eV, the response of the valence-electron excitations in the solid contributes to the reflectivity. Electronic valence-band-to-conduction-band excitations produce structure in $R(h\nu)$ down to the bandgap E_g, just below 3 eV. Then, below the electronic threshold, R levels off and there is a hiatus throughout the near-infrared regime in which the crystal is transparent. When the photon frequency is further reduced into the far infrared, the field oscillations become slow enough for the crystal lattice vibrations to respond to them. Since the atomic masses M exceed the electron mass m by a factor of about 10^5, while the interatomic forces and the forces upon the electrons are comparable in strength (both originate ultimately from the same Coulomb interactions), the spectral region exhibiting the phonon excitations lies lower in frequency than that of the electronic excitations by a factor of $(M/m)^{1/2} \approx 10^2-10^3$. Kramers–Kronig analysis shows that the optical absorption coefficient α reaches values of the order of 10^6 cm^{-1} in the ultraviolet electronic regime and 10^4 cm^{-1} in the far-infrared vibrational regime. In both regimes, the crystalline spectrum brandishes sharp features.

We may now use Fig. 6.5 as a template for the comparison to the corresponding amorphous solid. Of course, since As$_2$S$_3$ glass is, like the crystal, a nonconducting solid in which electronic and atomic motions can be excited by incident radiation, it too may be expected to exhibit two distinct regimes of optical response between far-ultraviolet and far-infrared wavelengths. Naturally this is so, but there are nevertheless quite significant and characteristic differences between the optical properties of crystal and glass. Most notably, we shall see that the amorphous solid is far more *democratic* in its response to radiation.

A comparison of the electron-excitation ultraviolet reflectivity spectra of the crystalline and amorphous forms of both As$_2$S$_3$ and As$_2$Se$_3$ is presented in Fig. 6.6. The spectra of the glasses are shown in the bottom panel; the other panels contain crystal spectra. First an obvious point. Amorphous solids, lacking any special directions associated with crystallographic axes, are *optically isotropic*. (This is one reason, incidentally, why SiO$_2$ glass is superior to crystal quartz as a window material.) Many crystals are optically anisotropic and display polarization-dependent spectra. This is the case for c-As$_2$S$_3$ and c-As$_2$Se$_3$, which are low-symmetry crystals having three distinct spectra corresponding to three possible orientations of crystal axes with respect to the electric-field direction of incident polarized light. Two polarizations are shown for each crystal in Fig. 6.6; only one was illustrated for c-As$_2$S$_3$ in Fig. 6.5.

More significant is the presence in the crystal spectra of sharp features (peaks, corners, edges) which, as shown in Fig. 6.6, become extremely sharp when observed at low temperatures. These sharp features are absent for glasses, even at low temperatures. As discussed in the previous section in connection with the crystalline/amorphous comparison of Fig. 6.1, this spectral fine structure is a band-structure consequence of **k** conservation in the crystal-

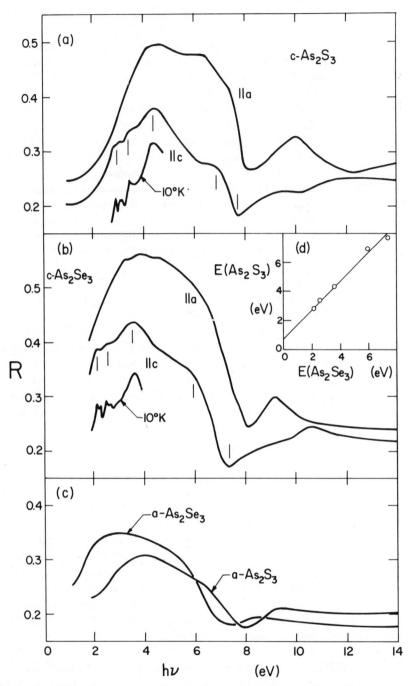

Figure 6.6 Reflectivity spectra of crystalline and amorphous As_2S_3 and As_2Se_3 in the electronic regime (Zallen, Drews, Emerald, and Slade, 1971). For each crystal, two independent polarizations are shown, part of one at low temperature (for clarity, the 10°K spectra are downshifted in R by 0.10). The glass spectra, even at low temperature, do not exhibit the sharp features characteristic of the crystal spectra.

line state. Translational periodicity demands that only **k**-conserving transitions ["vertical" transitions on an $E(\mathbf{k})$ diagram such as that of Fig. 5.9] contribute importantly to $R(E)$, and this symmetry-imposed selection rule causes $R(E)$ to show detailed structure similar to that present in the crystalline density of states $n(E)$. Lacking long-range order, amorphous solids do not show this fine structure in their comparatively bland spectra.

Although the glass spectra of Fig. 6.6 show no features on a scale of 0.1 eV to place in correspondence with the band-structure features of the crystal spectra, they *do* show definite features on an energy scale of *several* electron volts. Moreover, these features (roughly speaking, broad maxima near 5 and 10 eV separated by a well-defined minimum near 8 eV) very closely resemble the overall features of the spectra of the corresponding crystals. These features in $R(h\nu)$ reflect the presence, for both glass *and* crystal, of *two* main bands of electronic transitions in the ultraviolet spectrum. This situation for amorphous and crystalline As_2S_3 and As_2Se_3 is in contrast to that found earlier (Fig. 6.1) for amorphous and crystalline Ge, in which a single electronic band is seen in the ultraviolet.

The broad features sketched above, namely global similarity in the ultraviolet between crystal and glass, as well as an overall two-band format for As_2S_3 and one-band format for Ge, are comprehensible in terms of the chemical-bonding ideas illustrated earlier in Figs. 6.2 and 6.3. Since each of these pictures, Fig. 6.2 for a tetrahedrally bonded Ge-family semiconductor and Fig. 6.3 for a chalcogenide semiconductor, depends upon only the short-range order in the solid, it applies equally to crystal and glass. Moreover, the one-band shape for *c*- and *a*-Ge follows from Fig. 6.2 in which the valence band arises from bonding states, while the two-band shape for As_2S_3 follows from Fig. 6.3 in which there are separated bonding and nonbonding valence bands. In terms of the picture presented in Fig. 6.3, the lower valence band in As_2S_3 (or As_2Se_3) originates from bonding states associated with the As—S (or As—Se) covalent bonds, while the upper valence band originates from nonbonding states associated with lone-pair orbitals on the S (or Se) atoms.

For As_2S_3, the spectra of Figs. 6.5 and 6.6 indicate that the thresholds for the onset of nonbonding→antibonding and bonding→antibonding electronic transitions occur at photon energies of about 3 eV and 8 eV, respectively. Now let us focus down on the neighborhood of the fundamental electronic threshold near 3 eV. Figure 6.7 provides a closeup view of the vicinity of the optical absorption edge in both amorphous and crystalline As_2S_3. It shows the steep rise, with increasing photon energy, of the absorption coefficient α in the spectral region near the "optical bandgap," where the onset of the availability of electron excitations brings to an end the near-infrared and visible-spectrum transmission "window" of these solids. The variation in α is followed over several orders of magnitude; just to the right of Fig. 6.7, α attains values ($\approx 10^6$ cm^{-1}) characteristic of the main electronic regime.

The glass spectrum (Tauc, 1974) shown in Fig. 6.7 was taken at room temperature, while the crystal spectrum (Zallen et al., 1971) is shown at 10°K in order to emphasize the characteristic band-structure features: a strong direct-transition interband edge is seen at 2.9 eV [it corresponds to the first peak in the 10°K $R(h\nu)$ curve of Fig. 6.6a] and a weaker edge is seen at 2.8 eV. The absorption edge of amorphous As_2S_3, as expected, exhibits no sharp features. Its shape is unchanged at low temperature, although it does translate to higher energy by about 0.2 eV at 10°K. Note that when they are compared at the same temperature, the upper portions ($\alpha > 10^4$ cm^{-1}) of both edges are scarcely shifted ($\Delta h\nu < 0.1$ eV) with respect to each other, an impressive tribute to the overriding importance of their common short-range order in controlling the electronic structure (here, the bandgap E_g) of both solids.

In the absence, in Fig. 6.7, of any clearcut spectral marker on the absorption edge of the glass, the determination of a characteristic energy is not as unambiguous as it is in the crystal. An empirical definition of an optical bandgap E_o can be based on the observation (Tauc, 1974) that the upper portion of an amorphous solid's absorption edge is often well described by the relation

$$\alpha \approx \text{const.} \times (h\nu - E_o)^2. \qquad (6.1)$$

For As_2S_3 glass, Eq. (6.1), with E_o close to 2.4 eV, fits the data of Fig. 6.7 over the range 2.5–3.0 eV at absorption levels from 3×10^3 to 10^5 cm^{-1}.

A rough justification of the empirical relation (6.1) is provided by the following argument. In a crystal, a photon of energy $h\nu$ can induce a transition from a filled state of energy E to an empty state of energy $E + h\nu$ *only* if the initial and final states have the same wave vector **k** and satisfy certain selection rules. Thus, among all the pairs of electron states separated by energy $h\nu$, only a very few contribute to optical absorption. But in a glass, no such restrictions apply. Assuming that we are dealing with extended states, all such pairs of states (filled, at energy E, and empty, at energy $E + h\nu$) can contribute to optical processes. Assuming an approximation involving constancy of matrix elements (and of factors of ν) over a small range of energies near the electronic threshold, the imaginary (absorptive) part of the dielectric constant may be written

$$\epsilon_2(h\nu) = \text{const.} \times \int n_v(E)n_c(E + h\nu)dE. \qquad (6.2)$$

Here $n_v(E)$ and $n_c(E)$ are the valence band and conduction band densities of states. If we now assume that, as in crystals, the densities of extended states not too far from the band edges are given by $n_v(E) \sim (E_v - E)^{1/2}$ and $n_c(E) \sim (E - E_c)^{1/2}$, then Eq. (6.2) yields $\epsilon_2 \sim (h\nu - E_o)^2$ with $E_o = E_c - E_v$. Then, since α is proportional to ϵ_2 in the narrow spectral region near an edge, relation (6.1) follows. In any event, a plot of $\alpha^{1/2}$ versus $h\nu$ often yields a reasonably good straight-line fit to the absorption edge of a glass, and the extrapolated $h\nu$ at which $\alpha^{1/2} = 0$ provides a convenient experimental benchmark for the optical bandgap E_o. An edge which obeys Eq. (6.1) is sometimes called a *Tauc edge*. It is found in tetrahedrally bonded amorphous semiconductors, as well as in chalcogenide glasses.

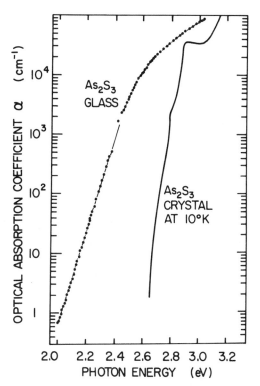

Figure 6.7 Comparison of the optical absorption edges of amorphous and crystalline As$_2$S$_3$ (Tauc, 1974; Zallen et al., 1971).

At low absorption levels ($\alpha < 10^3$ cm^{-1}), below the region of the Tauc edge, there is an exponential tail in $\alpha(h\nu)$, as seen for As$_2$S$_3$ glass in Fig. 6.7. The exponential tail persists at low temperatures, and is generally associated with the intrinsic disorder present in an amorphous solid. One possible explanation is the existence of disorder-induced local electric fields (such as the field corresponding to the potential contours of Fig. 5.15, or the field produced by the charged native defects to be discussed in Section 6.4).

Although the existence of this tail in the glass means that the absorption cutoff below E_o is not quite as sharp as it is in the corresponding crystal (Fig. 6.7), its exponential descent with decreasing $h\nu$ swiftly drives α to very small values. As$_2$S$_3$ glass is, in fact, used as an infrared window material in many applications, and it is even under development for use in infrared-transmitting fiber-optic lightguides. SiO$_2$-based glass is, of course, the fiber-optic material *par excellence*; its transparency thus far exceeds that of any other known solid. As mentioned in Chapter One, light of photon energy $h\nu = 0.8$ eV propagates through a one-kilometer length of this glass with an intensity loss of less than 10%. This remarkable low level of optical attenuation (<0.2

decibels/km) corresponds to an absorption coefficient α of less than 10^{-6} cm^{-1}, some 12 orders of magnitude below that of the ultraviolet absorption band. This low α at 0.8 eV is reached in a long exponential decline that begins near E_o, which is about 9 eV in SiO_2.

We now cross over to the other (far-infrared) side of the transparent regime, where we meet strong absorption processes associated with vibrational excitations, as shown for crystalline As_2S_3 at the left of Fig. 6.5. In a crystalline solid, the lattice-vibrational excitations are plane waves characterized by wave vector $\mathbf{k}$ as well as frequency ν. Each mode of excitation is termed a phonon, and the $\nu(\mathbf{k})$ phonon dispersion relations provide an energy-versus-momentum representation of the vibrational modes which is analogous to the $E(\mathbf{k})$ band-structure representation of the crystal's electron states. In an amorphous solid, the vibrational modes are no longer plane waves (and $\mathbf{k}$ has no meaning), but we will continue to use "phonons" as a convenient abbreviation for the vibrational elementary excitations of the solid. While $\nu(\mathbf{k})$ is not a valid concept in a glass, the concept of a vibrational density of states $g(\nu)$ retains its validity. Here $g(\nu)d\nu$ is the number of eigenvibrations, per unit volume, with frequencies between ν and $\nu + d\nu$.

The difference between crystal and glass, in the way that the elementary excitations show up in the optical spectrum, is far more striking in the case of phonon excitations than it is for electron excitations. In the case of electron transitions, the optical selectivity introduced by $\mathbf{k}$ conservation in the crystal (mandated by momentum conservation and the fact that the $\mathbf{k}$ vector of the absorbed photon is negligible on the scale of the Brillouin zone) imposes some fine structure on the ultraviolet absorption band, but the overall shape of the band is similar for crystal and glass because it reflects the same underlying chemical-bonding structure. Both crystal and glass exhibit a *continuum* of electronic excitations in the ultraviolet, a continuum which is closely connected to the electronic density of states $n(E)$.

In the case of phonon excitations, the optical consequences of $\mathbf{k}$ conservation in the crystal are far more restrictive than in the electron case. Of the 10^{24} phonon modes that exist in a typical macroscopic sample of a crystalline solid, only an elite few (of the order of 10^1 in number) possess the privilege of interacting with light in first-order optical experiments. The sharply structured infrared reflectivity shown for c-As_2S_3 at the left of Fig. 6.5 actually represents, not a continuum, but a *line spectrum*. It reveals the presence of six sharp spikes in optical absorption [i.e., six spikes in $\alpha(\nu)$ or $\epsilon_2(\nu)$]. The six crystal vibrations responsible for these lines are zone-center ($\mathbf{k} = 0$) phonons of the right symmetry to interact with light of the polarization corresponding to Fig. 6.5. Only zone-center modes can produce first-order (one photon annihilated, one phonon created) infrared absorption in a crystal. Crystal modes with $\mathbf{k}$ values away from the center of the Brillouin zone are optically "silent." Obviously, the full phonon density of states $g(\nu)$ is *not* accessible to first-order optical experiments in a crystal; neutron-scattering experiments are usually used to study $g(\nu)$ for crystalline solids.

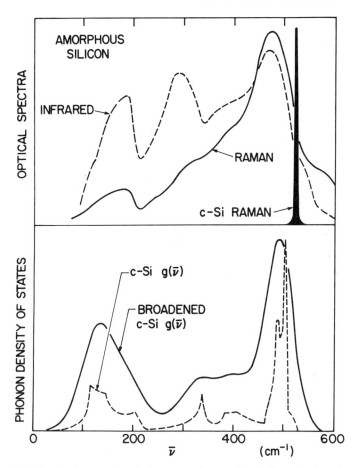

Figure 6.8 The fundamental optical spectra of amorphous silicon in the vibrational-excitation regime (Brodsky and Cardona, 1978). The upper panel exhibits the infrared absorption and reduced Raman-scattering spectra of amorphous silicon, as well as a comparison to the corresponding spectra of the crystal. The lower panel shows the phonon density of states of crystalline silicon, as well as a broadened version which reveals a rough similarity to the optical spectra of the amorphous form. In amorphous solids, *all* phonons participate in first-order optical processes.

In an amorphous solid, the elitism imposed by **k** conservation in the crystalline case is completely overthrown. Democracy reigns: *all* phonons may participate in first-order interactions with light. An amorphous/crystalline spectroscopic comparison for phonon–photon interactions has actually already been presented, not for As_2S_3 but for the closely related chalcogenide As_2Se_3, in Fig. 3.6 of Chapter Three. That figure shows Raman-scattering spectra rather than the infrared absorption spectra. (In first-order Raman scattering, an incident photon is annihilated and a phonon and scattered photon are created. The photon-energy shift of the scattered light, with respect to the incident light, locates the frequency of the phonon involved.) The line spectrum at the top of Fig. 3.6 corresponds to crystalline material, while the broad band at the bottom of Fig. 3.6 corresponds to amorphous material. A similar phenomenon occurs in infrared absorption, replacing the crystalline line spectrum (as seen in the c-As_2S_3 reflectivity of Fig. 6.5) by a broad band in the glass. The broad band is an image of $g(\nu)$. *In an amorphous solid, contributions from the entire phonon density of states appear in the first-order infrared and Raman spectra.*

This characteristic of a nonmetallic amorphous solid, namely that optical processes probe *all* of the solid's vibrational modes (rather than a tiny minority, as in crystals), allows optical experiments to yield a view of the main features of the phonon density of states $g(\nu)$. This is illustrated for amorphous silicon in Fig. 6.8 (Brodsky and Cardona, 1978). The two curves shown in the upper panel correspond to the infrared absorption and Raman scattering observed for a-Si. The spectral range covered, $0 < \bar{\nu} < 550$ cm^{-1}, corresponds to the fundamental lattice-vibrational regime in silicon. (In discussing phonons, it is usual to express frequency ν or energy $h\nu$ in terms of the equivalent wavenumber $\bar{\nu} = \nu/c$.) For comparison, the phonon density of states of crystalline silicon is shown in the lower part of Fig. 6.8 (the dashed curve), along with a broadened version (the solid curve) of the c-Si $g(\nu)$. The a-Si Raman and infrared spectra reveal a gross similarity to the broadened c-Si $g(\nu)$, which in turn indicates that the phonon densities of states of amorphous and crystalline silicon are not very different. Again the importance of the shared short-range order makes itself felt. The differences between the Raman and infrared spectra reflect matrix-element effects, which cause the two experiments to emphasize different parts of the vibrational spectrum.

Before leaving Fig. 6.8, it is enlightening to take note of the relative poverty of the first-order Raman and infrared spectra of crystalline silicon, which have also been included in the upper panel of that figure. The Raman spectrum of c-Si contains a single sharp Raman line; it corresponds to the $\mathbf{k} = 0$ optic mode which is (as shown in Fig. 6.8) the highest-frequency phonon in the crystal. The fundamental infrared spectrum of c-Si has even less content; *it is completely blank!* (This is spectroscopic elitism at its most extreme.) The infrared spectrum is empty because the only $\mathbf{k}$-conserving candidate, the zone-center optic mode, is *not* infrared active. Now, it is still widely (and mistakenly) assumed that the absence of one-phonon optical absorption in c-Si (and c-Ge and diamond) has to do with the fact that these are elemental solids. The direct interaction of infrared light with a single lattice fundamental requires that the phonon possesses an oscillating electric moment, and the absence of ionic charge in a homopolar solid composed of identical atoms has sometimes led to the assumption that elemental solids cannot display this direct photon–phonon ("reststrahlen") interaction. This is *wrong*. A vibrational mode in an elemental solid *can* possess a first-order (linearly-proportional-to-atomic-displacement) electric moment by the mechanism of displacement-induced charge redistribution. The reason that the zone-center optic phonon in c-Si fails to couple to infrared light is the very high degree of simplicity of the diamond structure, which permits symmetry to forbid a nonvanishing moment. Symmetry considerations reveal that an elemental crystal must satisfy a minimum-complexity condition—it must have at least three atoms per unit cell—if it is to display one or more infrared-active vibrational modes (Zallen, 1968). Examples of elemental crystalline semiconductors which display such modes are the chain-structure chalcogens Se and Te (Fig. 3.11). Moreover, as exemplified by *a*-Si in Fig. 6.8, *all* modes in *all* elemental *amorphous* semiconductors are infrared active.

We close our treatment of optical properties with two noteworthy crystal-line/amorphous comparisons for electron excitations in tetrahedrally bonded semiconductors. Figure 6.9 shows the imaginary part of the dielectric constant for amorphous and crystalline silicon, deduced by Kramers–Kronig analyses of their ultraviolet reflectivity spectra (Pierce and Spicer, 1972). As expected from our previous discussion, both solids manifest one main absorption band (of bonding → antibonding transitions) in the same general region of the spectrum, with the *c*-Si band being structure-ridden and the *a*-Si band being smooth. It happens that the detailed difference between the two materials in the way in which the absorption strength is *distributed* within the band has an interesting practical consequence.

Both crystalline and amorphous silicon are important solar-cell materials, with *c*-Si in use in applications (such as space probes) in which its high cost is not a problem, and *a*-Si (as mentioned in Chapter One) a promising candidate for large-area cells for power generation. Figure 6.9 reveals that the optical absorption of *c*-Si is much lower than that of *a*-Si in the region of 1–3 eV. (The sum rule which requires the integrated absorptions of the two solids to be very similar is satisfied by the higher absorption of the crystal in the 3–5 eV range.) The 1–3 eV region is just the region which contains most of the energy in the solar spectrum. Over much of this spectral region, the absorption coefficient of amorphous silicon is over an order of magnitude larger than that of crystalline silicon. Because of this difference, thin films (1 μm or less in thickness) of amorphous silicon can be used in solar-cell applications, while much thicker films (50 μm) are needed to adequately absorb the sun's photons in the case of crystalline silicon. The need for substantially less material in the *a*-Si case is a significant advantage for large-scale applications.

The optical bandgaps of crystalline and amorphous silicon are actually quite similar to each other; both are near 1 eV. It is a combination of spectroscopic elitism and band-structure characteristics that causes the crystal's optical absorption to be so feeble at photon energies in the 1–3 eV range. In the $E(\mathbf{k})$ band structure of crystalline silicon, the states near the top of the valence band have different $\mathbf{k}$'s than those at the bottom of the conduction band. A situation like this was illustrated in Fig. 5.9b of the last chapter. The strict selection rules that govern crystalline solids *forbid* first-order light-induced valence-band → conduction-band electron transitions between states of different $\mathbf{k}$. Only second-order photon-absorption electron-excitation processes, involving

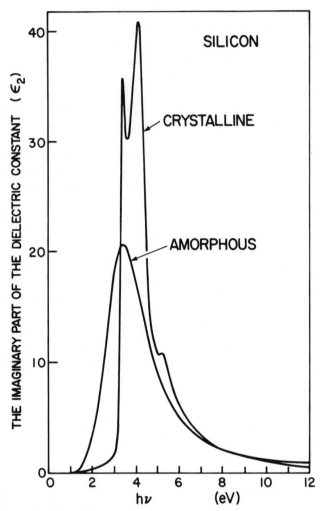

Figure 6.9 Comparison of the electronic-transition optical spectra of crystalline and amorphous silicon (Pierce and Spicer, 1972).

the simultaneous creation or destruction of a crystal phonon whose **k** can provide (which the photon cannot) the difference in **k** between the initial and final electron states, are permitted in this situation. Such second-order processes are orders of magnitude weaker than direct-transition processes ("vertical" transitions between states having the same **k** value), which accounts for the anemic absorption exhibited by crystalline silicon in the region just above its electronic threshold at 1 eV. Above 3 eV, direct transitions become available and (as seen in Fig. 6.9) the optical absorption rises strongly, but too late as far as the solar spectrum is concerned. Amorphous silicon, unencumbered by the burden of discriminatory selection rules, absorbs light strongly at photon energies starting just above its optical bandgap.

Note that, with respect to the two applications which have entered our survey of the optical properties of amorphous solids, one (the use of SiO_2-based glasses in fiber-optic communications) depends upon the high degree of *transparency* in the infrared window *below* the bandgap while the other (the use of amorphous silicon in solar cells) depends upon the high degree of optical *absorption* in the immediate region *above* the bandgap. Both reflect the fact that the transmission cutoff corresponding to the electronic threshold in amorphous solids is really quite clean.

Our final optical topic is illustrated in Fig. 6.10. This figure presents a set of electronic $\epsilon_2(h\nu)$ spectra obtained for an initially crystalline sample of the tetrahedrally bonded semiconductor GaAs which has been exposed to varying doses of bombardment by high-energy ions (Aspnes et al., 1982). What we have here is a single-sample crystalline/amorphous comparison analogous to that shown for As_2Se_3 in Fig. 3.6, except that the experiments represented in Fig. 6.10 were carried out in the opposite direction on a sample of GaAs as its surface region was converted to the amorphous form by the ion bombardment. The crystal features disappear as the crystallite size and the crystalline volume

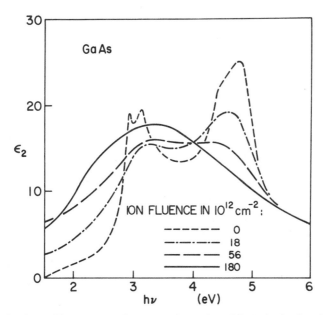

Figure 6.10 Crystalline-to-amorphous transformation of GaAs by ion bombardment. Ion implantation with high-energy As^+ ions converts a thin layer (about 0.1 μ deep) into the amorphous solid, as is beautifully demonstrated here in these ϵ_2 spectra derived from reflectivity measurements taken after different exposures to the ion flux (Aspnes, Kelso, Olson, and Lynch, 1982).

fraction decrease. The highest ion-implantation dose shown in Fig. 6.10 yields a top layer of amorphous GaAs which is indistinguishable from a-GaAs films produced by vapor quenching.

6.3   ELECTRICAL PROPERTIES

By now it is evident that nonmetallic glasses (i.e., amorphous insulators and semiconductors) have thus far totally dominated our survey of properties. There are several reasons for this. Much more is presently known about amorphous semiconductors, for example, than about amorphous metals. It is also true that the amorphous/crystalline comparison discloses more interesting differences for covalently bonded solids than for metallic solids. In the case of metals, perhaps the most striking feature is the *similarity* between the phenomena exhibited by both crystals and glasses. After all, metallic glasses which exhibit ferromagnetism (such as $Fe_{0.7}P_{0.2}C_{0.1}$) or superconductivity (such as $La_{0.8}Au_{0.2}$) eloquently attest to the inessentialness of long-range order in the solid state. Although amorphous ferromagnetism and superconductivity are not treated here, we attempt to redress the unbalance mentioned above by beginning our discussion of (normal) conductivity with the case of a metallic glass.

The temperature dependence of the electrical conductivity σ or electrical resistivity $\rho = 1/\sigma$ is the simplest and most informative electrical property of a conducting or semiconducting solid. Figure 6.11 presents a comparison of the $\rho(T)$ characteristics of crystalline and amorphous $Pd_{0.8}Si_{0.2}$ (Duwez, 1967). While the first splat-quenched (e.g., Fig. 1.4) metallic glasses to be prepared by Duwez and his colleagues in their pioneering experiments were gold–silicon alloys near the eutectic composition (e.g., quench path a in Fig. 1.5), amorphous $Au_{1-x}Si_x$ is stable only below room temperature. The first metallic glasses they extensively characterized by electrical measurements were in the palladium–silicon system, because amorphous $Pd_{1-x}Si_x$ is stable over a wide temperature range at compositions close to $x = 0.2$.

As usual, let us begin with a brief commentary on the crystalline case. In a crystalline metal, the Fermi energy falls within a band as in the example of Fig. 5.9a. The electron wave functions are extended-state Bloch functions, and in a partly filled band these plane-wave-like states give rise to a large electrical conductivity which is only *limited by scattering* introduced by the inevitable presence of *deviations* from perfect crystalline periodicity. "Defects" with respect to crystalline order, which scatter the electrons during their motion in an applied field (in this "nearly-free-electron" picture), include impurity atoms, surfaces, and—especially—the vibrational displacements of the atoms away from their equilibrium positions on the translationally periodic crystal lattice. The last-mentioned "defects," i.e., phonons, are always present, and they account for the characteristic resistivity-versus-temperature $\rho(T)$ shown for crystalline material in Fig. 6.11. The number of phonons present to scatter the electrons increases with increasing temperature, so that the electrical resistivity also in-

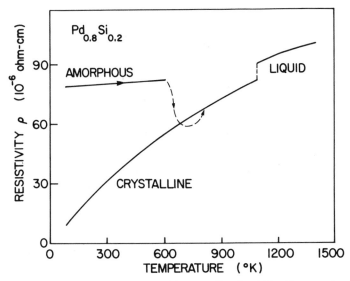

Figure 6.11 The electrical resistivity of the metallic glass $Pd_{0.8}Si_{0.2}$ can be observed up to 600°K (where it crystallizes). Here the resistivity-versus-temperature characteristic of the amorphous metal is compared to that of the crystalline form (Duwez, 1967).

creases as T increases. At high temperatures, the occupation number for a phonon mode of frequency ν is approximately $kT/h\nu$, and the $\rho(T)$ curve for a crystalline metal is also roughly linear.

For an amorphous metal, $\rho(T)$ is quite different. This is demonstrated by the data for palladium–silicon in Fig. 6.11. The resistivity of the metallic glass is much larger than that of its crystalline counterpart. Moreover, $\rho(T)$ is nearly flat; the resistivity of an amorphous metal changes very little as the temperature is changed. Both of these significant differences of the electrical properties of a metallic glass relative to the properties of its corresponding crystal, namely, much *higher resistivity* and much *lower sensitivity to temperature*, can be understood in terms of the topological disorder inherent in the atomic-scale structure of an amorphous solid (as discussed for random close packing in Chapter Two). In both the crystal and the glassy metal, conduction occurs via extended-state electrons and is limited by disorder-induced scattering processes. But because of the high degree of *static* disorder already present in the metallic glass, the additional *dynamic* disorder introduced by the presence of thermal phonons has little influence on the resistivity, which is already high as a result of the scattering caused by the intrinsic structural disorder. This is why $\rho(T)$ is high and flat for the glass.

We now turn our attention to semiconductors and insulators, which may be defined as materials whose conductivity vanishes at zero temperature. Referring to the schematic electron density-of-states diagrams of Fig. 5.14, the Fermi energy E_F falls between E_c and E_v, lying within the bandgap of the crystal or within the mobility gap of the glass. For the crystal, E_F lies at an energy devoid of states; for the glass, it lies within the region of localized states. In both cases, as discussed below, $\sigma(T \to 0) = 0$ since there is no electron motion unless thermal energy is supplied. Our treatment of electron transport in non-metallic amorphous solids will emphasize two interesting phenomena that arise as a specific consequence of strong disorder: *variable-range hopping* (Mott, 1969; Mott and Davis, 1979) and *dispersive transport* (Scher and Montroll, 1975).

Figure 6.12 shows a schematic illustration of electron transport mechanisms in amorphous semiconductors, using the format of an energy level diagram in real space. Let us first consider the mechanism of band transport. An electron that has been excited to the level labeled C, an extended state above the conduction-band mobility edge E_c, contributes to the conductivity in a way similar to that of a conduction electron in a metal. Its motion, within the band of extended states, is sporadically punctuated by scattering events. Also, because of the presence of localized states below E_c, the charge carrier's motion is

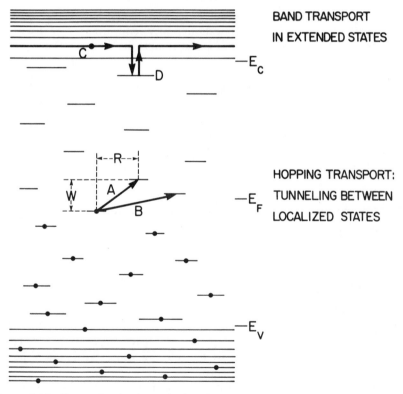

Figure 6.12 Energy-level schematic for electronic conduction mechanisms in an amorphous semiconductor. Energy is represented in the vertical direction, distance in the horizontal direction.

interrupted by trapping and release events as indicated in Fig. 6.12. Scattering and trapping severely limit the electron's drift mobility μ, the average velocity acquired per unit electric field. But by far the dominant limitation on the contribution of extended-state transport to the amorphous semiconductor's conductivity is (just as for a crystalline semiconductor) the small and temperature-sensitive carrier concentration n.

Assuming, for ease of discussion, that one sign of carrier dominates extended-state transport, the conductivity is given by $\sigma = ne\mu$, where n is the carrier density, e is the electronic charge, and μ is the mobility. For a metal, n is large and temperature insensitive. The relatively gentle temperature dependence of metallic conductivity (such as exhibited by the examples in Fig. 6.11) is controlled by the mobility, which gradually decreases with increasing temperature. For a semiconductor, it is the very steep variation of $n(T)$ which dominates $\sigma(T)$, which of course swiftly *increases* with increasing temperature. The behavior of $n(T)$ is itself dominated by the exponential function exp $(-\Delta E/kT)$, where the activation energy is $\Delta E = E_c - E_F$. E_F is near mid-gap in many amorphous semiconductors, so that $\Delta E \approx \frac{1}{2}E_g$ and $\sigma(T) \sim \exp(-E_g/2kT)$. This behavior, which resembles that seen in intrinsic crystalline semiconductors, is observed for many semiconducting chalcogenide glasses such as As_2Se_3.

However, in amorphous silicon and germanium, a quite different and unexpected behavior is observed for $\sigma(T)$ at temperatures below room temperature. Unexpected, that is, prior to Mott's classic back-of-the-envelope calculation [Eqs. (6.3) to (6.6), given below] which predicted this curious form of conductivity temperature dependence for amorphous semiconductors having a relatively high density of localized states in the gap. As discussed in the following section, the tetrahedrally bonded amorphous semiconductors (unless alloyed with hydrogen) are characterized by just such a high density of states within the mobility gap. The interesting behavior in question is exhibited in Fig. 6.13 for two samples of amorphous silicon at temperatures between 60 and 300°K (Hauser, 1973). As can be seen from the coordinates adopted for presenting these data, the observed temperature dependence corresponds to $\sigma(T) \sim \exp(-\text{const}/T^{1/4})$. This electrical behavior is associated with the transport mechanism called *variable-range hopping* (Mott, 1969; Mott and Davis, 1979). ·

The term *hopping* is an abbreviation for the phonon-assisted quantum-mechanical tunneling of an electron from one localized state to another. Two hopping transitions are schematically indicated by A and B in Fig. 6.12, two possible tunneling processes which take an electron in a filled state below the Fermi level to either of two nearby empty states above E_F. R denotes the distance the electron hops, while W denotes the energy separation of the final and initial states. In order for energy to be conserved, a phonon must be absorbed (for $W > 0$, as in A and B of Fig. 6.12) or emitted (for $W < 0$) during a hop. Since energy-lifting hops must occur for this tunneling mechanism to contribute to the macroscopic dc conductivity, and since phonons must be around to supply energy for such hops, the hopping conductivity is finite at finite temperature but vanishes at zero temperature.

Tunneling transport is sometimes observed to occur among impurities in doped crystalline semiconductors. It is found in a narrow temperature range in which T is low enough to quench band conduction and permit the tunneling mechanism to show up in the conductivity. Thus in Si : P, at doping concentrations somewhat below the insulator $\leftrightarrow$ metal transition of Fig. 5.16 (which corresponds to $T = 0$), hopping conductivity is seen at finite but low temperatures in samples in which some of the donor sites are empty because of the presence of some compensating acceptors. However, results such as those for amorphous silicon in Fig. 6.13 provide a far more striking manifestation of this electron transport mechanism. Hopping is found to dominate the conductivity of a-Si over a very wide temperature range, successfully competing with extended-state transport at temperatures up to room temperature.

Mott's intuitive argument for the "$T^{1/4}$ law" proceeds from the following approximate expression for the R, W, T dependence of the transition probability p of the occurrence of an energy-upward hop (such as A or B in Fig. 6.12) of distance R and energy W taking place at temperature T:

$$p \sim \exp[-2\alpha R - (W/kT)]. \tag{6.3}$$

This expression is the product of two exponential factors. The first factor, $\exp(-2\alpha R)$, is essentially the probability of finding the electron at distance R from its initial site. Here α is the inverse localization length that describes the exponential decay $\psi(r) \sim \exp(-\alpha r)$ of the electronic wave function at large distances. The second factor, $\exp(-W/kT)$, arises from the electron's need for energy-conserving phonon assistance in overcoming the energy mismatch W. In the low-temperature limit (which is assumed in this argument), the number of phonons of energy W is given by the Boltzmann factor $\exp(-W/kT)$. It is assumed that the localized electron states are randomly distributed in energy as well as in space, with a uniform distribution given by $n(E_F)$, the density of states per unit volume per unit energy at energies close to the Fermi energy. Because the initial state must be occupied and the final state must be empty for a hop to occur, Fermi–Dirac statistics [not yet included in the expression of Eq. (6.3)] limit these tunneling transitions to an energy range which brackets E_F and is about kT wide. Within this narrow energy range, $n(E)$ is taken to be constant, as is the localization length α^{-1}.

The essential point emphasized by Mott about Eq. (6.3) is that there exists a trade-off between hopping distance R and mismatch energy W which implies that the nature of the dominant (i.e., most likely) hops necessarily *changes* with temperature. Suppose that, at a given temperature, hops with an (R, W) combination such as that corresponding to A in Fig. 6.12 make a more important contribution to the hopping conductivity than hops such as B having larger R but smaller W. If we lower the temperature, the hopping probability p de-

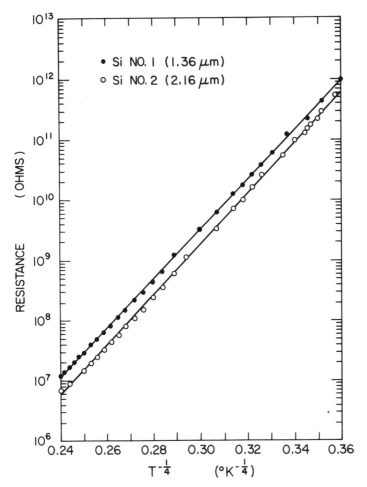

Figure 6.13 The temperature dependence of the electrical resistance of amorphous silicon, demonstrating the functional form predicted by Mott for variable-range phonon-assisted tunneling between localized states at the Fermi level (Hauser, 1973).

creases [because of the second term in Eq. (6.3)] for both A and B, but $p(A)$ drops much more rapidly than $p(B)$ because of A's greater energy mismatch W. At some lower temperature, $p(B)$ exceeds $p(A)$. It now becomes more favorable to make the longer hop, in spite of its smaller overlap factor $\exp(-2\alpha R)$, because this is more than compensated at low T by its larger phonon factor $\exp(-W/kT)$. The lower the temperature, the more difficult it is to overcome a given energy barrier W and the more desirable it is for the electron to hop further to find a lower W.

The enhanced probability of encountering a smaller W by permitting the electron to choose among the larger selection of final-state sites contained within a larger neighborhood, namely a sphere of (increased) radius R surrounding the initial site, is expressed by the relationship

$$(4\pi/3)R^3 W(R)n(E_F) = 1. \tag{6.4}$$

As is clear from its definition in Eq. (6.4), $W(R)$ is the energy for which the number of states spatially located within the sphere of radius R and energetically located within the range from zero to $W(R)$ is, on the average, equal to one. $W(R)$ is proportional to R^{-3} (in d dimensions, to R^{-d}), and is a reasonable measure of the minimum mismatch available for a hop of range R. Substituting $W(R)$ for W in Eq. (6.3) yields an exponent of the form $-aR - (b/R^3)$, and minimizing $aR + (b/R^3)$ to maximize p yields, for the most probable hopping distance,

$$R \approx [\alpha kT n(E_F)]^{-1/4}. \tag{6.5}$$

This shows how the hopping range changes with temperature, hence "variable-range hopping." It is a consequence of the temperature-dependent compromise struck in the bargaining between hopping distance and energy mismatch. Assuming that the most probable hops dominate the hopping conductivity, we substitute Eq. (6.5) into Eq. (6.3) to obtain

$$\sigma \sim \exp(-A/T^{1/4}), \tag{6.6}$$

where

$$A \approx [\alpha^3/kn(E_F)]^{1/4}.$$

In both Eq. (6.5) and the expression for A in Eq. (6.6), we have omitted the proverbial dimensionless factors of order unity.

Equation (6.6) is Mott's famous form for $\sigma(T)$. The power $1/4$ in the term $T^{1/4}$ enters as the reciprocal of an effective dimensionality characteristic of the problem; there are three space dimensions plus one energy dimension. An electron can only hop to a site which is close by in (x, y, z, E) four-space. [In d space dimensions, the $1/4$ is replaced $1/(d + 1)$.] Mott's result stimulated experimentalists to examine their electrical data for evidence of variable-range-hopping behavior via plots such as that shown for silicon in Fig. 6.13. For the tetrahedrally bonded amorphous semiconductors, slopes of plots of this type are typically compatible [via A of Eq. (6.6)] with values of the order of $\alpha^{-1} \approx 10$ Å and $n(E_F) \approx 10^{19}$ (cm^3 eV)$^{-1}$ for the localization length and the density of localized states at mid-gap. At $T = 100°$K, these parameters imply [via Eqs. (6.4) and (6.5)] values of $R \approx 80$ Å and $W \approx 0.05$ eV for the distance and energy of the dominant hops.

Refinements of the theory of this phenomenon, phonon-assisted tunneling among localized states at the Fermi level, recover the functional form of Eq. (6.6). In more detailed treatments, the transition probability of Eq. (6.3) is replaced by a transition rate which incorporates the Fermi–Dirac occupancy and vacancy factors for the initial and final states. Also, energy-downward transitions ($W < 0$, for which case no Boltzmann factor enters because phonon creation, not phonon annihilation, is involved) are explicitly included. The effect of an electric field is to bias energy-upward hops in the downfield direction by replacing W by $W - eFR$, where F is the field component parallel to R.

An elegant analysis has been given by Ambegaoker, Halperin, and Langer (1971) that involves an illuminating percolation argument. The percolation construct enters their model in the form of a random resistor network. Each node of the network represents a localized electron site of the amorphous solid, and each pair of nodes is connected by a wire whose conductance represents the tunneling transition rate between the corresponding sites. The conductances of different wires vary over many orders of magnitude, reflecting the enormous variation of hopping probabilities present in the solid. Suppose we imagine that all of the wires are removed from the network, and that we then put them back in the network one at a time, beginning with the highest conductance and then working on downward. At first there are isolated wires here and there, then some link up to form larger clusters, and eventually—at some critical conductance value—the percolation path appears. This is the microscopic conductance value (tunneling rate) which controls the macroscopic conductivity. Larger conductances do not matter because they form only isolated clusters and cannot span a macroscopic sample; the current path must still traverse wires having the critical conductance if it is to make it across the sample. Smaller conductances *can* form macroscopic paths, but contribute little because they are shorted out by paths of higher conductance. Important to this argument of a controlling critical conductance, which also leads to the $T^{1/4}$ law, is a great range of conductance values spanning several orders of magnitude. The vast majority of the conductances in the network are then either much larger or much smaller than the critical conductance, which strengthens the percolation argument.

The percolation process described above resembles a kind of bond percolation, but it is different from any of the cases discussed in Chapter Four. The closest percolation process about which information is known is the following. Consider sites distributed randomly in space. Sequentially connect the sites pairwise with bonds, starting with the closest pairs of sites and then proceeding to more distant pairs, until an infinite path appears. At this percolation threshold, each site has formed N_c bonds, on the average. N_c is known to be about 4.4 in two dimensions, about 2.7 in three dimensions, and it may be about 2.4 in higher dimensions (Zallen, 1977). Note that this stochastic-geometry problem, though quite similar, is not exactly the same as that stated in the preceding paragraph for the tunneling problem. This is because of the energy variable present in the hopping case, which permits some long bonds to appear before some shorter ones. The extent to which this modifies N_c is unknown.

Our final electrical topic concerns a remarkable and general aspect of stochastic transport in amorphous semiconductors. At issue is the highly "anomalous" behavior observed in transient photoconductivity experiments on highly disordered solids. Often in physics, experimental observations are termed "anomalous" before they are understood. Once theory succeeds in explaining and illuminating the observations, they are no longer "anomalous" and instead come to be regarded as "obvious." A crucial paper can trigger such an "anomalous → obvious" transition, and in the present case that key role was played by a 1975 paper by Scher and Montroll. That landmark paper has become basic to our understanding of a striking characteristic of carrier motion (now called *dispersive transport*) which is a common occurrence in amorphous semiconductors, though foreign to our experience with crystals.

Figure 6.14 *a* shows a highly idealized sketch of the setup used in transient photoconductivity experiments (also known as time-of-flight or drift-mobility experiments). The semiconductor sample is subjected to a large electric field produced by a voltage across blocking electrodes (electrodes which do not inject carriers into it) on opposite surfaces. The temperature is low enough and the

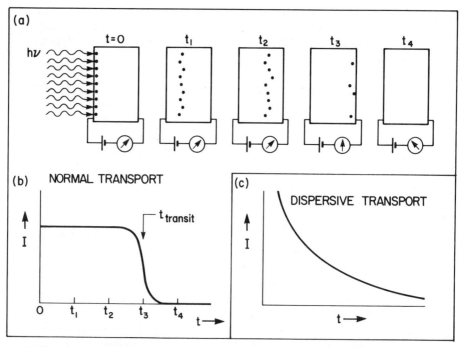

Figure 6.14 Transient photoconductivity or time-of-flight experiments. A short pulse of highly absorbed light injects carriers close to one electrode, and they drift across the sample under the influence of an electric field to be collected by the back electrode. In the normal situation seen in crystals, illustrated in *a*, the carriers move in a well-defined packet and produce a current-versus-time trace, illustrated in *b*, which has a flat top and a well-defined cutoff. Many amorphous semiconductors display an $I(t)$ trace, such as that illustrated in *c*, which is completely different from this and which is caused by the phenomenon of dispersive transport.

bandgap large enough so that negligibly few charge carriers are present and negligible current flows. A short pulse of light is flashed upon the sample through one electrode, which is semitransparent. The photon energy is chosen so that it exceeds the bandgap and so that the material's optical absorption coefficient is large enough to ensure that all of the light is absorbed within a shallow penetration depth very close to the illuminated front surface. The electron–hole pairs created by the absorbed light are swiftly pulled apart by the large field. Carriers of one sign are then quickly collected by the front electrode, while those of opposite sign (represented by the dots in Fig. 6.14) are pulled away and drift across the sample, under the influence of the field, to the back electrode. The situation resembles that which takes place during the imaging step of the xerographic process (Fig. 1.12a), except that now the experiment is done in the "small-signal limit" in which the small pulse of non-equilibrium photoinjected carriers scarcely perturbs the field, which does not change. In its motion across the sample, the drifting charge induces a current flow which is observed in the external circuit.

The situation depicted in Figs. 6.14a and 6.14b indicates what happens for the case of a crystalline semiconductor, for which the experiment is essentially equivalent to the classic Haynes–Shockley experiment for the hole drift mobility in germanium. A well-defined sheet of charge moves at constant speed across the sample, giving rise to a constant current. When the carriers arrive at the back electrode, the current ceases. This occurs at a definite transit time $t_{transit}$, which corresponds to the arrival time of the center of the charge packet. Because diffusion is superimposed upon the field-induced drift, the charge packet progressively spreads out about its mean position during its motion. The transient current characteristic $I(t)$ thus shows a rounded fall-off near $t = t_{transit}$, as in Fig. 6.14b.

What is observed for very many amorphous semiconductors is the quite different behavior indicated in Fig. 6.14c. The current *decreases continuously* from the earliest times observable, and the current flow extends to *very long times*. There is no apparent "marker" on the $I(t)$ curve to identify as a definite transit time.

The explanation of this behavior resides in the fact that carrier motion in strongly disordered solids is governed by a set of microscopic event times $\{t_i\}$ in which the values of the characteristic times t_i span an *enormous* range which covers many orders of magnitude. Microscopic mechanisms responsible for the vast spread of event times encompassed by the distribution $\{t_i\}$ will be discussed shortly, but for the moment the great breadth of the distribution will simply be assumed because *this assumption suffices* to account for Fig. 6.14c as well as other "peculiar" properties of transient transport in semiconducting glasses. The way these properties come about is illustrated in Fig. 6.15 (Scher and Montroll, 1975).

The curves of Fig. 6.15 correspond to calculated charge distributions shown at various times t after the light-pulse injection of carriers at the left. Figure 6.15a corresponds to normal transport, as in Figs. 6.14a and 6.14b.

Diffusion gives a Gaussian shape to the drifting charge packet, with the width increasing as $t^{1/2}$. If we use $x(t)$ to denote the mean position (center of gravity) of the drifting-and-diffusing packet at a particular time t, and if we use $\Delta x(t)$ to denote the spread (rms value of the deviation from the mean) of the packet at that time, then for normal transport

$$x(t) \sim t,$$

$$\Delta x(t) \sim t^{1/2}. \tag{6.7}$$

The linear $x(t)$ in Eq. (6.7) corresponds, of course, to a constant drift mobility and to the constant current of Fig. 6.14b at times shorter than the transit time. Note that $\Delta x(t)/x(t)$ decreases with time, so that the current cutoff observed near t_{transit} becomes relatively sharper for thicker samples. If L is the sample thickness and E is the applied electric field, the drift mobility μ is obtained from the observed transit time via the relation $\mu E = L/t_{\text{transit}}$.

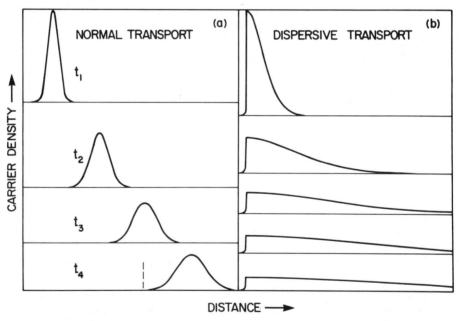

Figure 6.15 The evolution with time of the carrier-density profile of the injected-charge pulse in normal and dispersive transport (after Scher and Montroll, 1975). In a, the times used to label the sequence of pulse shapes roughly correspond to those used in a and b of Fig. 6.14, if the back electrode is considered to be at the place marked by the dashed line. The $I(t)$ trace of Fig. 6.14c corresponds to the dispersive-transport situation shown above in b, in which the current arises from the progressive smearing out of the charge distribution toward the back electrode. In marked contrast to the orderly parade illustrated in a, stragglers constitute the most prominent aspect of the disheveled march shown in b.

The situation depicted in Fig. 6.15*b* corresponds to *dispersive transport*, the term now widely adopted to describe the disorder-induced $I(t)$ behavior of Fig. 6.14*c*. First a word of warning about this term, which is placed in opposition to *normal transport* (Figs. 6.15*a* and 6.14*b*). It should be noted that normal transport is *not* nondispersive. Dispersion does occur in normal transport, in which diffusion disperses the drifting sheet by the spread Δx of Eq. (6.7). Its role here, however, is minor, and $\Delta x/x$ can be made very small by making the sample thickness L very large. By contrast, the effects of dispersion are of overriding importance in dispersive transport, and they cannot be evaded by any choice of L. Far from playing a minor role in the carrier motion, dispersion is now, in a sense, the whole story. We will find that one casualty of strong dispersion is the utility of the idea of a definite drift mobility.

In the Scher–Montroll theory of dispersive transport, the relations which replace Eq. (6.7) are

$$x(t) \sim t^{\eta},$$

$$\Delta x(t) \sim t^{\eta}. \qquad (6.8)$$

Here η is their dimensionless dispersion parameter which, as discussed below, characterizes the extended distribution $\{t_i\}$ of microscopic event times. It is assumed that

$$0 < \eta < 1. \qquad (6.9)$$

In Eq. (6.8), $x(t)$ and $\Delta x(t)$ denote the evolution with time of the mean position and rms spread of the charge distributions shown in Fig. 6.15*b*. For these distributions, it is clear that the motion of the centroid $x(t)$ does *not* reflect (as it does for normal transport) the motion of the peak. The peak, in fact, *does not move*; it remains close to the front electrode and is progressively reduced in magnitude as the distribution progressively spreads out toward the back electrode. The current flow (which is related to the centroid motion by $I \sim dx/dt$) arises from the continual *asymmetrical spreading out* of the charge. The mean position and the spread are proportional to each other, as revealed by Eq. (6.8) as well as by an examination of Fig. 6.15*b*. Dispersion and drift are synonymous with each other. Since $I \sim dx/dt$ and $x \sim t^{\eta}$, then $I \sim t^{\eta-1}$, and since $\eta < 1$, the current (and the apparent drift mobility) decreases with time. The current falls from the outset, well *before* appreciable charge arrives at the back electrode [which is not yet in the picture in the relations of Eq. (6.8)].

The model solved by Scher and Montroll is an asymmetric continuous-time random walk. The asymmetry in the walk, introduced to mimic the effect of the electric field, provides a bias favoring steps in the $+x$ direction over steps in the $-x$ direction. While the walk takes place on a regular lattice, so that the steps are equal in length, they are quite unequal in *time*. The temporal aspect is governed by an event-time distribution function $\psi(t)$. If the random walker arrives at a site at time $t = 0$, then $\psi(t)dt$ is the probability of his taking a step to an adjacent site within the time interval from t to $t + dt$. The theory thus folds the distribution $\{t_i\}$, the set of microscopic event times which corresponds to *all* of the localized sites upon which the injected carriers walk in their paths through the solid, onto the function $\psi(t)$ which is taken to apply to *each* site in the regular-lattice model. The event-time distribution is given an extended tail in the form of a slow (i.e., algebraic) falloff at long times:

$$\psi(t) \sim 1/t^{1+\eta} \tag{6.10}$$

The parameter η is the same as that of relations (6.8) and (6.9). Indeed, the assumption of the distribution of Eq. (6.10) in the continuous-time random walk gives rise to Eq. (6.8) and the charge-density profiles of Fig. 6.15b.

The $x(t)$ of Eq. (6.8), corresponding to a current falling as $1/t^{1-\eta}$, does not yet take into account the eventual arrival (and disappearance) of carriers at the back electrode. When they included this, Scher and Montroll obtained the following general form for the overall behavior of the pulse-induced transient current:

$$I(t) \sim t^{-(1-\eta)}, \quad \text{for } t < t_{\text{transit}}; \tag{6.11}$$

$$I(t) \sim t^{-(1+\eta)}, \quad \text{for } t > t_{\text{transit}}. \tag{6.12}$$

Equation (6.11) describes the behavior for times short enough so that $x(t) \ll L$, for which the back electrode is still effectively out of the picture and nearly all of the photoinjected carriers remain in transit. Equation (6.12) is the asymptotic behavior for times long enough so that $x(t) \gg L$, when few carriers are left. The characteristic "transit time" is the demarcation between the behavior of Eq. (6.11) and that of Eq. (6.12). From Eq. (6.8), we expect that $t_{\text{transit}} \sim L^{1/\eta}$.

Equations (6.11) and (6.12) imply the presence of two linear regimes on a log-log plot of current versus time. Double-logarithmic plots of $I(t)$ are now routinely used to analyze the transient photocurrents observed in time-of-flight experiments on amorphous semiconductors, because this predicted behavior is indeed found for many materials. Figure 6.16 displays data obtained on amorphous As_2Se_3 at room temperature and on amorphous selenium at low temperature (Pfister and Scher, 1978). The two linear regimes, in each of which the current exhibits a power-law time decay, are clearly seen. Moreover, the sum of the slopes of the two linear portions is -2, just as anticipated from the power-law exponents expressed in Eqs. (6.11) and (6.12). The observed behavior of Fig. 6.16 provides an empirical definition of the characteristic time; t_{transit} is taken to be the time given by the intersection of the extrapolated linear regimes. Thus this analysis yields a time "marker" which is invisible in a linear $I(t)$ display such as Fig. 6.14c.

The results of Fig. 6.16 display a striking *scaling* characteristic which also comes out of the Scher–Montroll theory. The data shown for hole transport in As_2Se_3 glass actually consists of six distinct experimental runs; measurements were made at three different applied fields on each of two samples of different thickness. For all six curves, I and t have been normalized to their values at the knee of the curve ($t = t_{\text{transit}}$, which differs from run to run and varies over three decades among the six runs). Thus normalized, *all of the data collapse onto a single curve.* This feature also applies to the selenium data shown in Fig. 6.16, which correspond to four sets of observations recorded at different fields. Such scaling behavior arises as a consequence of a scaling property possessed by the dispersive-transport charge-distribution curves of Fig. 6.15*b*; all of these curves can be renormalized to each other by simply rescaling the axes (with the origin a fixed point). From Eq. (6.8), a charge-density profile at time t_1 can be transformed to one at time t_2 by expanding its spatial extent by a factor of $(t_2/t_1)^\eta$ and decreasing its amplitude by the same factor.

The two branches of the Scher–Montroll curve [Fig. 6.16, Eqs. (6.11) and (6.12)] can be viewed in the following way. During the period spanned by the first branch [Eq. (6.11)], the current declines steadily because the electrons (or holes, as in the examples of Fig. 6.16) *slow down* in their passage across the amorphous solid. They slow down because of the enormous spread of microscopic event times represented by the distribution function $\psi(t)$ of Eq. (6.10). This extended distribution, physical mechanisms for which are addressed below, permits slow processes to become influential at long times. Note that the continually decreasing drift velocity means that *the conventional notion of a well-defined drift mobility loses its validity in this setting.* If we plug t_{transit} (located by the break in a log–log plot) into the relation $\mu E = L/t_{\text{transit}}$, we obtain a thickness-dependent mobility $\mu(L)$ which decreases with increasing L. Mobility, in this context, is not useful in terms of specifying an intrinsic characteristic of the solid.

The second branch of the Scher–Montroll curve [Eq. (6.12)] applies to times long enough so that a significant number of electrons are exiting the sample at the back electrode. Now there are *two* parallel processes which cause the current to go down with time. In addition to the carrier's steadily decreasing speed (the process operative in the first branch), now their *number* also decreases with time, so that the power-law falloff in $I(t)$ is faster than in the first branch. Note that even at these long times, when many carriers have successfully made it completely across the sample and have arrived at the back electrode, an appreciable fraction of the remaining carriers are *still immobilized* near the front electrode! (Look again at Fig. 6.15*b*.)

Why do the electrons get tired and slow down, i.e., what is the physical origin of the vast spread of microscopic event times that characterizes dispersive transport in amorphous solids? Two distinct mechanisms provide plausible candidates. If the transport takes place by tunneling from site to site within a manifold of localized states (as in the case of variable-range hopping), then a broad distribution of event times clearly arises from the exponential dependence of the transition probability on the intersite distance. The inevitable spread of site energies also contributes to the breadth of the time distribution. The other main candidate is extended-state transport interrupted by trapping events such as that indicated by D in Fig. 6.12. The event-time distribution

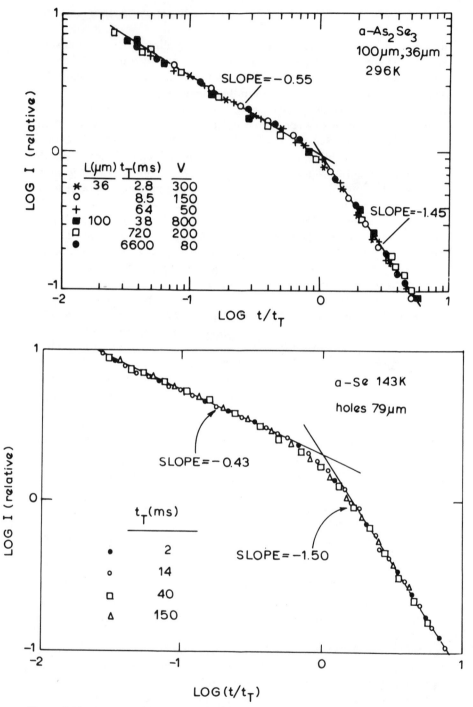

Figure 6.16 Log-log plots of transient-photoconductivity $I(t)$ traces for hole transport in amorphous As$_2$Se$_3$ and amorphous Se, showing the scaling behavior and the two branches of the Scher–Montroll curve (Pfister and Scher, 1978).

now reflects the distribution of release times out of the trapping localized states; these times span a wide range because the release rate decays exponentially with the depth of the localized-state energy relative to the mobility edge.

The two mechanisms mentioned above, both of which can give rise to dispersive transport, are *hopping* and *multiple trapping*. Both involve stepwise motion between localized states, but the localized states enter in completely different ways. In hopping, these states *enable* the transport to occur, since each transport step is by direct quantum-mechanical tunneling from one localized site to another. In multiple trapping, the localized states *impede* transport. Each step now corresponds to a sequence consisting of release from one site followed by extended-state transport and trapping at a second site. The time spent in bandlike motion is much shorter than the time spent in immobility on the trapping sites.

Despite their quite different microscopic natures, it is difficult to distinguish between hopping and multiple trapping on the basis of dispersive transport alone. Hopping has been identified as responsible for the dispersive transport observed in certain molecularly doped organic polymers (Mort et al., 1976). In these organic glasses the spacing of the relevant localized sites can be chemically controlled, providing an independent test of the hopping mechanism. Multiple trapping is believed to be important in the dispersive transport seen in the chalcogenide glasses.

6.4 NATIVE DEFECTS AND USEFUL IMPURITIES

The four-coordinated continuous random network discussed in Chapter Two for amorphous silicon and germanium, while enormously useful, is naturally an idealization of the structure of the actual solids. This point, while true to some extent for all glasses, is especially notable in the case of materials such as *a*-Si and *a*-Ge, which cannot be formed in bulk-glass form but instead are typically prepared as thin films by *very* rapid quenching from the vapor phase. Let us focus on amorphous silicon. It is known that *a*-Si contains microscopic voids (size on the order of several atoms on a side), a situation sometimes exaggeratedly referred to in terms of the "Swiss-cheese model" for this material. Evidence for the microvoids comes from small-angle X-ray and electron-scattering experiments and from the electrical and ESR behavior discussed below. Some silicon atoms adjoining such a void have an sp^3 electron orbital projecting into the void space, unbonded covalently to any other silicon. A single electron occupying such a covalently unsatisfied bond-prepared orbital is called a dangling bond. Two nearby dangling bonds on the periphery ("internal surface") of a void may react to form a not-quite-normal doubly occupied covalent bond in a manner similar to the "reconstruction" process that occurs on a free surface of crystalline silicon. Dangling bonds which do not reconstruct, and which remain as unpaired electrons, give rise to an electron spin resonance signal in *a*-Si. ESR measurements indicate roughly one spin per 10^3 Si atoms.

The void-associated dangling bonds and distorted reconstructed bonds are *native defects* intrinsic to amorphous silicon. Not associated with foreign atoms but instead consisting of unavoidable deviations from the ideal network ($z = 4$ crn), these native defects profoundly influence the electronic properties of a-Si. Both the dangling bonds and the weak reconstructed bonds introduce electron states within the energy gap of this amorphous semiconductor. That a dangling bond introduces a state in the gap may be seen from Fig. 6.2, since it can be considered to be equivalent to one of the sp^3 hybrids of the "prepared-for-bonding" isolated atom. That a reconstructed bond does so may be seen by considering it to be a "long bond" with an abnormally large Si—Si distance. A large bond length implies a small bonding–antibonding splitting and, again from Fig. 6.2, states within the gap. As a consequence, therefore, of its intrinsic native defects, amorphous silicon possesses a large density of states $n(E)$ within the pseudogap region, that is, between the mobility edges E_v and E_c in Fig. 5.14b. Roughly speaking, dangling bonds contribute midgap bumps to $n(E)$, while weak bonds contribute to the tails near E_v and E_c.

The large density of bandgap states in a-Si make this amorphous semiconductor electronically dirty: Its electrical resistivity is low, the Fermi energy E_F is "pinned" (insensitive to the addition of doping impurities) because $n(E_F)$ is large, and the large number of localized states energetically close to E_F reveal themselves via the variable-range tunneling conductivity behavior of Fig. 6.13. Thus, because of its significant concentration of native structural defects, *chemically pure amorphous silicon is electronically messy.* Consequently this material, the amorphous counterpart of the crystalline semiconductor which is perhaps the most technologically important material in existence, is itself rather unpromising from a technological viewpoint. An ironic discovery has overturned this state of affairs: The addition to this elemental glass of a hefty concentration of a chemical impurity (hydrogen) has the effect of cleaning out the bulk of the electronically obnoxious bandgap states! Hydrogenated amorphous silicon, usually written a-Si:H (although $Si_{1-x}H_x$ is really more appropriate, where x is typically 0.1–0.2), is the material mentioned in Chapter One as the silicon-based amorphous semiconductor of technological interest for solar-cell or photoreceptor applications.

Hydrogenated amorphous silicon is usually prepared by a technique known as glow-discharge decomposition or plasma deposition. Molecules of silane (SiH_4), at low pressure within a vacuum chamber, are torn apart by electron impact within an alternating electric field. The fragments react on the surface of a heated substrate, depositing a film of a-Si:H. The process has a certain resemblance to the formation of a-SiO$_2$ from $Si(OH)_4$ sketched earlier in Fig. 4.12, but in that gelation case the molecular dissociation and subsequent condensation reaction occur in solution while in the present case the molecular dissociation occurs in the gas phase and the subsequent condensation reaction takes place on a solid surface. Moreover, and this has turned out to be crucial, in the present instance *the elimination of hydrogen is quite incomplete.* The chemical incorporation of a large amount of hydrogen, typically in the range of 3 to 20 atomic percent, is responsible for the enormously beneficial effect upon the electronic properties of a-Si:H.

The beneficial action of H in $Si_{1-x}H_x$ is attributed to the strength of the Si—H covalent bond. Hydrogen terminates (bonds to) the dangling bonds occurring in amorphous silicon, removing the gap states associated with these native defects. This is demonstrated by the quenching of the dangling-bond ESR signal. In addition, hydrogen evidently opens up the weak reconstructed bonds associated with the voids and bonds to those silicon atoms as well, replacing each such long Si—Si bond by two Si—H bonds. Since the Si—H bond is very strong and its bonding–antibonding splitting is larger than that of Si—Si, the states introduced by the Si—H bonds lie at energies which are *outside* of the bandgap region of the a-Si "host." The effect of all of this is to wipe out most of the electron states in the gap which the network defects native to a-Si would normally introduce. Hydrogenated amorphous silicon is *chemically impure* (actually, there is so much H in amorphous $Si_{1-x}H_x$ that the material can be viewed as a silicon–hydrogen glassy alloy) but it is, vis-à-vis amorphous silicon, *electronically clean*.

In addition to mopping up the bandgap states derived from structural deviations from the ideal four-coordinated continuous random network, there is an additional aspect to the beneficial action of bonded hydrogen in a-Si:H. In the Polk model of Chapter Two for the $z = 4$ crn of a-Si, the primary loss of short-range order imposed by the constraints involved in building up the defect-free random network resides in the distribution of bond angles about the tetrahedral value. Hydrogen, in a-Si:H, enters into the network structure and changes it. The covalent coordination in $Si_{1-x}H_x$ is no longer four but is instead $z = 4 - 3x$. [Here z is the *average* coordination; a given atom still has either $z(Si) = 4$ or $z(H) = 1$.] With this reduction in coordination, the covalent network in $Si_{1-x}H_x$ is less overconstrained (Phillips, 1981) than is the network in pure amorphous silicon. As a result, the bond-angle strain in a-Si:H is less than it is in a-Si. (This lowering-of-constraints or increase-of-freedom explanation is essentially the same as the reason for the nearly-vanishing O—Si—O bond-angle strain in a-SiO$_2$.) Since the tailing of $n(E)$ into the pseudogap region in a-Si is attributed, at least in part, to the distribution of Si—Si—Si bond angles in the random network structure, the *narrowing* of this distribution in a-Si:H reduces the extent of the tailing and thus provides another mechanism by which the incorporation of hydrogen cleanses the solid of states in the gap.

$Si_{1-x}H_x$, with $x \approx 0.1$, reveals its electronic superiority to a-Si in its high electrical resistivity and in exhibiting clear photoconductivity and luminescence. With its energy-gap region now relatively free of states, the temperature dependence of its conductivity no longer resembles the variable-range-hopping behavior seen for a-Si (Fig. 6.13) but instead obeys a standard $\exp(-E/kT)$ semiconductor behavior with E about half the optical gap E_0. (E_0 is about 1.5–

1.9 eV for typical Si—H glasses, compared to 1.3 eV for a-Si.) Transport is dominated by extended states rather than by localized states as in a-Si (i.e., by processes such as C in Fig. 6.12, rather than processes such as A or B).

Most importantly, the semiconducting properties of a-Si: H can be controlled by *doping* the material either n-type (conduction by electrons, i.e., extended states above E_c) or p-type (conduction by holes, i.e., extended states below E_v) by chemically incorporating either column-five donors (such as phosphorus) or column-three acceptors (such as boron) in much the same way as in crystalline silicon. This is accomplished by mixing small amounts of gases such as PH_3 or B_2H_6 with the SiH_4 during the plasma deposition of the film (Spear and LeComber, 1975). While the doping process is not nearly as finely controlled as, for example, in the delicate example of phosphorus-doped crystalline silicon shown previously in Fig. 5.16, the fact that it can be done at all (enabling p-n junctions to be prepared) is extremely significant and is critically important for the electronic applications of this amorphous semiconductor. What makes it possible is that in a-Si: H the bandgap states inserted by the dopant impurities do not have to contend with an overwhelming density of states-in-the-gap native to the host solid, as they do in a-Si. Thus these impurity states are able to push the Fermi level around, and in particular to move it close to E_c or E_v, allowing a-Si: H to act as an *n*-type or *p*-type *extrinsic semiconductor*.

The fact that a-Si: H can be doped, for example, n-type via phosphorus incorporation shows that a fair fraction of the P atoms enter the covalent network with fourfold coordination (in effect, "substitutionally" for silicon), instead of the threefold coordination they would prefer according to the $8 - n$ rule. Evidently this happens because of the drastically fast vapor-quench techniques required for the preparation of amorphous silicon. It is in marked contrast to the behavior of semiconducting chalcogenide glasses, whose electrical properties are generally rather insensitive to changes (of even several percent) in chemical composition. During the relatively slow formation of these glasses by the gentle quenching of the melt, the proper bonding arrangements (needed to satisfy $z = 8 - n$) have enough time to establish themselves. Nevertheless, native defects also play a role in the properties of the chalcogenides, as we now describe.

Native defects occurring in the crn structure of chalcogenide glasses have already been mentioned earlier, in Section 3.5, using amorphous selenium as the example for simplicity. Chain-structure Se_N molecules form Se glass, so that the natural defects to expect are dangling bonds at the ends of chains, that is, singly coordinated Se atoms which terminate chains. Such defects are electrically neutral, and may be denoted $Se^0(z = 1)$, where the superscript denotes the net charge and the covalent coordination is given in parentheses. [In this notation, a normal atom of the crn network of this polymeric glass is denoted $Se^0(z = 2)$.] In Fig. 3.8, pairs of charged defects called *valence alternation pairs* (Kastner, Adler, and Fritzsche, 1976) were schematically illustrated. These consist of a positively charged three-coordinated selenium $Se^+(z = 3)$ and a negatively charged singly coordinated selenium $Se^-(z = 1)$. Such a pair of defects can be formed by the transfer of an electron from one dangling-bond selenium to another:

$$2Se^0(z = 1) + U_{eff} \rightarrow Se^+(z = 3) + Se^-(z = 1). \quad (6.13)$$

The chemical reaction expressed by Eq. (6.13) corresponds to the step which occurs between panels *c* and *d* of Fig. 3.8. $Se^+(z = 3)$ and $Se^-(z = 1)$ are represented in panels *e* and *f*, respectively. [In Fig. 3.8, the *s* electrons are counted among the lone pairs, so that four lone-pair electrons are shown on each $Se^0(z = 2)$, two on $Se^+(z = 3)$, and six on $Se^-(z = 1)$.] As discussed in Section 3.5, $z = 3$ and $z = 1$ are the natural $(8 - n$ rule) covalent coordinations of Se^+ and Se^-, respectively.

In the charge-transfer bond-switching reaction of Eq. (6.13), the energy needed to convert a pair of chain ends into a valence alternation pair is written as U_{eff}, an effective correlation energy. This is in analogy with the U of Fig. 5.12*c*, the electron–electron Coulomb energy of doubly occupying the electron orbital on a given site. In the present case, there is an e^2/r_{12} potential-energy cost which is the price to pay for bringing an electron (initially alone in a nonbonding orbital on another chain-end atom) to join its lone-pair partner on the chain-end atom which receives the electron and becomes Se^-. However, there is *also* an energy *benefit* which occurs as a consequence of the bond switching that takes place around the Se which supplied the electron and becomes Se^+, eager to form three bonds. The bond switching (actually, bond formation) converts two nonbonding electrons into bonding electrons, thus lowering their energy substantially. It is possible, therefore, that the net energy cost U_{eff} is *negative*.

This is believed to be the case for multicomponent chalcogenide glasses such as As_2Se_3 or Se—As—Ge. If so, dangling bonds are unstable, via Eq. (6.13), with respect to the formation of valence alternation pairs of charged defects having all electrons paired (in bonding and nonbonding orbitals). These defects would then be those lowest in energy relative to the ideal crn covalent network (note again that panel *d* in Fig. 3.8 contains the same number of bonds as the "perfect network" of panel *b*). Because of their low creation energy, a large concentration of these defects should be present in thermodynamic equilibrium at T_g, and these native defects would be frozen in during melt-quenching formation of the glass.

In the case of amorphous silicon, we have seen that the chemical addition of hydrogen has the effect of tying up the loose ends, as initially unpaired electrons in dangling bonds end up paired in strong covalent bonds. The above discussion suggests that in chalcogenide glasses a similar effect occurs *spontaneously*, via processes such as Eq. (6.13) in which dangling bonds annihilate each other in pairs. The resulting valence alternation pairs provide a means for understanding some important properties of chalcogenide semiconductors which are otherwise quite difficult to account for.

Electrical resistivity measurements on chalcogenide glasses generally find that the Fermi energy is strongly pinned near mid-gap. Also, these materials are extremely difficult to dope. A multicomponent glass such as $Se_{1-x-y}As_xGe_y$ is quite insensitive in its electrical properties to compositional changes on the order of several percent. But this electric evidence of a large density of states in the pseudogap is apparently contradicted by magnetic measurements: ESR and magnetic susceptibility studies fail to find substantial numbers of free spins in the chalcogenides (in marked contrast to the case for a-Si). Moreover, variable-range-hopping transport involving states near the Fermi level is *not* observed (again, in contrast to a-Si, with its densely populated pseudogap). Valence alternation pairs appear to neatly reconcile a high density of (electrically active) electron states near mid-gap with a low density of (magnetically active) unpaired electron spins. The states are there, but they are nearly all paired (and thus magnetically inert, since sites have no net spin) via Eq. (6.13) with $U_{eff} < 0$. The absence of electron hopping, as well as the tenacity of the Fermi-level pinning, are also consequences of the "negative U" (Anderson, 1975; Adler and Yoffa, 1976) which we define to be beyond the scope of this book, since it is getting late. One further point, however, should be mentioned in connection with an *optical* characteristic of these solids. The optical absorption edge in chalcogenide glasses typically exhibit an exponential tail (such as that shown for As_2S_3 in Fig. 6.7) which is broad and persists at low temperature. Since internal electric fields associated with charged defects have the effect of broadening an optical edge, the native defects described above may explain the exponential absorption tails seen in these glasses.

Having discussed native defects in both a tetrahedrally bonded amorphous semiconductor (a-Si) and in chalcogenide glasses, and having discussed the role of a "useful impurity" (hydrogen) in aiding electronic applications of the former, we now close with a quick discussion of a case of a useful impurity for the latter class of amorphous solid. One of the simplest chalcogenide glasses is amorphous selenium, and one of the most familiar applications of large-area chalcogenide photoconductors is in xerography. The first photoconductor in widespread use in xerography was, in fact, amorphous selenium. Pure selenium, however, had a problem. If heated to appreciably above room temperature, a-Se has a tendency to crystallize, and crystalline selenium is insufficiently insulating to hold an electrostatic charge in a xerographic application (Fig. 1.12a).

The fix, for this practical problem of thermal instability against crystallization, is chemical crosslinking. As shown earlier in the schematic of the covalent graph of $Se_{1-x-y}As_xGe_y$ (Fig. 3.7), incorporation into the glass of a sufficient concentration of three-coordinated (e.g., arsenic) or four-coordinated (e.g., germanium) atoms knits together the Se_N chains into a single three-dimensional network. This effectively stabilizes the glass against crystallization, solving the problem presented by pure selenium. Note that the structural role of the chemical impurities in this case, with respect to the continuous random network of the host glass, is opposite to the role of hydrogen in $Si_{1-x}H_x$. In the case of hydrogenated amorphous silicon, the addition of a low-coordination ($z = 1$) element reduces the number of constraints present by *lowering* the connectivity of the covalent graph; while in the case of $Se_{1-x-y}As_xGe_y$, the addition of high-coordination ($z = 3$ and $z = 4$) elements increases the constraints by *raising* the covalent connectivity of the system.

Assuming that arsenic is chosen as the cross-linking additive (the "useful impurity"), how much is needed, i.e., how large does x need to be in $Se_{1-x}As_x$ in order for the connectivity to be adequate to effectively curtail the tendency for thermally induced crystallization? The answer, in practice, is found to be: *very little*. One percent of arsenic is *more* than sufficient.* The reason for this has already been given in (surprise!) the percolation chapter, in Section 4.6. For an assembly of long chains of length N, the percolation threshold is of order $1/N$. Since $N \approx 10^5$ for selenium glass, $x \approx 10^{-2}$ is plenty. Which, having reached the topic of deciding "what is enough" brings us to: the end.

REFERENCES

General References on Amorphous Semiconductors

Brodsky, M. H. (editor), 1979, *Amorphous Semiconductors*, Springer-Verlag, Berlin.
Connell, G. A. N., and G. Lucovsky, 1978, *J. Non-Crystalline Solids* **31**, 123.
Connell, G. A. N., and R. A. Street, 1980, in *Handbook on Semiconductors*, Vol. 3, edited by S. P. Keller, North-Holland, Amsterdam, p. 689.
Hamakawa, Y. (editor), 1982, *Amorphous Semiconductor Technologies and Devices*, Ohmsha Ltd, Tokyo.
Knights, J. C., and G. Lucovsky, 1980, *CRC Critical Reviews in Solid State and Materials Sciences* **9**, 211.
Mott, N. F., and E. A. Davis, 1979, *Electronic Processes in Noncrystalline Materials*, Clarendon Press, Oxford.
Paul, W., and D. A. Anderson, 1981, *Solar Energy Materials* **5**, 229.
Weaire, D., and P. C. Taylor, 1980, in *Dynamical Properties of Solids*, Vol. 4, edited by G. K. Horton and A. A. Maradudin, North-Holland, Amsterdam, p. 1.

References

Adler, D., 1980, *J. Chem. Ed.* **57**, 560.
Adler, D., and E. J. Yoffa, 1976, *Phys. Rev. Letters* **36**, 1197.
Ambegaokar, V., B. I. Halperin, and J. S. Langer, 1971, *Phys. Rev. B* **4**, 2612.
Anderson, P. W., 1975, *Phys. Rev. Letters* **34**, 953.
Aspnes, D. E., S. M. Kelso, C. G. Olson, and D. W. Lynch, 1982, *Phys. Rev. Letters* **48**, 1863.
Brodsky, M. H., and M. Cardona, 1978, *J. Non-Crystalline Solids* **31**, 81.

*Actually, Cary Grant knew this in 1944.

Duwez, P., 1967, in *Phase Stability in Metals and Alloys*, edited by P. S. Rudman, J. Stringer, and R. I. Jaffee, McGraw-Hill, New York, p. 523.

Hauser, J. J., 1973, *Phys. Rev. B* **8**, 3817.

Kastner, M., 1972, *Phys. Rev. Letters* **28**, 355.

Kastner, M., D. Adler, and H. Fritzsche, 1976, *Phys. Rev. Letters* **37**, 1504.

Mooser, E., and W. B. Pearson, 1960, in *Progress in Semiconductors*, Vol. 5, Wiley, New York, p. 104.

Mort, J., G. Pfister, and S. Grammatica, 1976, *Solid State Commun.* **18**, 693.

Mott, N. F., 1969, *Phil. Mag.* **19**, 835.

Pfister, G., and H. Scher, 1978, *Adv. Phys.* **27**, 747.

Phillips, J. C., 1981, *J. Non-Crystalline Solids* **43**, 37.

Pierce, D. T., and W. E. Spicer, 1972, *Phys. Rev. B* **5**, 3017.

Scher, H., and E. W. Montroll, 1975, *Phys. Rev. B* **12**, 2455.

Spear, W. E., and P. G. LeComber, 1975, *Solid State Commun.* **17**, 1193.

Tauc, J., 1974, in *Amorphous and Liquid Semiconductors*, edited by J. Tauc, Plenum, London, p. 159.

Weinstein, B. A., R. Zallen, and M. L. Slade, 1980, *J. Non-Crystalline Solids* **35**, 1255.

Zallen, R., 1968, *Phys. Rev.* **173**, 824.

Zallen, R., 1977, *Phys. Rev. B* **16**, 1426

Zallen, R., and D. F. Blossey, 1976, in *Optical and Electrical Properties of Materials with Layered Structures*, edited by P. A. Lee, Reidel, Dordrecht, p. 231.

Zallen, R., R. E. Drews, R. L. Emerald, and M. L. Slade, 1971, *Phys. Rev. Letters* **26**, 1564.

Index

Abraham, F. F., 208–212, 251
Abrahams, E., 246, 248, 251
Adler, D., 30, 32, 105, 106, 133, 227, 251, 255, 292, 294–296
Agrawal, P., 192, 193, 203
Aklonis, J. J., 4, 32
Alben, R., 83–85
alkali atoms in oxide glasses, 100, 101
Ambegaokar, V., 281, 295
amorphous chalcogenides, *see* chalcogenide glasses
amorphous metals, *see* metallic glasses
amorphous semiconductors:
  applications of, 24, 28–31, 290–295
  chalcogenide, *see* chalcogenide glasses
  doping of, 292
  electronic structure of, 234, 235, 254–260
  electron transport in, 276–289
  mobility edges in, 234, 235
  optical properties of, 265–273
  preparation of, 8–10, 290
  tetrahedrally coordinated, 67–72, 76–84, 254–256, 258, 289–292
amorphous solid, 3
Anderson, P. W., 16, 231–233, 239–242, 246, 248, 251, 294, 295
Anderson Hamiltonian, 239–242
Anderson localization, *see* Anderson transition
Anderson transition, 205, 224–226, 231–251
Angell, C. A., 21, 32, 206, 251
antibonding–bonding splitting, 256–260, 265
antibonding–nonbonding splitting, 260, 265
antibonding orbitals, 255–257
applications of amorphous solids, 23–32, 273, 290–295
arsenic, 64, 93, 295
arsenic–selenium system:
  compositional freedom in, 97–99
  continuous random network of glasses in the, 65
  crystalline forms in the, 63, 64
  electrical properties of, 277, 286–288
  glass-forming compositions in, 101
  optical properties of, 263, 264
  photocrystallization of, 96
  Raman scattering of, 96, 97
  valence alternation in, 293
  in xerography, 24, 28, 294, 295

arsenic–sulfur system:
  annealing effects in, 95
  continuous random network of glasses in the, 65
  crystalline forms in the, 63, 64, 92
  electronic structure of, 258–260
  glass-forming compositions in the, 101
  glass transition in, 6
  macromolecular nature of, 90–93
  optical properties of, 258–267
  quenching rate for, 8
  short-range order in, 44, 93
  specific heat of, 17, 18
  structure of, 91–96
Aspnes, D. E., 273, 295
atomic polyhedron, *see* Voronoi polyhedron
average gap, 254–260

Band conduction, 274, 276, 277
bandgap, 235, 254–260, 271–273, 291
bandgap states, 290–292
band-structure effects in crystals, 225, 234, 254, 263–265, 271, 272
band theory of electrons in crystals, 223–225, 234, 254, 274
bandwidth, electron, 229–232, 241, 242
Bankstahl, H., 11, 32
Barker, J. A., 51, 52, 82, 85
Bell, R. J., 72, 73, 84, 85
Bennett, C. H., 81, 82, 85
Benoit, H., 113, 120, 133
Bernal, J. D., 50, 85, 95
Bernal model, 50, 52
Bethe lattice, 168, 171–174, 178–180
Bishop, G. H., 162, 203
bismuth, 10
Blachnik, R., 17, 32
Bloch functions, 223–226, 241, 274
blocked bonds in percolation, 138
blocked sites in percolation, 139
Blossey, D. F., 262, 296
boat configuration of a six-atom ring, 69, 71
bond-angle statistics, 65–69, 72, 291
bond circuits, *see* ring statistics
bonding–antibonding splitting, 256–260, 265
bonding orbitals, 255–257, 265
bonding schematics for electronic structure, 255–258
bond lengths in covalent glasses, 67, 69, 72

bond percolation, 138–141, 145, 168–171, 187, 198–201
bond reconstruction in chalcogenides, 95
borderline dimensionality, *see* marginal dimensionality
Boudreaux, D. S., 68, 85
bridging oxygens, 73, 182
bridging sulfurs, 93, 182
Brodsky, M. H., 269, 270, 295
Brownian motion, 115, 116, 119, 131, 132, 165

Cardona, M., 269, 270, 295
Cargill, G. S., 49, 80, 81, 85
Cayley tree, *see* Bethe lattice
chair configuration of a six-atom ring, 69, 71
chalcogen elements, 86
chalcogenide glasses, 17, 18, 65, 86–107, 256–260, 264–268, 277, 292–295
Chandrasekhar, S., 114, 133
Chaudhari, P., 7, 32, 49, 85
chemical-bonding view of electronic structure, 255–260, 265
chemical impurities, 103, 290–295
chemical order in glasses, 101–104
Chen, H. S., 11, 18, 32
close packing: *see* crystalline close packing; random close packing
clusters, *see* percolation clusters
cluster size:
    average 143, 154, 245
    distribution, 142, 158–162, 221
cobalt, 10
Cohen, M. H., 20, 32, 49, 85, 189, 203, 204, 215, 219, 221, 233, 236, 251
Cohen-Turnbull free-volume model, 215–218
communal entropy, 221, 222
compositional freedom in glasses, 97–101
computer simulations of:
    the Anderson transition, 250
    bond percolation, 145, 198–201
    continuum percolation, 188, 189
    the glass transition, 206–212
    random close packing, 51, 52
    real-space renormalization, 198–201
    site percolation, 160, 164
condensation reaction, 175–178
conductivity, 145, 190, 243–251, 274–280
connected bonds in percolation, 138
connected sites in percolation, 139
connectivity transition, *see* percolation transition
Connell, G. A. N., 82, 83, 85, 94, 134, 295
contiguity number, 56

continuous random network, 35, 60–73, 79, 81–85, 291
continuous-time random walk, 285, 286
continuum percolation, 186–191, 238, 239
coordination number, 33, 34, 38, 39, 291, 295
coordination shell, 40, 41
correlation energy, 228–230, 251, 293, 294
correlation-induced localization, *see* Mott transition
Cotton, J. P., 113, 120, 133
Coulomb interactions, 228–231, 293
covalent glasses, 63–73, 76–85, 101–104, 255–258, 289–295
covalent graphs, 62–64, 86–101, 295
covalent networks, 62–65, 67–73, 86–101, 182, 258, 295
Coxeter, H. S. M., 51, 85
critical concentration for percolation, 168–171
critical exponents:
    dimensionality dependence of, 166
    for localization, 245, 250, 251
    for percolation, 156–166, 200
    for polymer configurations, 123, 124
    universality of, 158, 165, 166, 197, 198
critical region, 156
critical volume fraction for percolation, 166, 185–191, 238
cross-linking in chalcogenide glasses, 98
crystalline close packing, 35–38, 54, 58, 164
crystalline order, 12, 35–38, 58
crystallinity, 12, 15
crystallization, 2, 5, 7, 18, 96

Dalton, N. W., 200, 203
dangling bonds, 104–106, 289–293
Davis, E. A., 252, 276, 277, 295
Dean, P., 72, 73, 84, 85
Decker, D., 113, 120, 133
DeConde, K., 242, 243, 251
decoration transformation, 63, 64, 72, 73, 93
defects in glasses, 104–107, 290–294
DeFonzo, A. P., 96, 133
de Gennes, P. G., 127, 133, 177, 180, 203
Delaunay division, 60
delocalization and kinetic energy, 46, 62, 229–231, 256
delocalization–localization transitions, 205–251
DeNeufville, J. P., 94, 133
dense random packing, *see* random close packing
density of states, 234, 235, 254–258, 290, 291
diamond structure, 61, 62
diffraction studies of structure, 42, 43, 73–85, 113, 119, 120

diffusive motion, 14, 15, 115, 213–217, 221, 247, 284
dihedral-angle statistics, 69, 70
DiMarzio, E. A., 20, 22, 32
dimensional invariants, 157, 170, 171, 186–188
dimensionality:
  of covalent networks, 87–94
  dependence of Anderson localization, 246–251
  dependence of close packing, 58, 59, 164
  dependence of critical exponents, 166
  dependence of the Flory exponent for polymers, 124–126
  dependence of Mott's variable-range-hopping exponent, 280
  Euclidean, 130, 132, 159
  fractal, 131–133, 158, 161–163
  Hausdorff–Besicovitch, 130
  marginal, 124, 126, 164–167, 239
  network, 87–94, 130, 182
  noninteger, 130–133, 161–167
  of percolation clusters, 161–163
  of polymer configurations, 130–133
  as a variable, 165–167
disorder, compositional, 13
disorder, diagonal, 240
disorder, topological, 13, 49–73, 184
disorder-induced localization, *see* Anderson transition
disorder-induced scattering, 275
disorder parameter, 232, 233, 241, 242, 247
dispersive transport, 282–289
Domb, C., 120, 123, 133, 200, 203
donors in silicon, 242–245, 251, 278, 292
Doolittle equation, 218
doping of *a*-Si:H, 292
Doyama, M., 82, 85
Drews, R. E., 264–267, 296
drift mobility, 277, 282–287
dual graph, 60, 61, 168
duality, 60, 61, 167
Duwez, P., 274, 275, 296

eclipsed bond configuration, 70
eight-minus-*n* rule, 101–104
Einstein, A., 115, 116, 133
elastic shear modulus of a gel, 180, 181
electrical networks, *see* resistor networks
electrical properties, 274–289, 292–294
electronic excitations, 252–254, 263–268
electronic structure, 234, 235, 254–260, 265
Emerald, R. L., 264–267, 296
empty sites in percolation processes, 139

end-to-end distance of a polymer coil, 116–119, 123
energy gap, 235, 236, 254–260
entropy, 17, 20–22, 221, 222
entropy, communal, 221, 222
entropy crisis at the glass transition, 21, 22
epsilon expansion, 165, 198
Euler-Poincaré relation, 51
eutectics, 10, 11
EXAFS, 43–45
excluded-volume effect on chain configurations, 120, 123, 124, 128, 129
extended-range percolation, 200, 202
extended-state transport, 274, 276, 277

face-centered-cubic lattice, 36, 37, 41, 54
Farnoux, B., 113, 120, 133
Fermi-level pinning, 290, 294
ferromagnetism, 24, 31, 32, 274
fiber-optic communications, 24–27, 267, 273
filled sites in percolation processes, 139
filling factor, 37, 49, 185–188
Finkman, E., 96, 133
Finney, J. L., 50–55, 81, 82, 85
Fischer, E. W., 113, 130, 133
Fisher, M. E., 124–127, 133, 165, 204
Fitzpatrick, J. P., 183, 203
fixed points, 195–197, 249, 250
Flory, P. J., 112, 113, 116, 121–127, 133, 177, 178, 203
Flory–Fisher argument for the self-avoiding-walk exponent, 124–127
Flory random coil model, *see* random coil model
Flory–Stockmayer theory of gelation, 177–180
flow diagrams, 194, 196, 248
fluidity, 14, 15, 218
Flynn, C. P., 186, 204
fractal dimensionality:
  for percolation clusters near threshold, 158–163
  for polymer configurations, 131–133
fractional exponents:
  for dispersive transport, 285, 286
  for percolation, 157
  for polymers, 113, 123, 124
  for self-avoiding walks, 113, 123, 124
  for variable-range-hopping, 280
free-volume model of the glass transition, 20, 206, 212–223
freezing point, 2, 5, 10, 11
Frisch, H. L., 177, 184, 191, 203
Fritzsche, H., 105, 106, 133, 233, 236, 251, 292, 296

froth, *see* Voronoi froth

gallium arsenide, 72, 273
Gaskell, P. H., 32
gelation, 175–182, 191–198, 206
gel macromolecule, 176–179
gel point, *see* sol-gel transition
germania, 6, 24, 27, 73
germanium, 10, 61, 72, 82–84, 252–256, 277
germanium–selenium system, 73, 97, 258
germanium–sulfur system, 73
germanium–tellurium system, 24, 30, 73
Gibbs, J. H., 20, 22, 32
Giessen, W. C., 7, 32
Gillis, J., 120, 123, 133
glass, 2, 3
glass-forming tendency, 8
glass–liquid transition, *see* glass transition
glass point, *see* glass transition
glass transition:
    definition of the, 2
    entropy aspects of the, 20–22, 221, 222
    free-volume model of the, 20, 206, 212–223
    kinetic aspects of the, 2–5, 19, 20, 223
    as a localization–delocalization transition,
        14, 205
    Monte Carlo computer simulations of the,
        206–212
    and the percolation model, 206, 219–223
    specific heat near the, 17–19
    temperatures for representative materials, 6
    thermodynamic aspects of the, 2, 5, 16–22,
        218–223
glassy metals, *see* metallic glasses
glassy semiconductors, *see* amorphous
    semiconductors
glow-discharge decomposition, 290
gold-silicon system, 6, 10, 11, 17, 18, 274
Graczyk, J. F., 76–78, 85
Grammatica, S., 289, 296
Grant, Cary, 295
graph concepts, uses of, 60–64, 138
graphite, 63, 64, 93
Grest, G. S., 203, 215, 219, 221, 251

Hales, Stephen, 56–58, 65
Halperin, B. I., 281, 295
Hamakawa, Y., 295
Hammersley, J. M., 139, 174, 177, 178, 184,
    191, 203
Harada, Y., 94, 134
hard-sphere model, 50, 207, 212, 213
Harrison, R. J., 162, 203
Hausdorff–Besicovitch dimensionality, 130

Hauser, J. J., 277, 279, 296
Haward, R. N., 108, 133
Hayes, T. M., 94, 134
Hayward, R., 62, 85
Henisch, H. K., 176, 204
Hess, H. F., 242, 243, 251
Higgins, J. S., 113, 120, 133
high-density percolation, 203, 219–222
Hoare, M. R., 51, 52, 82, 85
honeycomb lattice, 61, 63, 64, 168, 185
honeycomb, statistical, *see* Voronoi froth
Hoppe, A., 17, 32
hopping conduction, 276–281, 287, 289
hopping integral, 239–242, 246
hopping, variable-range, 277–281
Hutchinson, J. M., 4, 32
hydrogenated amorphous silicon, 24, 30, 31,
    290–292

Ibel, K., 113, 120, 133
Ichikawa, T., 75, 85
ideal chain configurations, 127–129
ideal glass, 103–105
impurities, 103, 290–295
infinite cluster, *see* percolation path
infrared absorption, 268–270
insulator–metal transitions, 137, 148, 186,
    190, 223–231, 243–245
intermolecular interactions, 86, 207–209, 260
ion implantation, 273, 274
iron, 10
iron–phosphorus system, 31, 76, 77

Jannink, G., 113, 120, 133
Jortner, J., 189, 204
Josephson scaling law, 164, 165

Kadanoff, L. P., 200, 204
Kastner, M., 105, 106, 133, 255, 258, 292, 296
Kauzmann, W., 21, 32
Kauzmann's paradox, 21, 22
Kelso, S. M., 273, 295
Kilgour, D. M., 50, 85
kinetic energy and delocalization, 46, 62,
    229–231, 256
kinetics of the glass transition, *see*
    glass transition
Kirkpatrick, S., 145, 155, 164, 198–201, 204
Kirste, R. G., 113, 120, 133
Knights, J. C., 295
Kovacs, A. J., 4, 32
Kruse, W. A., 113, 120, 133

Langer, J. S., 281, 295

laser glazing, 9
Leadbetter, A. J., 93, 134
Leath, P. L., 159, 160, 203, 204
LeComber, P. G., 292, 296
Lee, T. D., 220
Lehmann, Marianne, 171, 172
Leiser, G., 113, 120, 133
Lennard-Jones glass, 52, 208–212
Licciardello, D. C., 246, 248, 251
like-atom bonds, 72
liquid, supercooled, 7
liquid, undercooled, 7
liquid–crystal transition, *see* crystallization
liquid–glass transition, *see* glass transition
liquidity, 14, 15
Lobb, C. J., 189, 190, 204
localization, 155, 205, 226, 231–251
localization, correlation-induced, *see* Mott
    transition
localization, disorder-induced, *see* Anderson
    transition
localization, scaling theory of, 242–251
localization-delocalization transitions,
    205–251
localization length, 226, 250, 278, 280
localization and percolation, 233, 236–239,
    245
localized states, 226, 235, 276–280, 287, 289
lone-pair electrons, 257, 258, 293
lone-pair semiconductors, 258, 260
long-range order, 11, 12, 16, 223–225, 234,
    254
Lucovsky, G., 295
Lynch, D. W., 273, 295

macromolecules, 90–93, 176–179
magnetism, 24, 31, 32, 151–153, 157, 165, 274,
    294
Malt, R. B., 183, 203
Mandelbrot, B. B., 131, 132, 134, 161, 204
marginal dimensionality, 124, 126, 164–167,
    239
Masumoto, T., 76, 77, 85
Matsuoka, H., 82, 85
mean-field theory:
    for percolation, 164, 167, 172–174
    for polymer coils in solution, 124–127
    for self-avoiding walks, 127
melt quenching, 8, 9
melt spinning, 6, 7, 31
metal–insulator transitions, 137, 148, 186,
    190, 223–231, 243–245
metallic glasses:
    applications of, 24, 31

conductivity in, 274, 275
diffraction studies of, 75–77, 80–82
ferromagnetism in, 24, 31, 32, 274
glass transition in, 17, 18, 215
preparation of, 7–10, 215
random-close-packing model for, 35, 49–59,
    80–82
split second peak in the RDF of, 81, 82,
    211, 212
structure of, 35, 49–59, 75–77, 80–82
superconductivity in, 32, 274
metastability, 16
mobility, 277, 282–287
mobility edge, 233–239, 250, 251, 276
mobility gap, 236, 254, 276, 277
molecular solids, 86–97
Monte Carlo calculations, 198, 209–212
Montroll, E. W., 276, 282–288, 296
Mooser, E., 255, 296
Mort, J., 30, 32, 289, 296
Moss, S. C., 76–78, 85, 94, 133
Mott, N. F., 16, 102–104, 114, 116, 134, 227,
    233, 249, 251, 252, 276–280, 295, 296
Mott insulator, 229–230
Mott's derivation for variable-range hopping,
    278–280
Mott's use of the $8 - n$ rule, 101–104
Mott transition, 224, 227–231, 244
multiple trapping, 289

Nakanishi, H., 145, 193, 194, 204
native defects, 104–107, 290–294
Nemanich, R. J., 94, 134
network dimensionality, 87–94, 182, 258
network formers, 100, 103
network modifiers, 100, 103
nickel–phosphorus system, 6, 8, 80, 81
nonbonding–antibonding splitting, 260, 265
nonbonding electrons, 257, 258, 265
noncrystalline solid, 3

Ober, R., 113, 120, 133
odd-membered rings in silicon, 69, 71, 72
Olson, C. G., 273, 295
one-dimensional-network solids, 90, 91,
    107–113
optical:
    absorption coefficient, 261, 266
    absorption edge, 258–260, 265–268
    bandgap, 258–260, 265, 266, 271–273
    fine structure in crystals, 263–265
    properties, 24–26, 252–254, 258–273
    reflectivity, 252–254, 261–265
    response functions, review of, 261

optical (*continued*)
  transparency, 24–26, 267, 268
organic glasses:
  applications of, 24, 27, 107, 108
  Flory random-coil model for, 107–113,
    127–129
  glass transition in, 4, 6
  preparation of, 8
  structure of, 107–129
Ovshinsky, S. R., 30, 32, 94, 133, 233, 236, 251
oxide glasses, *see* germania; silica

packing fraction, *see* filling factor
pair connectedness, 154–156
pair correlation function, *see* radial
    distribution function
palladium–phosphorus system, 6, 8
palladium–silicon system, 274, 275
Papatheodorou, G. N., 94, 134
Paul, W., 82, 83, 85, 295
Pauling, L., 62, 85
Pearson, K., 114, 115, 134
Pearson, W. B., 255, 296
peas, 56–58
Peierls theorem, 250
percolation:
  applications of the theory of, 147–153,
    175–182, 186, 206, 219–223
  clusters, 140, 142–146, 156–163, 221
  critical concentration for various lattices,
    170
  critical exponents for, 153–167, 200
  critical volume fraction, 166, 185–191, 238
  fractal aspects of, 158–163
  and gelation, 177–181, 191–194, 197
  and the glass transition, 206, 219–223
  and localization, 155, 233, 236–239, 245
  and magnetism, 151–153, 157
  and metal–insulator transitions, 137, 148,
    186, 190
  path, 139–146
  and polymers, 163
  probability, 146, 173, 174
  scaling aspects of, 156–163, 193–201
  threshold, 136–138, 145, 156–159, 168–170,
    173, 187, 188, 199
  transition, 135–153
  and variable-range hopping, 281
  see also bond percolation; continuum per-
    colation; extended-range percolation;
    high density percolation; polychromatic
    percolation; site percolation; site-bond
    percolation
Perrin, J., 114–116, 131, 134

Pfister, G., 286, 288, 289, 296
phase transitions, *see:* Anderson transition;
    glass transition; metal-insulator transi-
    tions; Mott transition; percolation transi-
    tion; universality
Phelps, D. J., 186, 204
Phillips, J. C., 291, 296
phonons, 268–270
phosphorus-doped silicon, 242–245, 251, 278
photoconductivity, 24, 28, 29, 186, 282, 283,
    288, 294
photocrystallization, 96
Picot, C., 113, 120, 133
Pierce, D. T., 271, 272, 296
pinning of the Fermi level, 290, 294
plastics, 24, 27, 107
plumbing analogy for percolation, 139, 141,
    191
polarization catastrophe, 245
political analogy for renormalization, 193,
    195–197
political flow diagram, 196
Polk, D. E., 68, 69, 77, 79, 85
Polk model, 68–72, 79, 81–85, 291
polychromatic percolation, 200–203, 233
polymer chains, 90, 91, 93, 108–113, 116, 123,
    163
polymeric solids, 90–95, 107–113
polymerization index, 90, 93, 108
polymerization reaction, 175, 176
polystyrene, 6, 8, 24, 107–110
polyvinylacetate, 3, 4
Porter, Cole, 163, 204
Predel, B., 11, 32
preparation of amorphous solids, 5–10
pressure effect, 258–260
pseudogap, 235, 290

quenching rate, 3–6, 8, 9
Quinn, G. D., 162, 203

radial distribution function, 40–43, 74, 77,
    79–85, 156, 210–212
radius of gyration, 119
Ramakrishnan, T. V., 246, 248, 251
Raman scattering, 96, 97, 269, 270
random close packing, 35, 49–59, 80–82, 183,
    212, 214
random coil model, 35, 111–113, 127–129
random flights, *see* random walks
random potentials, 236–238
random resistor networks, 137, 145, 167, 169,
    181, 183, 200, 253
random walks, 113–127, 285

Rao, K. J., 206, 251
Rawson, H., 23, 32
Rayleigh, Lord, 114–116
real-space renormalization, 198
reconstructed bonds, 290, 291
Redner, S., 192, 193, 203
reflectivity spectra, 252–254, 262–265
Reich, G. R., 159, 160, 203, 204
renormalization-group fixed point, 195–197,
    249, 250
renormalization-group flow diagrams, 194,
    196, 248
renormalization-group theory, 193–201,
    247–250
resistivity, *see* conductivity
resistor networks, 137, 145, 167, 169, 181, 183,
    200, 253
Reynolds, P. J., 192–194, 203
ring statistics, 66, 71, 72, 177
Rogers, C. A., 37, 85
Rosenbaum, T. F., 242, 243, 251

scaling exponents, *see* critical exponents
scaling laws, 163–165
scaling properties:
    of the Anderson transition, 242–251
    of bond percolation, 198–201
    of Brownian trails, 129–133
    of dispersive transport, 286, 287
    of percolation clusters, 156–163
    of political elections, 193, 195–197
    of polymer configurations, 129–133
    of random walks, 129–133
    and renormalization, 193–201
    of self-avoiding walks, 129–133, 163
    of site-bond percolation, 193, 194, 197
scaling theory of localization, 242–251
scattering experiments for studying structure,
    42, 43, 73–85, 113, 119, 120
Scher, H., 185, 204, 236–238, 276, 282–288, 296
Scher-Montroll theory of dispersive transport,
    282–288
Scott, G. D., 50, 85
selenium:
    crystalline form, 90, 91
    electronic structure of, 254, 256–258
    glass transition of, 6
    photoconductivity of, 287, 288
    polymeric nature of, 90
    quenching rate for, 8
    structure of, 90–93
    valence alternation in, 105, 292, 293
    in xerography, 24, 28, 294
selenium–arsenic, *see* arsenic–selenium system

selenium–arsenic–germanium system, 97–99,
    293–295
selenium–germanium system, 73, 97, 258
self-avoiding random walks, 120–127, 163
self-repulsion effect on polymer coils, 120, 123,
    124, 128, 129
self-similarity, 115, 131, 132
Shapiro, B., 193, 194, 204
short-range order, 12, 34, 39, 67, 69, 72,
    101–104, 253–258, 265
Sichina, W., 21, 32
silane, 290, 292
silica:
    applications of, 23–27, 273
    continuous random network of, 72, 73
    glass transition of, 6
    network formers in, 100
    network modifiers in, 100
    quenching rate for, 8
    structure of, 72, 73, 84, 85, 176, 177
silica gel, 175–177, 182
silicate glasses, *see* silica
silicic acid, 175
silicon:
    bond-angle statistics in, 67–69, 291
    continuous random network of, 67–72
    crystalline form of, 61, 62
    dangling bonds in, 289, 290
    diffraction studies of, 76–81
    electrical properties of, 277–280
    role of hydrogen in, *see* hydrogenated
        amorphous silicon
    optical properties of, 269–273
    phosphorus-doped, 242–245, 251, 278
    preparation of, 9, 10, 290
    ring statistics in, 67, 68, 71, 72
    solar-cell applications of, 24, 30, 31,
        271–273, 290–292
    structure of, 67–72, 76–81, 289
    vibrational excitations in, 268–270
silicon–gold system, 6, 10, 11, 17, 18, 274
silicon–hydrogen system, *see* hydrogenated
    amorphous silicon
silicon–palladium system, 274, 275
simplex, 59
simplicial graph, 47, 60, 61
site-bond percolation, 139, 191–198
site percolation, 139, 141–144, 168–171, 186,
    187, 198
Slade, M. L., 94, 134, 259, 264–267, 296
Smith, L. N., 189, 190, 204
solar cells, 24, 30, 31, 271–273
sol–gel transition, 175–182, 191–194, 197, 206
solidity, 13–15

Solin, S. A., 94, 134
Spaepen, F., 183, 203
spanning cluster, 144, 199
spanning length, 154–157
Spear, W. E., 292, 296
specific heat, 17–19
Spicer, W. E., 271, 272, 296
splat quenching, 8, 9
split second peak in the RDF of metallic
    glasses, 81, 82, 211, 212
stable fixed points, 195, 196
staggered bond configuration, 70, 109
Stanley, H. E., 191–193, 203, 204
Stauffer, D. J., 177, 204
Steinhardt, P., 83–85
stochastic geometry, 35, 130, 138, 184, 185,
    189, 244, 281
Stockmayer, W. H., 177, 204
Straley, J. P., 169, 204
Street, R. A., 94, 134, 295
sulfur, 86–89, 94
sulfur–arsenic, *see* arsenic–sulfur system
sulfur–germanium system, 73
superconductivity, 32, 274
supercooled liquid, 7
Swiss cheese, 161, 289
switching in amorphous semiconductors, 30

Takahashi, T., 94, 134
Tauc, J., 96, 133, 252, 253, 265–267, 296
Tauc edge, 266, 267
Taylor, P. C., 295
tellurium, 94
tellurium–germanium system, 24, 30
Temkin, R. J., 82, 83, 85
tetrahedrally coordinated continuous random
    network, 67–72, 79, 81–85, 291
thermal expansion coefficient, 1, 4, 218
Thomas, G. A., 242, 243, 251
Thouless, D. J., 167, 204, 233, 246, 251
three-dimensional-network solids, 86–89
time-of-flight experiments, 282–289
topological defects in covalent glasses,
    104–107
topological disorder, 13, 49–73, 184
Toulouse, G., 164, 204
transfer energy, 239–241, 246
transit time, 283–288
transparency of glasses, 24–26, 267, 268
transport mechanisms, 276–280, 289
trapping, 277, 287, 289
tunneling transport, 276–281, 287, 289
Turnbull, D., 6, 7, 11, 18, 20, 32, 49, 85, 215,
    251

two-dimensional-network solids, 89–93

ultraviolet spectra, 252–254, 262–265
unblocked bonds in percolation, 138
unblocked sites in percolation, 139
unbounded cluster, *see* percolation path
undercooled liquid, 7
universality, 158, 165, 166, 197, 198
universality class, 197, 198
unstable fixed points, 195–197, 249, 250

valence alternation pairs, 104–107, 292–294
valence-electron excitations, 252–254,
    263–268
valence satisfaction in glasses, 101–104
van Hove singularities, 254
Van Vleck, J. H., 16
vapor-condensation techniques, 9, 10
variable-range hopping, 277–281
vibrational excitations, 268–270
viscosity, 23, 179, 218
vitreous solid, 3
Vogel-Fulcher equation, 23, 218
Voronoi froth, 46–48, 51–55, 60, 61, 219, 221
Voronoi polyhedron, 45–49, 53–58, 61, 214,
    216, 221
vulcanization, 182
Vycor glass, 176, 177

Waseda, Y., 76, 77, 85
water, 6, 10
Weaire, D., 83–85, 295
Webman, I., 189, 204
Weinstein, B. A., 259, 296
Wigner crystallization, 231
Wigner–Seitz cells, 45–49, 53–58
Wilmers, G., 120, 123, 133
Wilson, K. G., 165, 204
Wright, A. C., 93, 134
wrong bonds, 72

xerography, 24, 28, 29, 283, 294, 295

Yamakawa, H., 118, 134
Yamamoto, R., 82, 85
Yoffa, E. J., 294, 295
ytterbium, 224

Zachariasen, W. H., 35, 63–65, 73, 85, 91, 95,
    103, 134
Zallen, R., 88, 94, 134, 185, 200, 204, 236–238,
    251, 259, 262, 264–267, 271, 281, 296
zero-dimensional-network solids, 89, 90
Ziman, J. M., 236, 251